The Fundamentals of Nuclear Power Generation

Insert photo

Mill W Hubbell 2-1-13

By: M.W. Hubbell

AuthorHouse™
1663 Liberty Drive
Bloomington, IN 47403
www.authorhouse.com
Phone: 1-800-839-8640

First published by AuthorHouse   07/23/2011

ISBN: 978-1-4634-2441-1 (sc)
ISBN: 978-1-4634-2658-3 (ebk)

Library of Congress Control Number: 2011910044

Printed in the United States of America

Any people depicted in stock imagery provided by Thinkstock are models, and such images are being used for illustrative purposes only.

This photo is of Calvert Cliffs Nuclear Power Plant in Southern Maryland courtesy of the NRC.

Certain stock imagery © Thinkstock.

This book is printed on acid-free paper.

Because of the dynamic nature of the Internet, any web addresses or links contained in this book may have changed since publication and may no longer be valid. The views expressed in this work are solely those of the author and do not necessarily reflect the views of the publisher, and the publisher hereby disclaims any responsibility for them.

Table of Contents

Acknowledgements

The following are those people who have given me support, resources, questions, and/or answers in order to address most areas of the nuclear industry. This book, of course is **NOT** all encompassing in that this book is supposed to cause the reader to ask additional questions…this is how we learn. Questioning Attitude…

Judy Hubbell, Irene Hubbell, Sippie O'Kelley, Lea O'Gorman, Gustave Wolf, Mike Gahan, Bill Buchanan, Jason Schaefer, Ron Thompson, Jesse Davis, Joe Jaeger, Dave Burdin, Ann Yonkowski, Tiffany Nickels, Mark Simpson, Ken Allor, Mark Draxton, Chris and Michele Jones, Rodney Fleegle, Marc Reckner, Richard Best, Mike Orlando, Ed Schinner, Tim Beck, Nick Lavato, Lori Pyska, Parker Williams, Troy and Patricia Grigg, William (Bill) Wilson, J. B. Couch, Tom Morello, Nicole Walker, Jeffery Maddocks, Joe Gaffey, Steve Sanders, Dean Wood, Amanda Boswell, Joyce Turner, Bob Kreger, John M. Herron, Larry Smith

Thank You all for your continued support in this endeavor.

Introduction

This book is dedicated to everyone who is involved with the nuclear industry. The men and women who continue to make this industry one of the safest and most reliable resources for generating electricity. This book was created in order to help others understand more about nuclear energy and the processes.

In the nuclear industry, several words are used continuously in order to help those within the industry to remain focused on the task at hand. These words are: Safety, Integrity, Complacency, Lessons Learned, Training, Teamwork and Improvement.

You might ask "Why Questions and Answers?" Well the intention of the book is to not only make you more knowledgeable about the nuclear industry but to stimulate your thinking into why we do the things we do. Someone is out there, that will have the idea to revolutionize the way things are done. I want this book to stimulate that person so that the nuclear industry as well as the World can become a better place to live.

For we only have one World.

Simplified Steam Cycle

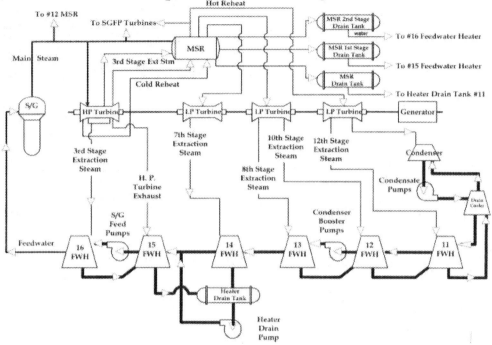

This simplified one line diagram should give you the basic idea of how we take water from the condenser and pump it through multiple heater exchangers, pumps, steam generator, moisture separator reheater (MSR), turbine and back to the condenser. No unit is exactly like another, so regardless if the units were built around the same time…there are differences. But this should help you better understand our flow through the secondary system. So let's get started.

What is the difference between fission and fusion? Nuclear fusion and nuclear fission are two different types of energy-releasing reactions in which energy is released from high-powered bonds between the particles within the nucleus. The main difference between these two processes is that fission is the splitting of an atom into two or more smaller ones while fusion is the fusing of two or more smaller atoms into a larger one.

Nuclear fusion is the reaction in which two or more nuclei combine together to form a new element with higher atomic number (more protons in the nucleus). The energy released in fusion is related to $E = mc^2$ (Einstein's famous energy-mass equation). On earth, the most likely fusion reaction is Deuterium–Tritium reaction. Deuterium and Tritium are both isotopes of hydrogen. $^2_1\text{Deuterium} + {}^3_1\text{Tritium} = {}^4_2\text{He} + {}^1_0n + 17.6$ MeV

Nuclear fission is the splitting of a massive nucleus into photons in the form of gamma rays, free neutrons, and other subatomic particles. In a typical nuclear reaction involving ^{235}U and a neutron: $\quad\quad {}^{235}_{92}\text{U} + n = {}^{236}_{92}\text{U}$

Followed by

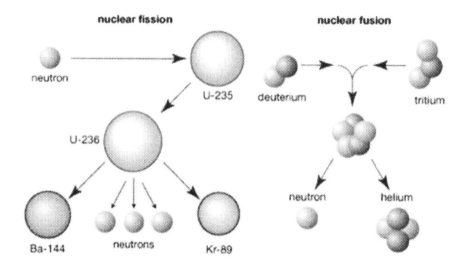

Fission vs Fusion

	Nuclear Fission	**Nuclear Fusion**
Nuclear Weapons	One class of nuclear weapons is a fission bomb, also known as an atomic bomb.	One class of nuclear weapons is the hydrogen bomb, which uses a fission reaction to "trigger" a fusion reaction.
Energy Ratios	The energy released by fission a million times greater than that released in chemical reactions; but lower than the energy released by nuclear fusion.	The energy released by fusion is 3 to 4 times greater than the energy released by fission.
Byproducts	Fission produces many highly radioactive particles.	Few radioactive particles are produced by fusion reaction, but if a fission "trigger" is used, radioactive particles will result from that.
Natural occurrence	Fission reaction does not normally occur in nature.	Fusion occurs in stars, such as our sun.
Definition	Fission is the splitting of an atom into two or more smaller ones.	Fusion is the fusing of two or more lighter atoms into a larger one.
Conditions	Critical mass of the substance and neutrons are required.	High density, high temperature environment is required.
Energy Requirements	Takes little energy to split two atoms in a fission reaction.	Extremely high energy is required to bring two or more protons close enough that nuclear forces overcome their electrostatic repulsion.

Time - Line

5000 B.C. – Evidence of the use of gold in Bulgaria.

4000 B.C. – The element Silver is discovered. Iron is discovered.

3750 B.C. – Carbon was used in "charcoal" as a domestic smokeless fuel.

3700 B.C. – Egyptians develop bronze.

3500 B.C. – The element lead and tin was known to be used. The wheel was invented. Lubricating grease was first used.

3000 B.C. – The element Antimony is discovered. Shape metal into molds. Candles are invented.

2700 B.C. – Merit Ptah, an Egyptian physician is the first woman recorded by name in the history of science.

2500 B.C. – Brazing, Soldering, and Welding is discovered. Wrought iron is discovered.

2000 B.C. – Sanskirt writings outlined purifying water through coarse sand / charcoal, gravel filtration, boiling and straining.

1500 B.C. – The element Mercury was found in Egyptian tombs. Egyptians discovered the principle of coagulation using alum for suspended particle settlement. Invention of steel hardening iron with carbon.

800 B.C. – Sundial was created.

750 B.C. – The pulley is invented by the Assyrians.

600 B.C. – Thales of Miletus writes about amber becoming charged when rubbed – he was describing what is now known as static electricity.

550 B.C. – Crane was invented by the Greeks.

513 B.C. – Cast iron is produced in China.

500 B.C. – Leucippus is credited along with his pupil Democritus for postulating the theory of atoms and voids. Hippocrates used a bag filter to trap sediment thus removing tastes and odors from the water, his device was called a "Hippocratic sleeve".

470 B.C. – Athenian teacher Anaxagoras suggests that the Sun, Moon and stars are composed of the same matter as the Earth.

450 B.C. – Greek philosopher Anaxagoras stated that matter cannot be created nor destroyed.

440 B.C. – Leucippus of Miletus introduces the concept of the atom, an indivisible unit of matter.

460 – 370 B.C. - Democritus (c.460-c.370 BC.) was the first physicist to come up with a theory of matter, which is rather similar to our idea of matter today. He believed that matter cannot be divided infinitely and therefore there exist the smallest elements, the atoms, which are joined to each other by hook-like "protuberances". The word atom means undividable in Greek.

300 B.C. – Rome built the aqueducts.

276 – 194 B.C. – Eratosthenes of Cyrene made several accurate measurements of the Earth. He realized that at noon the sunlight would shine to the bottom of a well in Aswan, Egypt but in Alexandria miles away that the sunlight at noon would not go all the way to the bottom of that well. Even though he was not the first to measure the Earth he was the first to be extremely accurate in his measurements.

260 B.C. – The lever is first described by Archimedes.

250 B.C. – The screw-pump (screw conveyor) was invented by Archimedes. The compound pulley is introduced by Archimedes.

200 B.C. – The windmill is invented by China and Persia to harness wind power.

130 B.C. – Hero of Alexandria creates the first ever device to be powered by steam called the Aeolipile.

100 B.C. – The belt drive is first used by the Romans. The water mill is first used by the Greeks.

79 A.D. – First known use of uranium. Roman artisans produced yellow colored glass in mosaic mural near Naples.

40 A.D. – Roman Nemi ships were found to have ball bearings.

105 – Tsai Lun invents paper.

850 – Moors in Spain prepared pure copper by re-acting its salts with iron, a forerunner to electroplating.

1094 – The chain drive is first used by the Chinese.

1182 – The modern magnetic compass is invented.

1206 – The crankshaft and the camshaft were invented by Islamic scholar Al-Jazari.

1250 – Albertus Magnus discovered Arsenic.

1269 – Petrus Peregrinus mapped out the magnetic field on the surface of a spherical magnet.

1300 – The earliest known example of a rotary to reciprocating motion is the waterwheel powered pump invented by Al-Jazari.

1348 – 1350 – Black plague sweeps Europe killing up to 60% of its population.

1500s – Commercial use of zinc.

1551 – Taqi al-Din in Ottoman, Egypt invented a steam turbine with practical applications.

1564 – Graphite is discovered.

1590 – Zacharias Janssen invents the microscope.

1593 – Galileo invented the water thermometer.

1600 – William Gilbert first coined the term "electricity" from the Greek word for amber. Gilbert wrote about the electrification of many substances in his "Demagnete, magneticisique corporibus". He also first used the terms electric force, magnetic pole, and electric attraction.

1601 – Giovanni Battista della Porta performs experiments to create positive and negative (vacuum) pressures.

1620 – William Oughtred created the slide rule that was often considered the first analog computer.

1622 – The slide rule is invented by William Oughtred.

1623 – Wilhelm Schickard first to invent the mechanical calculator.

1627 – Sir Francis Bacon had the first record of experimentation with water filtration; even though unsuccessful in his experiment it led future sciences to improve water purification.

1629 – Giovanni Branca develops a primitive steam turbine which rotates driven machinery allowing a stamping mill to be developed.

1630 – Vincenzo Cascariolo discovers fluorescence.

1636 – W. Gascoigne invents the micrometer.

1638 – Galileo attempts to measure the speed of light by a lantern relay between distant hilltops.

1640 – Otto von Guericke invented the vacuum pump.

1643 – Evangelista Torricelli invents the barometer used to measure atmospheric pressure.
- Blaise Pascal built the first adding machine.

1650 – Otto von Guericke invented the dasymeter, used to determine the density of gases.

1652 – Fluid pressure laws are determined by Blaise Pascal.

1660 – Otto von Guericke invented a machine that produced static electricity.

1666 – Newton discovered that white light is composed of all the colors of the spectrum.

1669 – Hennig Brand discovered Phosphorus by distilling urine.

1670 – Robert Boyle discovered that the reaction between a metal and an acid produces hydrogen.

1671 – Isaac Newton experiments with prisms to show that white light is a mixture of all the colors.

1674 – Gottfried Wilhelm von Leibniz described the binary number system and co-invented differential calculus and designed a machine that would perform four basic arithmetic functions.
  - Anton van Leeuwenhoek invents the compound microscope.

1675 – Robert Boyle discovered that electric force could be transmitted through a vacuum and observed attraction and repulsion. Boyle also published experiments and notes about the production of electricity.

1676 – Robert Hooke created his law which states that the strain produced is directly proportional to the stress. He also invents the universal joint.

1679 – The earliest and simplest safety valve was the steam digester invented by French physicist Denis Papin…(a pressure cooker is a fine example of this today).

1684 – Street lighting introduced to London.

1689 – The centrifugal pump is invented by Denis Papin.

1698 – Thomas Savery builds a steam powered water pump for pumping water out of mines.

1700's – The first water filters for domestic application were applied…wool, sponge and charcoal.

1702 – Guillaume Amontons suggests that there may be a state of absolute zero, later taken up by Kelvin.

1704 – Isaac Newton proposes a mechanical universe with small solid masses in motion.

1705 – Francis Hauksbee put some mercury into a glass globe, pumped out the air and then spun it. When he did this in the dark and rubbed the globe with his bare hand, it glowed. (He invented the first fluorescent light.)

1709 – Daniel Gabriel Fahrenheit invented the alcohol thermometer. Abraham Darby revolutionizes iron making by coke-based iron smelting.

1712 – Thomas Newcomen builds a piston and cylinder steam powered water pump for pumping water out of mines.

1714 – Daniel Gabriel Fahrenheit invents the mercury thermometer.

1716 – Jacob Hermann suggests that gas pressure is proportional to density and to the square of the average velocity of the gas particles in motion.

1722 – C. Hopffer patents the fire extinguisher.

1729 – Stephen Gray discovered the conduction of electricity. He shows that charges on electrified objects resides on their surfaces.

1730 – First compound magnet produced by Savery.

1732 – The pitot tube is invented by Henri Pitot.

1733 – Charles Francois du Fay discovered that electricity comes in two forms which he called resinous (-) and vitreous (+).
- Benjamin Franklin and Ebenezer Kinnersley later renamed the two forms as negative and positive.

1739 – Georg Brandt discovered Cobalt.

1741 – Charles Wood discovered Platinum.

1742 – Anders Celsius develops the Celsius temperature scale.

1745 – Georg von Kleist discovered that electricity was controllable.
- Pieter van Musschenbroek invented the "Leyden Jar" the first electrical capacitor which stores static electricity.
- Doctors in Geneva begin to treat patients with electric shocks.

1746 – Andreas Marggraf discovered Zinc.
- William Watson suggested conservation of electrical charges.

1747 – Benjamin Franklin experiments with static charges in the air and theorized about the existence of an electrical fluid that could be composed of particles.
- William Watson discharged a Leyden jar through a circuit, that began the comprehension of current and circuit.
- Henry Cavendish started measuring the conductivity of different materials.
- William Watson passes electricity more than 2 miles down a wire.

1748 – Benjamin Franklin first coined the term "battery" to describe an array of charged glass plates.

- William Watson uses an electrostatic machine and a vacuum pump to make the first glow discharge. His glass vessel is three inches in diameter and approximately three feet long…the first fluorescent light bulb.
- Jean Antoine Nollet developed a primitive electroscope and put forward early theories about electrochemistry.

1750 – John Michell discovers that the two poles of a magnet are equal in strength and that the force law for individual poles is inverse square.
- Abraham Darby produces the first wrought iron.

1751 – Axel Fredrik Cronstedt isolated Nickel.

1752 – Benjamin Franklin invented the lightning rod – he demonstrated lightning was electricity with his kite flying experiment.
- Johann Sulzer puts lead and silver together in his mouth, performing the first recorded "tongue test" of a battery.

1753 – Claude Geoffroy identified Bismuth.

1755 – Joseph Black discovers carbon dioxide.
- The bimetallic strip was invented by John Harrison.

1759 – Francis Ulrich Theodore Aepinus shows that electrical effects are a combination of fluid flow confined to matter and action at a distance.

1761 – Joseph Black describes latent heat and specific heat.

1764 – Johannes Wilcke invents an electro-phorous machine that generates electrical current.

1765 – James Watt invented the steam engine with a separate condenser.

1766 – Henry Cavendish discovered Hydrogen.

1767 – Joseph Priestley discovered that electricity followed Newton's inverse-square law of gravity.

1769 – James Watt patents his first improved steam engine.

1772 – Daniel Rutherford discovered Nitrogen.

1774 - Joseph Priestley identified oxygen.
- Carl Wilhelm Scheele discovered Manganese.
- Carl W. Scheele discovered Chlorine and it was identified by Sir Humphry Davy in 1810.

1775 – Henry Cavendish invents the idea of capacitance and resistance.
- Alessandro Volta invents the electrometer.
- Joseph Priestley discovered hydrochloric and sulphuric acids.

1776 – United States of America becomes independent from England.

1777 – Joseph Louis Lagrange invents the concept of the scalar potential for gravitational fields.
- Antoine Lavoisier put forth the idea of chemical compounds, composed of more than one element.

1778 – Carl W. Scheele isolated Molybdenum.
- Carl Scheele and Antoine Lavoisier discover that air is composed mostly of nitrogen and oxygen.

1781 – Joseph Priestly creates water by igniting hydrogen and oxygen.
- Compound steam engine is invented by Jonathan Hornblower.

1782 – Franz Joseph von Reichenstein discovered Tellurium.

1783 – Juan Jose and Fausto d'Elhuar y de Suvisa discovered Tungsten.

1785 – Charles Coulomb – uses a torsion balance to verify that the electrical force law is inverse square. He also discovers that the electric force near a conductor is proportional to its surface charge density and makes contributions to the two fluid theory of magnetism.

1786 – Luigi Galvani demonstrated what is now understood to be the electrical basis of nerve impulses when he made a frog muscle twitch by jolting them with a spark from an electrostatic machine.

1787 – Martin Heinrich Klaproth discovered Zirconium.

1789 – Adair Crawford identified Strontium.
- Johan Gadolin identified Yttrium.
- Coulomb's Law of electrostatic force is first described.

- Martin Klaproth realized that a pitch-blende contained an unknown element…uranium later discovered in 1841.

1790 - Lavoisier clarifies the concept of chemical element.

1791 - Reverend William Gregor discovered Titanium.
- John Barber invented the first true gas turbine used to power a horseless carriage.

1792 – William Murdoch invents gas lighting.

1793 – Alessandro Volta invented the first electric battery. Volta proved that electricity could travel over wires.

1794 – The internal combustion engine is invented by Robert Street.

1797 – Louis Nicolas Vauquelin discovered Chromium.
- Henry Maudslay invents the first metal or precision lathe.

1798 – Louis Nicolas Vauquelin discovered Beryllium.
- Henry Cavendish measures the gravitational constant.
- Rumford discovers the link between heat and friction.

1799 – The high-pressure steam engine is invented by Richard Trevithick.

1800 – William Nicholson and Anthony Carlisle discovered that water may be separated by electrolysis into hydrogen and oxygen by the action of Volta's pile.
- Alessandro Volta announces his invention of the electric battery.
- Sir Humphry Davy first got the idea of a light bulb when he ran current through a platinum wire…it glowed but was not bright enough, it is at this point where numerous scientist improve the light bulb up to today.
- William Herschel discovers a point below the frequency of red light which he terms infrared.
- Jonathan Hornblower invented the double beat valve.

1801 – Andres Manuel del Rio discovered Vanadium.
- Charles Hatchet discovered Niobium.
- The first steam-powered pumping station is built near Philadelphia to supply power.
- Johann Wilhelm Ritter discovers light beyond the violet end of the spectrum which he terms ultraviolet.

1802 – Johann Wilhelm Ritter develops a dry cell battery.
- Anders Gustav Ekeberg discovered Tantalum.

1803 – William Hyde Wollaston discovered Rhodium and Palladium.
- Jons Jakob Berzelius, Whilhelm Hisinger and Martin Klaproth all discovered Cerium.
- Smithson Tennant discovered Osmium and Iridium.
- John Dalton developed the first useful atomic theory of matter.
- John Dalton publishes a table of comparative atomic weights.
- John Dalton "Thou knowest no man can split the atom."

1804 – Robert Thom designed the first municipal water treatment plant in Scotland using a slow sand filtration system.

1805 – Luigi Brugnatelli invents modern electroplating.
- Gay-Lussac proves that water is composed of two parts hydrogen and one part oxygen by volume.

1807 – Humphrey Davy – shows that the essential element of Volta's pile is chemical action since pure water gives no effect. He argues that chemical effects are electrical in nature.
- Humphrey Davy isolated Sodium and Potassium.
- Isaac de Rivas makes a hydrogen gas powered vehicle.

1808 - Dalton pointed out that the atoms of a given element are the same and different materials are made up of a few types of atoms.
- Davy demonstrates the first electric carbon arc light.
- Humphrey Davy, Joseph-Louis Gay-Lussac and Louis Jacques Thenard isolated Boron.
- Humphrey Davy isolated and indentified Magnesium, Calcium and Barium.
- John Dalton puts forward the modern atomic theory with his book 'A new system of chemical philosophy'.

1809 – Gas powered street lights first appeared in London.

1810 – Sir Humphry Davy uses the first carbon arc lamp to improve the light bulb.

1811 – Bernard Courtois discovered Iodine.
- Amedeo Avogadro stated that equal volumes of gases at the same temperature and pressure contain the same number of molecules regardless of their chemical nature and physical properties, now known as Avogadro's Law.

1812 – Michael Faraday is hired to work for Humphrey Davy as a scientific assistant at the Royal Institution.

1814 – George Stephenson builds the first practical steam powered railroad locomotive.

1815 – Prout's hypothesis states that hydrogen is an atom, and all other elements are built from multiple hydrogen units.

1816 – The first energy utility is founded in the United States.
- Robert Stirling invented the Stirling engine.

1817 – Johan August Arfwedson discovered Lithium.
- Jons Jakob Berzelius discovered Selenium.
- Friedrich Strohmeyer discovered Cadmium.

1820 – The relationship of electricity and magnetism is confirmed by Hans Christian Oersted who observed that electrical currents affected the needle on a compass orienting itself perpendicular to the wire, discovering electromagnetism.
- Marie Ampere discovered that a coil of wire acted like a magnet when a current is passed through it. She also discovers that parallel currents attract each other and that opposite currents attract.
- British researcher John Frederich Daniell developed an arrangement where a copper plate was located at the bottom of a wide-mouthed jar. A cast zinc piece commonly referred to as a crowfoot, because of its shape, was located at the top of the plate, hanging on the rim of the jar. Two electrolytes, or conducting liquids, were employed. A saturated copper sulphate solution covered the copper plate and extended halfway up the remaining distance toward the zinc piece. Then a zinc sulphate solution, a less dense liquid, was carefully poured in to float above the copper sulphate and immerse the zinc. As an alternative to zinc sulphate, magnesium sulphate or dilute sulphuric acid was sometimes used. The Daniell Cell was one of the first to incorporate mercury, by amalgamating it with the zinc anode to reduce corrosion when the batteries were not in use.

1821 – Invention of the first electric motor by Faraday.
- Humphrey Davy shows that direct current is carried throughout the volume of a conductor for long wires. He also discovers that resistance is increased as the temperature rises.
- Pierre Berthier discovered that iron-chromium alloys are corrosion resistant.

1822 – Thomas Seebeck invents the thermocouple and discovers the thermoelectric effect by showing that a current will flow in a circuit made of dissimilar metals if there is a temperature difference between the metals. This is called thermoelectricity. Thermo is a Greek word meaning "heat".
- William Sturgeon invents the electromagnet.

1824 – Jons Jakob Berzelius isolates Silicon.
- Joseph Aspin is given a patent for Portland cement.

1825 – Michael Faraday discovers benzene.
- Oersted observes that some undefinable magnetic effect is associated with charged particles in motion.

1826 – Ohms Law is created by Georg Simon Ohm that states "conduction law that relates potential, current, and circuit resistance." (V=IR) Ohm did develop the idea of voltage as the driver of electric current.
- Antoine-Jerome Balard discovered Bromine named after the Greek word bromos meaning "stench".
- Samuel Morey invents the internal combustion engine.

1827 – Joseph Henry's electromagnetic experiments lead to the concept of electrical inductance. Joseph Henry built one of the first electrical motors. Insulated wire is invented by Joseph Henry.
- F. Savery after noticing that the current from a Leyden jar magnetizes needles in alternating layers, conjectures that the electric motion during the discharge consists of a series of oscillations.
- Hans Christian Oersted discovered Aluminum.
- Ohm's law of electrical resistance is established.

1828 – Jons Jakob Berzelius discovered Thorium.
- Friedrich Wohler discovered Beryllium.
- Mark Seguin developed the first multi-tube boiler engine.

1829 – The thermocouple is invented by Thoms Seebeck.

1830 – The thermostat is invented by Andrew Ure.
- The pyrometer is invented by John Frederic Daniell.

1831 – Michael Faraday discovers the principles of electromagnetic induction, generation and transmission. Faraday shows that changing currents in one circuit induce currents in a neighboring circuit in which can be explained by the idea of changing magnetic flux.
- Charles Wheatstone and William Cooke created the first telegraph machine.
- Joseph Henry proposes and builds an electric telegraph.
- Joseph Henry invents the multiple coil magnet.

1832 – Michael Faraday studied the effect of electricity on solutions, coined the term "electrolysis" as a splitting of molecules with electricity, developed the laws of electrolysis.
- Faraday also coined the terms electrode, anode, cathode, ion, cation, anion, and electrolyte.
- Joseph Henry independently discovers induced currents.
- Samuel Morse conceived the idea of using a system of dots and dashes to represent letters and would eventually be known throughout the world as the Morse code.
- Fourneyron produces the first water turbine.

1833 – Faraday begins work on the relation of electricity to chemistry.
- The magnetometer is invented by C.F. Gauss.

1834 – Faraday discovers self-inductance.
- Jean Charles Peltier discovers the opposite of Seebeck's thermoelectric effect. He finds that current driven in a circuit made of dissimilar metals causes the different metals to be at different temperatures.
- Emil Lenz formulates his rule for determining the direction of Faraday's induced currents.
- Wheatstone measures the speed of electricity using revolving mirrors and several miles of wire.

1835 – Samuel Morse developed the Morse Code.
- James Bowman Lindsay improves the incandescent light bulb.
- Joseph Henry invents the electro-mechanical relay.

1836 – John Daniel invented the Daniel Cell that used two electrolytes: copper sulfate and zinc sulfate. The Daniel Cell was somewhat safer and less corrosive then the Volta cell.
- Nicholas Callan invented the induction coil which was the first widely used transformer.
- Samuel Morse demonstrated the first device to send signals over wires.

1837 – The first industrial electric motor is created.
- Faraday discovers the idea of the dielectric constant.
- The invention of the telegraph.

1838 – Faraday shows that the effects of induced electricity in insulators are analogous to induced magnetism in magnetic materials.
- Faraday also discovers Faraday's dark space, the dark region in a glow discharge near the negative electrode.

1839 – William Grove developed the first fuel cell which produced electricity by combining hydrogen and oxygen.
- Carl Gustaf Mosander discovered Lanthanum.
- Bacquerel discovered the photovoltaic effect.
- Charley Peckham patented the first gate valve.
- Daguerre discovers photography which later becomes the basis for personnel dosimety and discovery of radioactivity in uranium.
- Vulcanization is invented by Charles Goodyear.
- Isaac Babbitt invented a low-friction tin-based metal alloy used on bearings.

1840 – Warren De la Rue invents the vacuum tube enclosure.
- John Herschel invents the blueprint.

1841 – J.P. Joule's law of electrical heating published. He shows that energy is conserved in electrical circuits involving current flow, thermal heating, and chemical transformations.
- Eugene-Melchior Peligot isolated and identified Uranium.
- Frederick de Moleyns uses a powdered charcoal filament to improve the light bulb.

1843 – Carl Gustaf Mosander discovered Terbium and Erbium.
- Joule describes the mechanical equivalent of heat.
- Variable resistor was invented by Samuel Christie but was named after his compatriot Charles Wheatstone.

1844 – Samuel Morse constructed the first practical telegraph system and built a line from Baltimore to Washington, D.C.
- K.K. Klaus discovered Ruthenium.

1845 – Faraday discovers that the plane of polarization of light is rotated when it travels in glass along the direction of the magnetic lines of force produced by an electromagnet.
- Wilhelm Roentgen is born.
- Portland Cement is formulated by Joseph Aspdin.

1846 – Faraday discovers diamagnetism. He sees these effects in heavy glass, bismuth and several other materials.

1847 – Thomas Alva Edison is born.
- H. von Helmholz states that energy may be converted to other forms but may not be destroyed or lost.

1848 – Kelvin develops his temperature scale.

1850 – William Thomson (Lord Kelvin) invents the idea of magnetic permeability and susceptibility.
- Joseph Swan improved the light bulb by using carbonized paper filament.
- H. Thompson and J.T. Way treated various clays with ammonium sulfate or carbonate in solution to extract the ammonia and release calcium. (ion exchange).
- First commercial use of uranium in glass by Lloyd & Summerfield in England.

1851 – Kelvin proposes absolute zero.

1852 – The American Society of Civil Engineers is created.
- Henri Becquerel is born.

1854 – Faraday clears up the disagreement about the measured speed of signals along the transmission lines by showing it is crucial to include the effect of capacitance.
- John Snow discovered that chlorine could be used to purify the water and reduce the cholera epidemic that was attacking England. This if the first indication of using chemicals to purify water.

1855 – Heirrich Gobel improves the light bulb using a carbonized bamboo filament.
- Johann Geissler produces a mercury vacuum pump and with it he produces the first good vacuum discharge tube.

1856 – Karl Kronig suggests that gas molecules in equilibrium travel in straight lines unless they collide with something.
- Stephen Wilcox patent water tube boiler.
- John Ramsbottom invented a tamper proof spring safety valve.
- Joseph John Thomson, first person to identify the existence of subatomic particles is born.

1857 – William Kelly invents the blast furnace for steel production.

1859 – J. Plucker demonstrated that magnetic fields bend and builds one of the first gas discharge tubes (cathode ray tube).
- Gaston Plante developed the first practical storage lead-acid battery that could be recharged. This is the type of battery which is primarily used in vehicles today.
- William John Macquorn Rankine proposed the Rankine temperature scale.
- Spectrometer invented by Robert Bunsen and Gustav Kirchhoff.

1860 – Kirchhoff's Law published.
- Robert Bunsen and Gustav Kirchhoff discovered Cesium.
- Uranium is first used in homeopathic medicine for treatment of diabetes.

1861 – Robert Bunsen and Gustav Kirchhoff discovered Rubidium.
- Sir William Crookes discovered Thallium.
- The discovery of osmosis.
- Auguste Mouchout granted a patent for a motor running on solar power.
- The American Civil War Begins

1862 – Geissler and Topler developed the mercury vacuum pump.
- Gustav Kirchhoff coined the term "black body" radiation and two sets of independent concepts in both circuit theory and thermal emission are named after him.

1863 – Ferdinand Reich discovered Indium.

1864 – Karol Olszewski and Zygmunt Wroblewski predicted that ultra-cold temperatures would drop resistance levels in electrical conductors.
- Antoine Becquerel suggests an optical pyrometer.

1865 – The American Civil War Ends

1866 - The first cell developed by Georges Leclanché in France was a wet cell having its electrodes immersed in a liquid. Nevertheless, it was rugged and easy to manufacture and had a good shelf life. He later improved the battery by substituting a moist ammonium chloride paste for the liquid electrolyte and sealing the battery. The resulting battery was referred to as a **dry cell**. It could be used in various positions and moved about without spilling. Carbon-zinc dry cells are sold to this day in blister packages labeled "heavy duty" and "transistor power". The anode of the cell was zinc, which was made into a cup or can which contained the other parts of the battery. The cathode was a mixture of 8 parts manganese dioxide with one part of carbon black, connected to the positive post or button at the top of the battery by a carbon collector rod. The electrolyte paste may also contain some zinc chloride. Around 1960 sales of Leclanché cells were surpassed by the newer alkaline-manganese batteries. www.allaboutbatteries.com/history-of-batteries.html

1867 – Michael Faraday passes away.
- Becquerel devised and demonstrated a fluorescent lamp.
- Babcock & Wilcox is founded as a partnership.
- Reinforced concrete is developed by Joseph Monier.

1868 – Pierre Janssen discovered Helium.
- The first traffic light is introduced in London to ease traffic issues.
- Robert Mushet invents tungsten steel.

1869 – E. Goldstein coins phrase "cathode rays".
- First high-voltage, smooth, direct-current generator created by Zenobe Theophile Gramme.

1871 – Gramme introduced the first commercially significant electric motor.
- Ernest Rutherford is born.

1873 – James Maxwell wrote equations that described the electromagnetic field and predicted the existence of electromagnetic waves traveling at the speed of light.
- Amedee Bollee of Le Mans builds first steam powered car.
- Zenobe Theophile Gramme invented the first powerful electrical motor to be used in industries by accident while experimenting with a direct current dynamo.
- William Coolidge is born.

1874 – G.J. Stoney proposes that electricity is made of discrete negative particles he calls electrons. Stoney is credited with coining the term "electron".
- Kelvin puts forth the Second Law of thermodynamics.
- Henry Woodward and Mathew Evans improve the light bulb by containing the filament within a glass cylinder filled with nitrogen.

1875 – The first detection of electric currents emanating from the brain.
- Paul-Emile Lecoq de Boisbaudran discovered and identified Gallium.

1876 – Henry Rowland performs and experiment inspired by Helmholtz which shows for the first time that moving electric charge is the same thing as an electric current.
- Alexander Graham Bell invents the telephone and uses electricity to transmit speech for the first time.
- Nikolaus Otto designs a four-stroke internal combustion engine.

1878 – Edison Electric Light Company (U.S.) and American Electric and Illuminating (Canada) are founded.
- London street lighting goes electric.
- Jean de Marignac discovered Ytterbium.
- Solar powered steam engine exhibited in Paris.
- Edison files his first patent for the improved light bulb.

1879 – William Crookes discovers properties of cathode rays such as traveling in straight lines from the cathode; causing glass to fluoresce; imparting a negative charge to objects they strike; being deflected by electric fields and magnets to suggest a negative charge; causing pinwheels in their path to spin indicating they have mass.
- The first commercial power station opens in San Francisco which uses Charles Brush's generator and arc lights. The first commercial arc lighting system is installed in Cleveland, Ohio.
- Thomas Edison demonstrates his incandescent lamp in Menlo Park, New Jersey.
- Edwin Hall performs an experiment that had been suggested by Henry Rowland and discovers the Hall Effect, including its theoretical description by means of the Hall term in Ohm's law.
- Lars Fredrik Nilson discovered Scandium.
- Paul-Emile Lecoq de Boisbaudran discovered Samarium.
- Per Teodor Cleve discovered Holmium and Thulium.
- Albert Einstein is born.
- Otto Hahn is born.
- Lord Kelvin suggested to measure time using atomic vibration.
- Karl Benz invents the four-stroke engine.

1880 – The first power system isolated from Edison.
- Grand Rapids, Michigan: Brush arc light dynamo driven by a water turbine used to provide theater and storefront illumination.
- The American Society of Mechanical Engineers (ASME) was founded by Alexander Holley, Henry Rossiter, George Babcock and John Sweet.
- William Wheeler develops domestic lighting using light pipes running from an arc lamp.
- William Fisher invents the constant pressure pump governor in Marshaltown, Iowa.
- John Milne invented the modern seismograph.

1881 – In Niagra Falls, New York, Brush's dynamo is connected to a turbine in Quigley's flour mill...lighting city street lamps.
- Carl Gassner invented the first commercially successful dry cell battery.

- Godalming, Surrey, hosts the first public electricity supply with power generated by a waterwheel.
- Alexander Graham Bell invents the first crude metal detector.

1882 – Edison's Pearl Street Station becomes operational.
- The world's first hydroelectric station opens in Appleton, Wisconsin on the Fox River.
- Hans Geiger is born.
- Nikola Tesla identified the concept of the rotating magnetic field.
- Schuyler Skaats Wheeler invented the electric fan.

1883 – Edison introduces the "three wire" transmission system.
- Horace Lamb and Oliver Heaviside analyze the interaction of oscillating electromagnetic fields with conductors and discover the effect of skin depth.
- Pierre Curie discovers piezoelectricity.
- First electric railway built at Brighton by Magnus Volks.
- Nikola Tesla invented a practical two phase induction motor.
- Charles Fritts invents the first solar cell using selenium wafers.
- Thomas Edison first observed thermionic emission in a vacuum known as the Edison Effect.
- ASEA was created which in 1988 would merger with BBC to form ABB.

1884 – Charles Parsons builds a turbine that is connected to a dynamo that generated 7.5 kW of electricity and which would become the technology widespread in the power generation.
- The American Institute of Electrical Engineers (AIEE) is created in New York. This institute laid the foundations for all work on electrical standards done in the U.S.
- Daimler produces an internal combustion engine that runs on petrol.

1885 – Heinrich Hertz shows that Maxwell was correct and generates and detects electromagnetic waves.
- William Stanley an engineer for Westinghouse built the first commercial A.C. transformer.
- Hungarian engineers Zipernowsky, Blathy and Deri from the Ganz Company in Budapest created the efficient "ZBD" closed core model based on Gaulard and Gibbs design. Their patent application made the first use of the word "transformer".
- Carl Auer von Welsbach isolated and identified Praseodymium.
- Carl Auer von Welsbach discovered Neodymium.
- Electric Arc Welding invented by Nikolai Benardos and Stanislaus Olszewski.

1886 – Dmitri Mendeleyev devised the first logical arrangement of 63 known elements, which he called "The Periodic Table of Elements". Because of the precise organization of this table, he was able to predict the existence of many unknown elements. The element Mendelevium (101) is named in his honor.

- William Stanley develops the transformer and Alternating Current electric system. Frank Sprague builds first American transformer and demonstrates use of step up and step down transformers for long distance AC power transmission in Great Barrington, Massachusetts.

- The Westinghouse Electric Company is created.

- In this year it is reported that between Canada and the U.S. there are between 40 and 50 water-powered electric plants on line or under construction.

- Henri Moissan isolated Fluorine.

- Clemens Winkler discovered Germanium.

- Paul-Emile Lecoq de Boisbaudran and Jean de Marignac discovered Gadolinium.

- Paul-Emile Lecoq de Boisbaudran discovered Dysprosium.

- Charles M. Hall and Paul L.T. Heroult discovered an inexpensive method of obtaining pure aluminum.

1887 – The High Grove Station in San Bernadino, California is the first hydroelectric plant in the West.

- Svante Arrhenius deduces that in dilute solutions electrolytes are completely dissociated into positive and negative ions.

- Heinrich Hertz finds that ultraviolet light falling on the negative electrode in a spark gap facilitates conduction by the gas in the gap.

- Heinrich Hertz invented radar.

- Philip Diehl invents and patents the ceiling fan.

1888 – Nikola Tesla invents the rotating field AC alternator.

- Nikola Tesla patents the induction motor.

- The electric meter is invented by Oliver Shallenberger.

- Wind turbine generator created by Charles Brush.

1889 – In Oregon City, Oregon, the Willamette Falls Station becomes the first AC hydroelectric plant. It's power is sent as single phase power that is transmitted 13 miles to Portland at 4,000 volts and stepped down to 50 volts for distribution.

- J.J. Thomson shows that Canton's effect in which a red hot poker can neutralize the electrification of a small charged body is due to electron emission causing the air between the poker and the body to become conducting.

- Russian engineer Mikhail Dolivo-Dobrovolsky developed the first three phase transformer.

- Almon Strowger patents the direct dial telephone.

- Otto Blathy patent kilowatt-hour electric meter.

1890 – Turbine driven electric power generators began to appear.

- Dr. Fuller that rapid sand filtration improved when it was preceded by coagulation and sedimentation techniques in water purification.

1891 – The 60 Hz (cycle) AC system is developed in the U.S.

- La Cour windmill first used to generate electricity.

- Vladimir Shukhov and Sergei Gavrilov invented the process called thermal cracking which is used to produce diesel.

- (BBC) Brown Boveri and Cie is formed, in 1988 they would merge with ASEA to form ABB.

1892 – Through mergers of Thomson-Houston and Edison General Electric, the General Electric Company is formed.

- Henri-Louis Chatelier builds the first optical pyrometer.

- Rudolf Diesel develops the diesel engine.

1893 – Westinghouse demonstrates the "universal system" of generation and distribution at Chicago exposition.

- Austin, Texas the first dam designed specifically for hydroelectric power built across the Colorado River is completed.

1894 – Philip Lenard studies the penetration of cathode rays through matter.

- Lord Rayleigh and Sir William Ramsay identified Argon.

- The first steam turbine powered ship called the Turbinia is put into operation.

- Heinrich Hertz died at the early age of 37.

1895 – Wilhelm Roentgen discovers x-rays produced by bremsstrahlung in cathode ray tubes. The world immediately appreciates their medical potential. Within five years the British Army is using a mobile x-ray unit to locate bullets and shrapnel in wounded soldiers in the Sudan.

- Ernest Rutherford discovered alpha and beta radiation.

- Pierre Curie experimentally discovers Curie's law that relates some magnetic properties to change in temperature. The Curie point is about $1,060^0$F.

- William Ramsay discovered Helium.

- The first electric hand drill is invented by Fein.

- Three 4-ton 100 kW Parsons radial flow generators were installed in Cambridge Power Station and used to power the first electric street lighting scheme in the city.

- Photographic emulsions and electroscopes are primary instruments used when radiation is discovered.

1896 – Henri Becquerel discovers the emission rays from uranium, when the invisible rays from uranium ore darken a photographic plate. He found different types of decay and using an electric or magnetic field they could be split up into three beams. It was obvious from the direction of the electromagnetic forces that alpha rays carried a positive charge, beta rays carried a negative charge and gamma rays were neutral.

- J.J. Thomson discovers that materials through which X-rays pass are rendered conducting.

- Hydroelectric power generators designed by Tesla became operational at Niagara Falls, which shortly supply electricity to New York.

- Clarence Madison Dally a glass blower at Thomas Edison's Menlo lab is the first person known to have been killed by x-ray exposure.

- Harvey Hubbell patents the pull-chain electrical light socket.

1897 – J.J. Thomson discovers the electron.

- Guglielmo Marconi sends a radio message from the Isle of Wight to Poole (20 miles away). Later he sends a message across the Atlantic.

1898 – Pierre and Marie Curie discover the first radioactive elements: radium and polonium – the elements that constitute most of the radioactivity in uranium ore.

- Sir William Ramsay discovered Neon, Krypton and Xenon.

- Wien identified a positive particle later known as the proton.

- Marie Curie use the term "radioactivity" for the first time.

- Leo Szilard is born.

1899 – Ernest Rutherford distinguishes two kinds of rays from radium and its products. Some are stopped by a thin aluminum foil (20 micron). These rays he named alpha and beta.

- Waldmar Jungner invented the first nickel-cadmium rechargeable battery.

- The first electrically operated railway of Europe of the Burgodorf-Thun course was created (40km long, 750 volts and 40 Hz.)

- Andre Debierne discovered Actinium.

1900 - Villard discovers the radiation of highest penetrability: the gamma radiation.

- Max Planck of Germany develops quantum theory, which deals with matter and energy on the subatomic level. He used the idea of quanta (discrete units of energy) to explain hot glowing matter.

- Frederick Soddy observes spontaneous disintegration of radioactive elements into variants he calls "isotopes" or totally new elements. He also discovers "half-life" and makes initial calculations on energy released during decay.

- The voltage on the transmission system reaches its highest point yet at 60 kilovolts.

- Marie and Pierre Curie show that beta rays and cathode rays are identical.

- Friedrich Ernst Dorn discovered Radon.

1901 – Henri Alexandre Danlos and Eugene Bloch place radium in contact with a tuberculous skin lesion.

- Max Planck publishes Laws of Radiation.

- Thomas Edison invented the alkaline storage battery.

- Eugence-Anatole Demarcay isolated Europium.

- Peter Cooper Hewitt develops the fluorescent lamp.

- First report of death due to X-rays is published.

- Peter Cooper Hewitt invented the mercury vapor lamp.

1902 – Rutherford and Soddy publish theory of radioactive decay. The atoms of a radioactive element emit charged particles (alpha or beta) and in doing so change into atoms of a different element.

- A five megawatt turbine begins operation for the Fisk St. Station in Chicago.

- Giuseppe Mercalli developed the first earthquake intensity scale.

- Willis Carrier invents the air conditioning unit.

- Georges Claude invented the neon lamp.

- Egidius Elling built the first gas turbine that was able to produce more power than needed to run its own components; it put out 11 hp.

- Peter C. Hewitt develops the mercury vapor lamp.

- Calcium hypo chlorite and ferric chloride were added to the drinking water supply in Belgium resulting in both coagulation and disinfection.

- Rollins experimentally shows X-rays can kill higher life forms.

1903 – Rutherford and Soddy establish the theory of nuclear reactions.

- Alexander Graham Bell suggests placing sources containing radium in or near tumors.

- Rutherford found that the time behavior of the activity of the daughter nucleus of thorium is exponential and therefore the element transforms during the decay.

- Becquerel shares Nobel Prize for Physics with Pierre and Maire Curie for 1898 discovery of natural radioactivity.

- Nagaoka postulates a "Satumian" model of the atoms with flat rings of electrons revolving around a positively charged particle.

- The first successful gas turbine is created in France.

- The World's first all turbine station begins operation in Chicago.

- Shawinigan Water and Power installs the world's largest generator (5,000 watts) and the world's largest and highest voltage line at 136 Km and 50 Kilo volts to Montreal.

- The first electric vacuum cleaner is created along with the first electric washing machine.

- The Hagen – Rubens connections between the conductivity of metals and their optical properties are established.

- William Coolidge invented the ductile tungsten used in lightbulbs.

- Charles Curtis and William Emmet develop the steam turbine generator and the steam turbine, respectively.

- Water softening was invented as a technique for water desalination. Cations were removed from the water by exchanging them with sodium or other cations, in ion exchangers.

- The Wright Brothers first to fly a fixed wing aircraft (airplane).

1904 - J.J. Thomson pointed out that the number of electrons must be in the order of magnitude of the atomic weight. Based on this he developed his theory in 1904, which stated that the atom is a positively charged sphere and the electrons are embedded in this sphere. This theory was known as "plum pudding model".

- Rutherford discovers that alpha rays are heavy positively charged particles.

- Abegg discovered that inert gases had a stable electron configuration which lead to their chemical inactivity.

- John Ambrose Fleming invented the diode rectifier vacuum tube.

- Hendrik Lorentz gives his electron – collision theory of electrical conduction.

- Sandor Just and Ferenc Hanaman were granted a Hungarian patent for a tungsten filament lamp, improving the light bulb.

- Arthur Wehnelt patents a diode.

- Richard Mollier first published his enthalpy-entropy diagram for steam.

- Bertram Borden Boltwood when working with radium discovered that one element can decay into another.

- Colormetric dosimetry system devised by Saboroud and Noire.

- Rutherford coins the term "half-life".

- Harvey Hubbell invented the AC electrical plugs and sockets.

1905 – Albert Einstein develops the theory about the relationship of mass and energy…$E=mc^2$.

- Sault Ste. Marie, Michigan puts in service the first low head hydroelectric plant with direct connected vertical shaft turbines and generators.

- Arrhenius predicts global warming as a consequence of carbon dioxide emissions.

1906 – Hans Geiger developed an electrical device to "click" when hit with alpha particles.

- In Ilchester, Maryland the first fully submerged hydroelectric plant is built inside the Ambursen Dam.

- Reginald Fessenden proved that he could transmit music and voice over the radio and that it was not just for Morse code. On Christmas Eve he transmitted the first music and voice program that originated in Massachusetts and could be received as far away as Virginia. Fessenden is also the creator of Amplitude Modulation (AM) radio.

- Henri Moissan produced fluorine in its pure form.

- Nernst's Third Law of thermodynamics is published suggesting that changes in body at absolute zero incur no change in entropy.

- Ozone was first applied as a disinfectant in France.

- Pierre Curie is killed by a horse drawn carriage.

1907 – Lee De Forest invented the electric amplifier.

- Carl Auer von Welsbach and Georges Urbain discovered Lutetium.

- Sven Wingquist from Sweden was given the first patent for ball bearings.

- Robert Goddard concluded that nuclear propulsion would be essential for space travel if only a means could be found to release the energy of the atom.

- Mendeleev died

- William Thomson (Lord Kelvin) died.

- Boltwood conceives of radioactive dating

- Frederick G. Cottrell invented the electrostatic precipitator.

1908 – Rutherford receives Nobel prize for discovering that alpha particles are helium nuclei and beta particles are electrons.

- William Bailley of the Carnegie Steel Company invents a solar collector with copper coils and an insulated box.

- Hans Geiger and Ernest Rutherford invent the Geiger counter.

- Jean-Baptiste Perrin calculates the approximate size of the atom.

- Hugh and Stanley Rockwell invented the Rockwell hardness test.

1909 – The first pumped storage plant in Switzerland begins operation.

- Soren P.L. Sorenson suggested that the hydrogen ion concentration be used to categorize solutions in which he devised a scale. It is the pH scale that we continue to use today.

- Ernest Rutherford and Thomas Royds demonstrate that alpha particles are doubly ionized helium atoms.

- Liquid chlorine was developed for disinfection of water supplies.

1910 – Robert Milikan accurately measures the electric charge ($e=1.602x10^{-19}$ coulomb) and the mass ($m=9.11x10^{-28}$ gram) of an electron in his famous oil drop experiment.

- Irving Langmuir of General Electric experiments with gas filled lamps using nitrogen to reduce evaporation of the tungsten filament, thus raising the temperature of the filament and producing more light. To reduce conduction of heat by the gas, he makes the filament smaller by coiling the tungsten.

- J.J. Thomson while working with neon confirms the existence of isotopes.

- Jesuit Father Theodor Wulf measures radiation at ground level and at top of the Eiffel Tower. Radiation increases at higher elevations.

1911 – George von Hevesy conceives the idea of using radioactive tracers. This idea is later applied to, among other things, medical diagnosis.

- Rutherford made a decisive discovery. He investigated the inside of the atom by bombarding it with alpha-particles stemming from radioactive decay, discovering the atomic nucleus. Rutherford develops the atom model which is similar to a miniature solar system. The attractive force here is of course not gravity but Coulomb force.

- Marie Curie receives a second Nobel Prize, this time for Chemistry, for the isolation of radium and polonium and for her investigation of their chemical properties.

- R.D. Johnson invents the differential surge tank and the Johnson hydrostatic penstock valve.

- Heike Kamerlingh Onnes discovers superconductivity when he cooled down mercury to 4.2^0K and discovered that the resistance in the mercury disappeared.

- The first Solvay Congress of physicists in Brussels on radioactivity and quantum theory, with Einstein, Planck, Lorenz, Poincare, Marie Curie, Langevin and Rutherford attending.

1912 – The Institute of Radio Engineers (IRE) is created.

- Harry Brearley invents stainless steel.

- Combustion Engineering was created through the merger of the Grieve Grate Company and the American Stoker Company, headquartered in New York City.

- Max von Laue studied X-ray diffraction patterns and found that crystals are composed of regular repeating units of atoms.

- April 14, RMS Titanic sinks after hitting an ice berg.

1913 – Niels Bohr publishes theory of atomic structure combining nuclear theory with quantum theory. Bohr proposed that electrons traveled in circular orbits and that only certain orbits were allowed.

- Frederick Proescher publishes the first study on the intravenous injection of radium for therapy of various diseases.

- The first radiation detector is invented.

- The electric refrigerator is invented.

- Southern California Edison puts into service a 150,000 volt line to bring electricity to Los Angeles. Hydroelectric power is generated along the 233 mile long aqueduct that brings water from Owens Valley in the eastern Sierra.

- Kasimir Fajans and O.H. Gohring discovered Protactinium.

- Frederick Soddy coins the term "isotope".

1914 – H.G. Moseley using x-ray tubes, determines charges on the nuclei of most atoms.

- The periodic table was reorganized to be based upon atomic number instead of atomic mass.

- H.G. Wells publishes a novel, The World Set Free, in which an atomic war in 1956 destroys the major cities of the world.

- Ernest Rutherford suggests that positively charged atomic nucleus contains protons.

- James Franck and Gustav Hertz observe atomic excitation.

- Niels Bohr discovered that electrons travel around the nucleus in fixed energy levels.

- Ernest Marsden, an assistant of Rutherford reports an odd result when he bombards nitrogen gas with alpha particles; something is thrown back with a much greater velocity. (first report of nuclei fissioning).

1915 – Albert Einstein publishes general theory of relativity.

- British Roentgen Society proposes standards for radiation protection workers; including shielding, restricted work hours, medical exams, no limits because of lack of units for dose or dosimeters, voluntary controls. This is believed to be the first organized step toward radiation protection.

1916 – Idea of covalent bonding put forward by Lewis.

1917 – In major Russian town, particularly St Petersburg a boost to scientific research where over 10 physics institutes were established.

- The Hydracone draft tube is patented by W.M. White.

- The first long distance high voltage transmission line is established by American Electric & Power (AE & P). The line originates from the first major steam plant to be built at the mouth of a coal mine the Windsor Plant in West Virginia on the Ohio River and travels 55 miles away to Canton, Ohio.

- Edwin Howard Armstrong invents the superheterodyne circuit.

1918 – General Electric started their gas turbine division.

- The rad was first proposed that quantity of X rays which when absorbed will cause the destruction of the biological cells.

- Mass spectrometer was invented by Arthur Dempster.

1919 – Rutherford discovered that alpha particles ionize hydrogen gas. When he bombarded nitrogen gas with alpha particles, he found that it knocked off a proton and created oxygen. 4He + 14N -> 17O + 1H:

- Francis Aston discovers existence of isotopes through the use of a mass spectrograph.

- Elmer Samuel Imes accurately measures the inter - atomic distances in molecules.

- Arc welder invented.

1920 - The name proton was suggested by Rutherford.

- The first U.S. station to only burn pulverized coal begins operation.

- The Federal Power Commission (FPC) is created.

- Westinghouse is granted the first U.S. broadcasting license for its station KDKA and the first scheduled public broadcast takes place.

- American Roentgen Ray Society (ARRS) establishes standing committee for radiation protection.

1921 – Rutherford and Chadwick achieve transmutation of all elements except carbon, oxygen, lithium and beryllium.

- The first robot is developed and built.

1922 – Neils Bohr receives the Nobel Prize in physics for his theory of electron orbits. He developed an explanation of atomic structure that underlies regularities of the periodic table of elements. His atomic model had atoms built up of successive orbital shells of electrons.

- Connecticut Valley Power Exchange (CONVEX) starts, pioneering interconnections between utilities.

- G. Pfahler recommends personnel monitoring with film.

- First radio broadcast took place.

1923 – In supporting Einstein's theory, Louis de Broglie discovers that electrons have a dual nature that is similar to both particles and waves.

- The first U.S. 220 kv line goes into operation.

- John Logie Biard invents the television.

- Dirk Coster and George Karl von Hevesy discovered Hafnium.

- William Conrad Roentgen died from carcinoma of the intestine.

- The decibel was created to replace the "TU" transmission unit and to honor Alexander Graham Bell. The unit bel proved to be to inconveniently large so the decibel was created.

- A. Mutscheller puts forth first "tolerance dose" (0.2R/day).

- Dr. Robert Andrews Millikan – "There is no likelihood man can ever tap the power of the atom…Nature has introduced a few foolproof devices into the great majority of elements that constitute the bulk of the world, and they have no energy to give up in the process of disintegration".

1924 – Georg de Hevesy, J.A. Christiansen and Sven Lomholt perform the first radiotracer (lead-210 and bismuth-210) studies in animals.

- The first reheat generating unit, the Philo Plant, Ohio goes into operation.

1925 – The first cloud chamber photographs of nuclear reactions.

- The Pauli exclusion-principle was created by Wolfgang Pauli which states that two particles of a certain class can never be in the same energy state.

- Otto Berg and Wilhelm Noddack discovered Rhenium.

- John Biard transmits the first television signal.

1926 – G.N. Lewis proposes the name "photon" for a light quantum.

- The U.K. National Grid introduced following Electricity Supply Act.

- Lise Meitner physicist became the first female full time professor in Germany.

- Schroedinger introduces wave mechanics.

- Edith Quimby devises film badge dosimeter with energy compensating filters.

1927 – Herman Blumgart, a Boston Physician, first uses radioactive tracers to diagnose heart disease.

- Heisenberg described atoms by means of a formula connected to the frequencies of spectral lines. He also proposed the Principle of Indeterminancy – you cannot know both the position and velocity of a particle. (the uncertainty principle)

- In the U.S. electric cables designed to carry over 130,000 volts are laid.

- Heitler and London prove the electrical nature of chemical bonding.

- The first zeolite material column was used to remove interfering calcium and magnesium ion from solution to determine the sulfate content of water (ion exchange).

- Dutch Board of Health recommends tolerance does equivalent to 15R/year.

- H. Muller shows genetic effects of radiation.

1928 – Paul Dirac combines quantum mechanics and special relativity to describe the electron.

- Construction of the Boulder Dam begins.

- The Federal Trade Commission begins investigation of holding companies.

- Paul Dirac proposes the existence of antiparticles.

- Philip F. Labre invented the three-prong plug.

1929 – Nuclear fusion as a source of stellar radiation is first investigated in a theoretical paper by Atkinson & Houtermans.

- Ernest Lawrence conceives idea for the first cyclotron (atom smasher).

- John Cockcroft and Walton develop a high voltage apparatus for accelerating protons and are awarded the Nobel Prize in 1931 for their studies in atomic transmutation. They bombarded lithium with protons to produce alpha particles.

- The first triple compound generating unit, Philo Plant, Ohio goes into operation.

- Dr. Robert A. Millikan, "The energy available through the disintegration of radioactive or any other atoms may perhaps be sufficient to keep the corner peanut and popcorn man going in our large towns for a long time, but that is all".

1930 - Schrödinger views electrons as continuous clouds and introduces "wave mechanics" as a mathematical model of the atom. Schrödinger and Heisenburg developed probability functions to determine the regions or clouds in which electrons would most likely be found.

- Paul Dirac first proposed anti-particles:

- Anderson will discover the anti-electron (positron) in 1932 and Segre/Chamberlain will detect the anti-proton in 1955.

- Vannevar Bush built the first general purpose analog computer.

- Around the 1930's Edwin Armstrong improved the sound quality of radio transmissions in developing Frequency Modulation (FM) radio. It would take numerous years before FM would overtake the AM radio as king of the air waves.

- Paul Dirac introduces electron hole theory.

- The International Radium Standard Commission made the "curie" the equilibrium quantity of any decay product of radium. However, workers began to apply the definition of a curie to any radioactive substance. In 1948, The Committee on Standards and Units of Radioactivity of the National Research Council (U.S.) recommended that this unofficial definition be made the official definition of the curie.

1930 thru 1932 – Walther Bothe and Herbert Becker in Germany, Irene and Frederic Joliot-Curie in France, and James Chadwick in the United Kingdom conduct a series of experiments which culminate in Chadwick's discovery of the neutron a product when beryllium is bombarded by alpha particles. 9Be + 4He -> 12C + neutron

- Heisenberg and Tamm found differences in mass and coined the term mass defect.

- The neutron apparently came from the Latin root for neutral and the Greek ending – on to imitate the electron and proton.

1931 – Albert Einstein urges all scientists to refuse military work.

- Harold C. Urey of the United States and associates discover deuterium (heavy hydrogen) which is present (0.014%) in all natural hydrogen compounds including water.

- Frank Whittle patented the design for a gas turbine for jet propulsion.

- Wolfgang Pauli postulates the existence of a subatomic particle Enrico Fermi dubs "neutrino", a mass less uncharged particle that carries energy and momentum.

1932 – Ernest O. Lawrence and M. Stanley Livingston publish the first article on "the production of high speed light ions without the use of high voltages." It is a milestone in the production of usable quantities of radionuclides.

- Carl Anderson will discover the anti-electron and gave it it's name (positron) a positively charged electron.

- Walter Heisenberg presents the proton-neutron model of the nucleus and uses it to explain isotopes.

- Ernst Ruska builds the first electron microscope.

- Mark Oliphant discovered helium 3 and tritium and that heavy hydrogen nuclei could be made to react with each other.

- Charlie Lauritsen invents the pocket dosimeter.

1933 – Hideki Yukawa combines relativity and quantum theory to describe nuclear interactions by an exchange of new particles (mesons called "pions") between protons and neutrons. From the size of the nucleus, Yukawa concludes that the mass of the conjectured particles (mesons) is about 200 electron masses. This is the beginning of the meson theory of nuclear forces.

- The U.S. government established the Tennessee Valley Authority (TVA) which introduced hydroelectric plants to the South's troubled Tennessee River Valley.

- Gilbert Lewis obtains heavy water by electrolysis.

1934 – Irene and Frederic Joliot Curie discover artificial radioactivity. Irene is Marie Curie's oldest daughter.

- Fermi bombards heavier elements with neutrons in order to produce transuranic elements and the first nuclear fission without realizing it.

    238U + neutron -> 239U -> beta decay 239Np -> beta decay 239Pu

- The coiled-coil electric light bulb is invented, increasing the amount of radiated light.

- Leo Szilard realizes that nuclear chain reactions may be possible.

- Leo Szilard is legally recognized as the inventor of the atomic bomb.

- Leo Szilard filed the first patent application for the method of producing a nuclear chain reaction aka nuclear explosion

- Enrico Fermi formulates his theory of beta decay.

- Ernest Lawrence and Stan Livingston invent the cyclotron.

- Beckman invents the pH meter.

- George de Hevesy and Hoffer use heavy water in a biological tracer experiment to estimate the rate of turnover of water in the human body.

- Norsk Hydro, Norway built the first commercial heavy water plant at Vemork, Tinn.

- Fermi discovers slow neutrons.

- Marie Curie dies of Leukemia on July 4th.

- Evans at MIT starts whole body counting.

- Pavel Alekseyevich Cherenkov discovers Cherenkov radiation.

1935 – The first night baseball game in major leagues takes place.

- The Public Utility Holding Company Act is established.

- The Federal Power Act is established.

- Securities and Exchange Commission and the Bonneville Power Administration is established.

- The clothes dryer was invented.

- Charles Richter invents a logarithmic scale to measure the intensity of earthquakes.

- Hoover Dam begins operation with its first of 17 generators.

- Sir Henry Doulton invented the ceramic cartridge for removing bacteria from water.

- Electric pH meter is invented by Glen Joseph.

- Arthur Hardy invents the spectrophotometer.

1936 – John H. Lawrence, the brother of Ernest, makes the first clinical therapeutic application of an artificial radionuclide when he uses phosphorus-32 to treat leukemia.

- The principle of boron neutron capture therapy (BNCT) was developed as early as in 1936 by an American scientist Locher.

- The highest steam temperature reaches 900^0F vs 600^0F in the early 1920's.

- A 287 Kilovolt line runs 266 miles to Boulder (Hoover) Dam.

- The Rural Electrification Act is established.

- The Osram Company presents at the world exhibition in Paris the first fluorescent lamp.

- The first American TV network (NBC) National Broadcasting Corporation is created.

- Eugene Wigner develops the theory of neutron absorption by atomic nuclei.

- Louis Neel successfully explained anti-ferromagnetism.

1937 – John Livingood, Fred Fairbrother and Glenn Seaborg discover iron-59.

- A particle of 200 electron masses was discovered in cosmic rays. While at first physicists thought it was Yukawa's pion, but was later to be discovered as a muon.

- The first million lb/hr high pressure boiler (1,250 psi) the Logan Plant, West Virginia goes into service.

- The first hydrogen cooled turbine begins operation in the U.S.

- Emilio Segre and Carlo Perrier discovered Technetium.

- Frank Whittle and Hans von Ohain built the first successful jet engine.

- Niels Christensen invented the O-ring.

- Ernest Rutherford dies.

- Lauritsen electroscope used to measure dose.

- May 6, airship Hindenburg explodes into flames and crashes to the ground in Manchester Township, New Jersey.

1938 – John Livingood and Glenn Seaborg discover iodine-131 and cobalt-60 all isotopes currently used in nuclear medicine.

- Three scientists, Otto Hahn and Fritz Strassman and Lise Meitner demonstrate nuclear fission. Lise Meitner being both a women and a Jew fleeing Nazi persecution did not have her name on their seminal paper.

- Hans Bethe and C. Critchfield make the first complete analysis of the proton-proton reaction. Later that year, Bethe's further research led him to describe the carbon-nitrogen-oxygen (CNO) cycle.

- E. Stuckelberg observes that protons and neutrons do not decay into any combination of electrons, neutrinos, muons, or their antiparticles. The stability of the proton cannot be explained in terms of energy or charge conservation; he proposes that heavy particles are independently conserved.

- Konrad Zuse built the first binary computer.

- Lise Meitner conducts experiments verifying that heavy elements capture neutrons and form unstable products which undergo fission. This process ejects more neutrons continuing the fission chain reaction.

1939 – Halban, Frederic Joliot Curie and Kowarski demonstrate that fission of uranium can cause a chain reaction. They take the first patent on the production of nuclear energy.

- Emilio Segre and Glenn Seaborg discover technetium-99m an isotope currently used in nuclear medicine.

- Albert Einstein writes a letter to President Roosevelt which discusses German nuclear research and possibility of building an atomic bomb. After bombing Hiroshima, Einstein states "I could burn my fingers that I wrote that letter to Roosevelt".

- NBC held its first broadcast at the World's fair.

- Marguerite Perey discovered Francium.

- Lise Meitner and Otto Frisch announce the theory of nuclear fission.

- Hans Bethe recognizes that the fusion of hydrogen nuclei to form deuterium releases energy. He suggests that much of the energy output of the Sun and other stars results from energy-releasing fusion reactions in which four hydrogen nuclei unite and form on helium nucleus.

- William Arnold coins the term "fission".

- Otto Frisch detects fission fragments in an ionization chamber.

- Szilard writes to Fermi describing the idea of using a uranium lattice in graphite to create a chain reaction. This is the first proposal of the graphite moderated reactor concept.

- Robert Oppenheimer hearing the discovery of fission immediately grasps the possibility of atomic weapons.

- Fermi immediately saw the possibility of emission of secondary neutrons and a chain reaction.

- Hitler invades Poland starting WW II in Europe

1940 – The Rockefeller Foundation funds the first cyclotron dedicated for biomedical radioisotope production at Washington University in St.Louis.

- Peierls and Frisch had initially predicted that almost every collision of a neutron with a U-235 atom would result in fission and that both slow and fast neutrons would be equally effective. Later that year it was found that slower neutrons were much more effective at causing fission than fast neutrons.

- German scientists fail to observe neutron multiplication in the reactor in Hamburg.

- Edwin M. McMillan and Philip H. Abelson first produce Neptunium.

- Dale R. Corsun, K.R. Mckenzie and Emilio Segre create Astatine.

- The University of California begins building a giant cyclotron under the direction of Ernest Lawrence.

- Edwin McMillan bombards uranium with neutrons and produces the first "transuranic" element, neptunium (Atomic Number 93).

- First contract is signed with Columbia University to develop bomb material.

- Photomultiplier tube is developed by Larson and Salinger which makes scintillation radiation detectors much more useable.

- Joseph John Thomson dies.

- Brazilian President Getulio Vargas initiates a national nuclear program. He allows the U.S. to mine Brazil's large uranium reserves in return for American nuclear technology.

1941 – One of the most uncertain pieces of information was confirmed, the fission cross section of U-235.

- C. Moller and Abraham Pais introduce the term "nucleon" as a generic term for protons and neutrons.

- The first very high pressure (2,300 psi) natural circulation generating unit the Twin Branch Plant, Indiana goes into operation.

- Glenn T. Seaborg discovered Plutonium.

- Enrico Fermi invented the neutronic reactor.

- Pearl Harbor is bombed by the Japanese.

- The Office of Scientific Research and Development is created under the direction of Vannevar Bush to develop atomic energy.

- The day before the bombing of Pearl Harbor, President Roosevelt authorizes the Manhattan Engineering District to begin working on a secret U.S. project to build an atomic bomb, later to be called the Manhattan Project and it is put under the direction of the Office of Scientific Research and Development.

1942 – The Manhattan Project is formed to secretly build the atomic bomb before the Nazis.

- Construction begins on Chicago Pile 1 (CP-1).

- Fermi demonstrates the first self-sustaining nuclear chain reaction in a lab using a graphite pile at the University of Chicago (Chicago Pile 1). Leo Szilard and Fermi originate the method of arranging graphite and uranium which makes the reaction possible. Soon afterwards many top secret nuclear research/production facilities are built for the Manhattan Project.

- Los Alamos becomes site for an atomic laboratory and Robert Oppenheimer is selected as director.

- From Fermi's demonstration of the chain reaction comes a term "SCRAM" meaning "Safety Control Rod Axe Man". Meaning that during the demonstration a man was standing with an axe that would cut the rope causing the control rod to drop into the reactor, shutting down the core, this term is still used today.

- John Atanasoff and Clifford Berry built the first electronic digital computer.

- McMillan and Glenn Seaborg synthesize plutonium.

1943 - Richard Courant developed a technique using a form of Finite Element Analysis (FEA) to find approximate solutions to vibrational systems.

- The world's first operational nuclear reactor is activated at Oak Ridge, Tennessee. Oak Ridge X-10 Clinton reactor, first to generate electricity with a model steam engine. It is the first true plutonium production reactor.

- Thirty scientists assemble at Los Alamos, New Mexico for an introductory series of lectures on the theory and practicalities of designing and building an atomic bomb using uranium-235 or plutonium-239.

- The Military Policy Committee of the Manhattan Project develops the idea to use the atomic bomb on Japan rather than on Germany. The committee is chaired by General Leslie Groves.

1944 - The purpose of Oak Ridge was to produce the Uranium 235 for the bomb. Both gaseous diffusion and electromagnetic isotope separation plants were built at Oak Ridge. The plant was finished and in production.

- The first reactor with higher power was the "Pile B" in Hanford, WA. It was capable of producing plutonium. The structure of Pile B was somewhat similar to that of the Chicago Pile 1 (Fermi's reactor) but in this case the system had to be cooled. The reactor was cooled by water flowing through the core at a flow rate of 5 m^3/s. The moderator - the material that slows neutrons down - was graphite. The mass of the graphite was 1200 tons, while the fuel was 200 tons of metal uranium. The power of Pile B was about 250 MW and it produced 6 kg of plutonium in a month.

- Fermi and Szilard applied for a patent on reactors, was delayed 10 years because of wartime secrecy.

- Glenn T. Seaborg and other scientists created Americium.

- Glenn T. Seaborg, Ralph A. James and Albert Ghiorso create Curium.

- A second uranium reactor is built at Clinton, Tennessee for manufacturing plutonium for an atomic bomb.

- Barely sixteen months after the feasibility of achieving a self-sustaining nuclear chain reaction was established by Enrico Fermi in Chicago -- a tightly held secret known only to a very limited number of individuals in the U.S., UK and Canada -- Homi Jehangir Bhabha initiates efforts to start nuclear research programms in India.

- Joseph Rotblat , Polish refugee and physicist, resigns from the Manhattan Project since he believed that Nazi Germany would not succeed in developing an atomic weapon. He later explains, "I felt there was no need to make a bomb. The only reason I started in 1939 was to stop Hitler using it against us." Rotblat was thereafter barred from entering the United States for 20 years. In 1957 he helped start the Pugwash Conferences on Science and World Affairs, of which he was the first president. In 1995 Rotblat and Pugwash jointly were awarded The Nobel Peace Prize for their work towards the abolition of nuclear weapons.

- Martin Deutsch and Robley D. Evans develop the electron spectrometer.

1945 – Trinity test of the first atomic explosive device at Alamogordo, New Mexico.

- The United States drops atomic bombs on Hiroshima and Nagasaki. Japan surrenders ending World War II days later.

- Edwin McMillan and Vladimir Veksler devise the synchrotron.

- Harry Daghlian is the first American to die of acute radiation sickness. He accidentally created a supercritical mass when he dropped a tungsten carbide brick onto a plutonium core. He quickly removed the piece, but was fatally irradiated in the incident.

- The ZEEP reactor (1st outside of the U.S.) achieves first self-sustaining fission chain reaction in Canada near Chalk River.

1946 – Samuel M. Seidlin, Leo D. Marinelli and Eleanor Oshry treat a patient with thyroid cancer with iodine-131, an "atomic cocktail".

- President Truman signs the Atomic Energy Act of 1946, putting the fledgling nuclear energy industry under civilian control and creating the powerful Joint Congressional Committee on Atomic Energy.

- U.S. performs an underwater detonation of an atomic bomb at Bikini Atoll.

- The first Soviet nuclear reactor started operation on 25 September 1946 near Moscow.

- I. Rabi introduces the term "lepton" to describe objects that do not interact too strongly.

- Purcell and Block develop nuclear magnetic resonance.

- United Nations Atomic Energy Commission was created.

- Percy Spencer invents the microwave oven.

1947 – Benedict Cassen uses radioiodine to determine whether a thyroid nodule accumulates iodine, helping to differentiate benign from malignant nodules.

- The U.S. Atomic Energy Commission first investigates the possibility of peaceful uses of atomic energy, issuing a report the following year.

- It was determined that a meson that does interact strongly is found in cosmic rays, and is determined to be the pion.

- Physicists develop procedures to calculate electromagnetic properties of electrons, positrons, and photons. Introduction of Feynman diagrams.

- The transistor is invented by John Bardeen, Walter Brattain and William Shockley at Bell Laboratories.

- The first extra-high-voltage (EHV) line testing up to 500,000 volts occurred at the Tidd Project in Ohio.

- J.A. Mirinsky, L.E. Glendenin and C.D. Coryell discovered Promethium.
  Willard Libby introduces carbon-14 dating.

- Hyman Rickover is sent to Oak Ridge Tennessee to work on the possibility of nuclear ship propulsion, this leading to the USS Natutilus in 1955.

1948 – Nuclear chain reaction achieved in ZOE, the first French atomic pile in Fort de Chatillon.

- Abbott Laboratories begins distribution of radioisotopes.

- The U.S. performs atomic tests at Eniwetok Atoll. The Berkeley synchro-cyclotron produces the first artificial pions.

- John Bardeen and Walter Houser discovered the transistor effect which led to the miniaturization of electrical devices.

1949 – The National Bureau of Standards (NIST) announces the world's first atomic clock using the ammonia molecule as the source of vibrations from Isidor Rabi a physic professor at Columbia University.

- Discovery of K^+ via its decay.

- Fermi and Yang suggest that a pion is a composite structure of a nucleon and an anti-nucleon.

- The first use of high pressure (2.000 psi) and high temperature ($1,050^0$F) combination unit, the Twin Branch Plant, Indiana goes into service.

- Glenn T. Seaborg, Stanley Thompson and Albert Ghiorso created Berkelium.

- Radioactive carbon (C-14) dating is developed by Willard Libby at the University of Chicago.

1950 – K.R. Crispell and John P. Storaasli use iodine-131 labeled human serum albumin (RISA) for imaging the blood pool within the heart.

- President Truman orders the Atomic Energy Commission to develop the hydrogen bomb and initiates above ground test in Nevada.

- Westinghouse was awarded a huge contract to begin the development and ground was broken in the spring of 1950 at the Bettis Atomic Power Laboratory outside of Pittsburgh, Pennsylvania. Not to be left out, General Electric was later awarded a similar contract at the Knolls Atomic Power Laboratory (KAPL) near Schenectady, New York. Given that this project was first of kind, the Westinghouse and GE teams would face many technological challenges.

- The neutral pion is discovered.

- The Philip Sporn Plant, West Virginia is the first to have a heat rate below 10,000 Btu/kwh.

- Stanley Thompson, Kenneth Street Jr., Albert Ghiorso and Glenn T. Seaborg created Californium.

- Julius and Ethel Rosenberg indicted in atom spy case.

- The tokamak reactor was created by Andrei Sakharov and Igor Tamm.

1951 – First usable electricity created from nuclear energy first produced in the US (100 kilowatts) in Experimental Breeder Reactor 1 (EBR1) at the National Reactor Station, later called the Idaho National Engineering Laboratory.

- The US Food and Drug Administration (FDA) approve sodium iodine-131 for use with thyroid patients. It is the first FDA approved radiopharmaceutical.

- Using a field ion microscope, scientists are able to observe a single atom for the first time.

- Color television pictures are first transmitted.

1952 – Operations begin at the Savannah River Plant in Aiken, South Carolina with the startup of the heavy water plant.

- First British atomic detonation, Monte Bello Islands, Australia.

- The U.S. explodes the first fusion device, a 65 ton behemoth code named Mike, was exploded on Eniwetok in the Marshall Islands.

- The NIST completes the first accurate measurement of the frequency of the cesium clock resonance which is named NBS-1.

- Donald Glaser invents the bubble chamber.

- The Brookhaven Cosmotron, a 1.3 GeV accelerator begins operation.

- The Atomic Energy of Canada Limited (AECL) was created by the Canadian government to promote peaceful use of nuclear energy.

- Chalk River in Ontario, Canada suffered a major hydrogen/oxygen explosion destroying the reactor core due to the loss of control over the reactor's power level.

- Albert Ghiorso and other scientists created Einsteinium and Fermium.

1953 – In his Atoms for Peace speech, President Eisenhower proposes joint international cooperation to develop peaceful applications of nuclear energy.

- Gordon Brownell and H.H. Sweet build a positron detector based on the detection of annihilation photons by means of coincidence counting.

- EBR-1 achieved the first demonstration of the breeding principle in a reactor.

- First Soviet fusion device exploded on a tower in Siberia.

- The Nautilus first starts its nuclear reactors the Mark 1 prototype naval reactor started up for submarine applications in Idaho.

- The beginning of a "particle explosion" – a true proliferation of particles.

- The first 345 Kilo volt transmission line is created.

- The first nuclear power station is ordered.

- Microwave oven first produced by the Raytheon Corporation.

- Vanadium-silicon superconductor is discovered.

- The American Electric Power Company (AEP) commissions a 345 kV system that interconnects the grids of seven states. The system reduces the cost of transmission by sending power where and when it is needed rather than allowing all plants to work at less than full capacity.

- The Atomic Industrial Forum (AIF) was created was an American industrial policy organization for the commercial development of nuclear energy.
- CP-1 experimental reactor finally decommissioned
- Joseph Stalin dies
- Russia announces that it possesses the hydrogen bomb…beginning the Cold War arms race.

1954 – The first nuclear submarine, USS Nautilus, is launched.

- The world's first nuclear power plant begins operation in the Russian city of Obninsk at the Institute of Physics and Power Engineering (5 MWe's).

- David Kuhl invents a photo-recording system for radionuclide scanning. This development moves nuclear medicine further in the direction of radiology.

- President Dwight Eisenhower signs Atomic Energy Act, opening the door to private use of nuclear energy giving the civilian nuclear energy program further access to nuclear technology.

- Edward Salpeter describes the triple-alpha process. The "BRAVO" H-bomb test.

- The first high voltage DC (HVDC) line (20 MW / 1900 Kilovolts, 96 Km).

- Gerald Pearson, Calvin Fuller, and Daryl Chapin invented the first solar battery.

- Bell Laboratories develop silicon solar cells, with a 6% efficiency.

- Enrico Fermi died.

- The Society of Nuclear Medicine holds its first meeting.

- An experimental sodium-cooled reactor utilized aboard the USS Seawolf, the U.S.'s second nuclear submarine, was scuttled in 9,000 ft of water off the coast of Delaware/Maryland. The reactor was plagued by persistent leaks in its steam system and was later replaced with a more conventional model.

1955 – Arco, Idaho becomes the first U.S. town to be powered by nuclear energy, an experimental boiling water reactor BORAX III.

- Rex Huff measures the cardiac output in man using iodine-131 human serum albumin.

- USS Seawolf enters service as the second US nuclear submarine.

- The Atomic Energy Commission announces the beginning of a cooperative program between government and industry to develop nuclear power plants.

- The first international conference on the peaceful uses of nuclear energy is held in Geneva, Switzerland, sponsored by the United Nations.

- The Soviet Union test first fusion device; development was led by Andrei Sakharov.

- In Russia the BR-1 (bystry reaktor) (fast reactor) fast neutron reactor began operating.

- Albert Ghiorso and other scientists created Mendelevium.

- Owen Chamberlain, Emilio Segre, Clyde Wiegand and Thomas Ypsilantis discover the antiproton.

- General Electric develops the boiling water reactor.

- Cowan and Reines discover the neutrino.

- Albert Einstein dies.

- The Health Physics Society is formed.

1956 – G1, first French nuclear power plant begins operation in Marcoule (5MWe) a gas-graphite design similar to Magnox.

- British developed a series of reactors designed to use natural uranium metal, moderated by graphite, and gas-cooled. The first of these 50 MWe Magnox type, Calder Hall 1 started up in 1956 and ran until 2003. The first reactor to provide electricity to a national grid opens in Calder Hall, England.

- Experimental nuclear powered aircraft engine (X-39) were created and tested successfully…though the engine was never put into service because the lack of…radiation protection for the aircraft crew.

- Frederick Reines and Clyde Cowan detect anti-neutrinos.

- Chien-Shiung Wu discovered test method for beta decay.

- National Academy of Sciences and ICRP recommend lower basic permissible dose for radiation workers to 5 rad/year.

- Irene Joliot-Curie dies

1957 – The United States sets off first underground nuclear test in a mountain tunnel in the remote desert 100 miles from Las Vegas.

- Radiation is released when the graphite core of the Windscale nuclear reactor in England catches fire. Since then no air cooled reactors have been built.

- The first power from a civilian nuclear unit is generated by the Sodium Reactor Experiement at Santa Susana, California and provided power until 1966.

- President Eisenhower signs into law the Price-Anderson Act, legislation to protect the public, utilities and contractors financially in the event of an accident at a nuclear power plant.

- The first U.S. large scale nuclear power plant begins operating in Shippingport, Pennsylvania. Twenty one days later the unit reaches full power, generating 60 MWs of electricity and operates until 1982.

- Electron beam used for fabrication of safer nuclear fuel rods.

- First British H-bomb exploded at Christmas Island.

- A G.E.-Vallecitos, (5 MW) boiling water reactor near Pleasanton, California connects to the electrical grid and becomes the first privately funded plant to supply power in megawatt amounts to the electric utility grid.

- The first use of supercritical pressure (4,500 psi) and super high temperature steam $(1,150^0F)$ in the Philo Plant, Ohio goes into operation.

- The first use of a double reheat steam system in the Philo Plant, Ohio goes into operation.

- Windscale, Cumbria, United Kingdom nuclear reactor caught fire and burned out of control while trying to anneal the graphite moderator. This accident released radiation out onto the countryside.

- Superconductivity was first observed in 1911 but it was not explained until 1957 when John Bardeen, Leon Cooper and Robert Schrieffer present a theory known as BCS.

- The International Atomic Energy Agency (IAEA) is created.

- A serious accident occurred near the town of Kyshtym in the Urals. A defect in a nuclear liquid waste storage reservoir overheated causing an explosion of nitrate-acetate salts. The explosion caused about 7.4×10^{17}Bq of radioactive products to be ejected. Hundreds died from radiation sickness.

- Russia launches first artificial satellite Sputnik.

1958 – Hal Anger invents the "scintillation camera", an imaging device that made it possible to conduct dynamic studies.

- USSR approves potato and grain irradiation.

- The second Atoms for Peace Conference in Geneva is held, where fusion scientists around the World shared fusion research for the first time, laying down the foundation for future collaboration.

- The commercial cesium clocks become available, costing approx. $20,000 each.

- Robert Noyce and Jack Kilby successfully developed the integrated circuit.

- Albert Ghiorso and other scientists identified Nobelium.

- Australia's first nuclear reactor goes into operation at Lucas Heights, Sydney.

- American, British and Soviet scientists began to share previously classified fusion research as their countries declassified controlled fusion work as a part of the Atoms for Peace conference in Geneva.

1959 – The Dresden 1 (250 MWe) designed by G.E. nuclear power station in Illinois achieves a self-sustaining nuclear reaction. It's the first U.S. nuclear power plant built entirely without government funding.

- First fast breeder reactor begins operation in the Obninsk Scientific Center, Russia (12 MWe).

- The first power nuclear merchant vessel, Savannah, at Camden, New Jersey is launched and operates for 12 years traveling around the World.

- There were two commercial reactors operating in the U.S.

- Both U.S. and Russia launched their first nuclear powered surface vessels.

- The first reactor in Hungary was built at the KFKI[*].

- The Budapest Reactor was a 2 MW thermal power research reactor and served physics research purposes. The reactor has been reconstructed and now the maximum power is 10 MW.

- The research topics are related to isotope production, solid state physics, radiochemistry, nuclear physics etc.

- After a six-year long revolution President Fulgencio Batista flees Cuba and Fidel Castro assumes power after proclaiming victory in Santiago. This begins the Cuban Missile Crisis until 1963.

- UCLA became the first to demonstrate the reverse osmosis (RO) process.

1960 – Louis G. Stang Jr. and Powell Richards advertise technetium-99m and other generators for sale by Brookhaven National Laboratory. Technetium-99m had not yet been used in nuclear medicine.

- Canada approves potato irradiation.

- The Atomic Energy Commission announces the successful development of a 220 pound nuclear reactor to provide electric power for space vehicles.

- The Atomic Energy Commission publishes its 10 year plan for nuclear energy.

- The first fully commercial Pressurized Water Reactor (PWR) U.S. nuclear power plant, Yankee Nuclear Power Station (250 MWe) in Rowe, Massachusetts achieves a self-sustaining nuclear reaction.

- Small nuclear power generators are first used in remote areas to power weather stations and to light buoys for sea navigation.

- France conducts first nuclear test in the Sahara desert.

- The NBS-2 is put into service in Boulder, Colorado; it can run for long periods unattended and is used to calibrate secondary standards.

- The first heat rate below 9,000 Btu/kwh took place in the Clinch River Plant, Virginia.

- The first large supercritical pressure generating unit began operation was the Breed Plant, Indiana.

- The first working laser is built by Theodore Maiman.

- The Roentgen Meter is created to measure x-rays and gamma rays.

- American Board of Health Physics begins certification of health physicists.

- The Alco PM-2A the first portable nuclear reactor to generate electrical power (2MW) for Camp Century.

1961 – The U.S.S. Thresher (SSN-593) a nuclear powered attack submarine, launched in 1960 was lost to sea during deep sea diving test, there were no survivors and the vehicle was never recovered.

- The U.S., U.K. and Soviet Union observe an informal suspension on nuclear tests.

- President Kennedy advises Americans to build fallout shelters.

- The U.S. Navy commissions the world's largest ship, the U.S.S. Enterprise. It is a nuclear-powered aircraft carrier with the ability to operate at speeds up to 30 knots for distances up to 400,000 miles (740,800 kilometers) without refueling.

- The first U.S. nuclear accident that killed three men, SL-1 Stationary Low-Power Reactor in Idaho Falls, Idaho. The cause was rapid removal of one single control rod causing the reactor to go super-critical at light-speed.

- Albert Ghiorso, Tl Sikkeland, A.E. Larsch and R.M. Latimer created Lawrencium.

- Reactor LOCA on the first USSR nuclear missile submarine, the K-19. Fourteen crew members allegedly died from radiation exposure rigging a provisional cooling system using a reserve tank and pipes cut off one of the torpedoes. Over thirty sailors receive doses from 100 to 5,000 rem. Eight officers and sailors died within days, six more died within the next several years.

1962 – David Kuhl introduces emission reconstruction tomography. This method later became known as SPECT and PET. It was extended in radiology to transmission X-ray scanning, known as CT.

- President John F. Kennedy asks the Atomic Energy Commission to report on the role of nuclear energy in the economy.

- EBR-1 is the world's first reactor to produce electricity with a plutonium core.

- Experiments verify that there are two distinct types of neutrinos (electron and muon neutrinos).

- Researchers at the Naval Ordnace Laboratory in White Oak, Maryland discovered that nickel-titanium alloy has shape memory properties, meaning that the metal can undergo deformation yet "remember" its original shape.

1963 – The United States and Soviet Union sign the Limited Test Ban Treaty, which prohibits underwater, atmospheric, and outer space nuclear tests. More than 100 countries have ratified the treaty since this time.

- The FDA exempts the "new drug" requirements for radiopharmaceuticals regulated by the Atomic Energy Commission.

- Henry Wagner first uses radiolabeled albumin aggregates for imaging lung perfusion in normal persons and patients with pulmonary embolism.

- U.S. FDA approves irradiated bacon, wheat, and wheat flour and potatoes.

- Jersey Central Power and Light Company announce its commitment for the Oyster Creek nuclear power plant, the first time a nuclear plant is ordered as an economical alternative to a fossil fuel plant.

- The U.S. and the Soviet Union sign "hot line" agreement between the White House and the Kremlin.

- Canada's CANDU reactors using natural uranium in fuel tubes surrounded by heavy water used as a moderator and a coolant.

- The search for a clock with improved accuracy and stability results in NBS-3.

- The Clean Air Act is established.

- The first natural draft cooling tower is built in the Western Hemisphere at the Big Sandy Plant, Kentucky.

- The IEEE is created through the merger of the American Institute of Electrical Engineers (AIEE) and the Institute of Radio Engineers (IRE).

- John F. Kennedy was assassinated in Dallas, Texas.

- Maria Goeppert-Mayer was awarded the Nobel Prize in physics for developing the nuclear shell model of how protons and neutrons are arranged in the nucleus of an atom.

1964 – U.S. FDA approves flexible packaging materials of food as contaminants during irradiation processing.

- The world's first nuclear powered lighthouse, the "Baltimore Light", on the Chesapeake Bay in Maryland, goes into operation. A 60 watt radioisotope nuclear generator, 345" high, weighing 4,600 pounds, supplies a continuous flow of electricity for 10 years without refueling.

- President Johnson signs the Private Ownership of Special Nuclear Materials Act, which allows the nuclear energy industry to own the fuel for its own units.

- After June 1973, private ownership of the uranium fuel is mandatory.

- Three surface ships powered by nuclear energy – Enterprise, Long Beach and Bainbridge – complete a round the world cruise without any logistical support.

- The Atomic Energy Commission issues Oyster Creek nuclear power plant's construction permit.

- China test first nuclear bomb.

- The first two Soviet nuclear power plants were commissioned.

- A 100 MWe boiling water graphite channel reactor began operating in Beloyarsk (Urals).

- In Novovoronezh (Volga region) a small (210 MWe) pressurized water reactor known as a VVER (veda-vodyanoi energetichesky reaktor) (water cooled power reactor) was built.

- EBR-2 is a sodium cooled reactor generating 19 MWe as an Integral fast reactor (IFR).

- EBR-1 is decommissioned.

- Gell-Mann and Zweig discover the quark.

- Rolf Sievert founded the International Radiation Protection Association

- International Radiation Protection Association (IRPA) is formed.

1965 – The first nuclear reactor operates in space, SNAP-10A.

- The Atomic Energy Commission gives the Liquid Metal Fast Breeder reactor highest priority and decides to build the Fast Flux Test Facility which began operation in 1982.

- The first major electrical blackout occurs in the Northeast United States after a lightning bolt strikes a 345 kV line.

- A-4E aircraft loaded with one nuclear weapon rolls off deck of USS Ticonderoga in North Pacific 70 miles from Japan; sinks in 16,000 ft of water, bomb, pilot and plane never recovered.

- Humboldt Bay 3, a BWR in Eureka, CA grossly failing stainless steel clad fuel is replaced with zircaloy-clad fuel.

1966 – The large number of utility orders for nuclear power reactors makes nuclear power a commercial reality in the United States.

- French tests first H-bomb in Tuamoto Islands.

- The first major combination of pumped storage and run of the river hydroelectric development at the Smith Mountain Project, Virginia.

- The first use of control room simulator to train power plant operators took place at the Cardinal Plant, Ohio.

- Fermi 1 Atomic Power Plant in Lagoona Beach, Michigan goes on-line.

- Fermi 1 Atomic Power Plant suffers meltdown a month after going on-line.

1967 – The Outer Space Treaty bans nuclear weapons being placed in orbit.

- The Treaty of Tiatelolco is signed.

- China test first fusion device with a yield of 3 megatons.

- The 13th General Conference on Weights and Measures defines the second on the basis of vibrations of the cesium atom; the world's timekeeping system no longer has an astronomical basis.

- The first tidal power station in the world in France goes into operation producing (240 MWs).

- The electroweak theory is created to unite separate nuclear and electromagnetic forces.

1968 – Nuclear Nonproliferation Treaty (NPT) calling for halting the spread of nuclear weapons capabilities is signed. By 1970, more than 50 countries had ratified the NPT. By 1986, more than 130 countries had ratified it.

- The U.S.S. Scorpion (SSN-589) a nuclear powered submarine launched in 1959 was lost at sea in 1968. There were no survivors and the vehicle was never recovered.

- NBS-4, the world's most stable cesium clock is completed. This clock was used into the 1990's as part of the NIST time system.

- The North American Electric Reliability Council (NERC) is created.

- The first 1,200 foot stack is built at the Mitchell Plant, West Virginia.

- A serious accident aboard an experimental Soviet nuclear submarine, the K-27, allegedly kills five crew members, the rest are hospitalized. After lengthy repair attempts, the sub is scrapped near Novaya Zemlya, USSR, along with its nuclear fuel.

1969 – C.L. Edwards reports the accumulation of gallium-67 in cancer.

- The FDA announces that it will gradually withdraw the exemption granted to radiopharmaceuticals and start regulating them as drugs. The change would be completed by 1977.

- The U.S. has 15 commercial reactors in operation.

- The National Environmental Policy Act is created.

- Albert Ghiorso and other scientists created Rutherfordium.

- Apollo 11 lands on the moon.

- Apollo 12 deploys SNAP-27 nuclear generator on the lunar surface.

1970 – The first Earth Day is celebrated. Electricity "brownouts" hit the Northwest during a heat wave.

- The Environmental Protection Agency (EPA) is established to control environmental pollution.

- The Water and Environmental Quality Act is created.

- Albert Ghiorso and other scientists created Dubnium.

- Woodward and Hoffmann devised the rules which determine the probable locations of electrons around a central nucleus.

- USSR nuclear submarine K-8 sinks in the Bay of Biscay, allegedly killing the captain and 53 crewmen.

1971 – The American Medical Association officially recognizes nuclear medicine as a medical specialty.

- President Nixon announces a national goal of completing the Liquid Metal Fast Breeder unit by 1980.

- Twenty-two commercial nuclear power plants are in full operation in the United States. They produce 2.4 percent of U.S. electricity at this time.

- The training reactor of the Budapest University of Technology was put into operation and has been operating since then. Its maximum thermal power is 100 kW. Besides training, research topics such as radiochemistry, archeology and medical applications are investigated.

- Te Fujita-Pearson tornado intensity scale is created used to determine the strength of a tornado.

- The first microprocessor is created; it has a processing speed of 740 kHz and contains 2300 transistors.

- Ralph Lapp first used the term "China Syndrome" describing a core meltdown that would burn through the bottom of the reactor vessel penetrating the earth below and if hot enough could reach China on the opposite side of the earth.

1972 – Computer axial tomography, commonly known as a CAT scan is introduced. A CAT scan combines many high definition, cross sectional x-rays to produce a two dimensional image of a patient's anatomy.

- In Shevchenko in Kazakhstan, the world's first commercial prototype FBR is started up, the BN-350 (bystry neutron – fast neutron) producing 120 MWe and heat to desalinate Caspian seawater.

- The NBS-5, an advanced cesium beam device, is completed and serves as the primary standard.

- The Clean Water Act is created.

1973 – Phenix, first French fast breeder power plant in operation in Marcoule (250 MWs).

- H. William Strauss introduces the exercise stress test myocardial scan.

- President Nixon proposes to replace the Atomic Energy Commission with the Energy Research and Development Administration and the Nuclear Regulatory Commission.

- The Organization of Petroleum Exporting Countries (OPEC) agrees to use oil as a foreign policy weapon, cutting exports 5% until Israel withdraws from Arab territory occupied during the Yom Kippur War. Days later Saudi Arabia cuts oil production by 25% and joins many other oil producing nations in embargoing oil shipments to the United States.

- The U.S. utilities order 41 nuclear power plants, a one-year record.

- The first large RBMK (1,000 MWe) high power channel reactor started up at Sosnovy Bor near Leningrad.

- The first of four small (12 MWe) boiling water channel type units in the eastern Arctic town of Bilibino for the production of both heat and power.

- The first wide scale, up to the minute supervisory system for measuring air quality near coal fired units transmitting data electronically to a computer center is developed.

- Akira Hasegawa and Fred Tappert propose the use of solitary waves to carry information in optical fibers.

- ICRP 22 "Implication of Commission Recommendations that Doses be Kept as Low as Readily Achievable" published.

- Oil embargo and price rise bring first "energy crisis".

1974 – The first 1,000 MW nuclear plant goes into service – Commonwealth Edison's Zion 1 plant.

- President Ford abolishes the Atomic Energy Commission and creates in its place the Energy Research and Development Administration and the Nuclear Regulatory Commission to begin regulating the nuclear industry.

- The Joint Congressional Committee on Atomic Energy is also abolished. India sets off a low yield device (10-15kt) under the Rajasthan desert.

- A major power outage affected more than a million customers for up to one week in Nova Scotia, New Brunswick and Prince Edward Island, Canada.

- Karen Silkwood testifies to AEC on Kerr-McGee safety violations. Karen Silkwood is killed in a car crash, documents allegedly substantiating Kerr-McGee mishandling of plutonium missing from Silkwood's car.

1975 – Energy Research and Development Administration begins operating.

- Laser separation of uranium isotopes are created.

- The NBS-6 begins operation; an outgrowth of NBS-5, it is one of the world's most accurate atomic clocks, neither gaining nor losing one second in 300,000 years.

- Browns Ferry nuclear unit 1 reactor building caught fire during a cable penetration test using a candle. The fire started outside the containment wall and was limited but the negative draft from within the containment building caused the fire to spread destroying cables supplying power to the unit's equipment causing equipment to start/stop and open/close valves without the operators having control of this equipment.

- The Becquerel was defined to succeed the curie.

- The gray was defined to succeed the rad as the measure of absorbed dose. Named after Louis Harold Gray.

- Martin Peri discovered the tau lepton, an elementary particle which exists for a fraction of a billionth of a second.

- NRC orders 23 BWR nuclear reactors shut-down because of cracking in cooling pipes.

- The Biblis plant in Hesse, Germany has two pressurized water reactors. The combined power output is 2,500 MWs electrical. Biblis was the biggest nuclear power plant in the World and a milestone of power generation.

1976 – John Keyes develops the first general purpose single photo emission computed tomography (SPECT) camera.

- Ronald Jaszczak develops the first dedicated head SPECT camera.

- Joint Expert Committee on Food Irradiation (JEFCI) approves several irradiated foods and recommends that food irradiation be classified as a physical process.

- The first major research in the U.S. on pressurized fluidized bed combustion takes place.

- The Resource Conservation and Recovery Act (RCRA) is passed to protect human health and the environment from the potential hazards of waste disposal.

1977 – President Carter bans the recycling of used nuclear fuel from commercial reactors.

- The Voyager 2 spacecraft is launched carrying a 12 inch copper phonograph record containing greetings in every language. The spacecraft's electricity is generated by the decay of plutonium-238 pellets.

- President Carter combines the Energy Research and Development Administration with the Federal Energy Administration, creating the Department of Energy.

- New York City experiences a blackout.

- Hideki Shirakawa, Alan MacDiarmid, and Alan Heeger discover organic polymers which conduct electricity, leading to the production of LED's and solar cells.

- Davis Besse nuclear plant suffers a PORV pressurizer valve stuck open, notices the problem and fixes the situation without any harm to the plant. This would be the same incident that occurs a year and a half later at Three Mile Island but the operators handled the scenario differently leading up to the accident. Communication not being shared among the nuclear industry.

1978 – David Goldenberg uses radiolabeled antibodies to image tumors in humans.

- The U.S. stops development of the neutron bomb.

- The Public Utilities Regulatory Policies Act (PURPA) is passed, ends utility monopoly over power generation.

- The Power Plant and Industrial Fuel Use Act limits use of natural gas in electric generation, but repealed in 1987.

- France expenses a blackout that covers 80% of its territory.

- A laser cooling process for atoms is developed by Steven Chu.

1979 – Three Mile Island nuclear power plant near Harrisburg, Pennsylvania suffers a partial core meltdown. Minimal quantity of radioactive material is released and no one is injured.

- The U.S. nuclear energy industry creates the Institute of Nuclear Power Operations (INPO) to address issues of safety and performance.

- In the U.S. seventy two commercial nuclear reactors are in operation.

- The sievert is created to mean the derived unit for dose equivalent or organ equivalent dose.

- U.S. Council for Energy Awareness (USCEA) is created.

- Cronin and Fitch discover asymmetry of elementary particles.

- The movie "The China Syndrome" debuts a few weeks before the Three Mile Island accident.

- NRC establishes the Accident Sequence Precursor (ASP) analysis program.

1980 – JEFCI approves all irradiated foods treated with a maximum average dose of 10kGy.

- Nuclear energy generates more electricity than oil.

- The Low-level Radioactive Waste Policy Act was passed, making states responsible for the disposal of their own low-level nuclear waste, such as from hospitals and industry.

- The first U.S. wind-farm is in operation.

- The Pacific Northwest Electric Power Planning and Conservation Act establish regional regulation and planning.

- Last known atmospheric nuclear test is conducted by China.

1981 – J.P. Mach uses radiolabeled monoclonal antibodies for tumor imaging.

- President Regan's administration lifts the ban on reprocessing used nuclear fuel and announces a policy that anticipates the need for a high level radioactive waste storage facility.

- Israel destroys Iraq's Osirak reactor.

- PURPA was ruled unconstitutional by a Federal judge.

- The first application of sliding pressure technique on supercritical pressure generating unit to maintain uniform efficiency from full load to minimal load took place at the Gen. James M. Gavin Plant, Ohio.

- Peter Armbruster and Gottfried Munzenberg created Bohrium.

- The NRC set a requirement that all nuclear power plants must develop training simulators for their operators in order to enhance their abilities to reduce nuclear accidents.

- An auxiliary operator, working his first day on the job without proper training, inadvertently opened a valve which led to the contamination of eight workers when more than 100,000 gallons of radioactive coolant leaks into the containment building of the Tennessee Valley Authority's Sequoyah 1 plant.

- EPA establishes 25 millirem/year whole body limit to general public from nuclear fuel cycle activities and 100 millirem lifetime dose limit.

1982 – Steve Larson and Jeff Carrasquillo use iodine-131 labeled monoclonial antibodies to treat cancer patients with malignant melanoma.

- After 25 years of service, the Shippingport Power Station is shut down. Decommissioning is completed in 1989.

- The U.S. Supreme Court upholds legality of PURPA in FERC v. Mississippi (456 US 742).

- Peter Armbruster and Gottfried Munzenberg created Meitnerium.

- The U.S. FDA approves the use of Potassium Iodide as a thyroid protective measure against radiation exposure.

- Rupture of central fuel assembly at Chernobyl 1 (USSR) due to operator errors; radioactivity vented to Pripyat and surrounding area.

- The Space Shuttle begins its commercial operation.

1983 – U.S. FDA and Canadian Health & Welfare Department approve spice irradiation.

- President Reagan signs into law the Nuclear Waste Policy Act, authorizing the creation of a high level nuclear waste repository.

- Funding for the Clinch River Breeder Reactor project is killed by Congress.

- Research at CERN shows evidence of weakons (W and Z particles); this validates the link between weak nuclear force and electromagnetic force.

- Nuclear energy generates more electricity than natural gas.

- A storm destroyed power lines resulting in almost a total blackout of Sweden.

- The television movie "The Day After" aires, showing survivors dealing with a nuclear holocaust.

1984 – The atom overtakes hydropower to become the second largest source of electricity after coal.

- Eighty-three nuclear power reactors provide about 14 percent of the electricity produced in the United States.

- Annapolis, N.S., tidal power plant which is the first of its kind in North America (Canada).

- Peter Armbruster and Gottfried Munzenberg created Hassium.

- The movie "SL-1" was a documentary about the first disastrous nuclear reactor explosion in the United States in 1961.

- USS Kitty Hawk collides with a Soviet attack submarine; both are armed with nuclear weapons.

1985 – U.S. FDA approves irradiation of pork to control trichinosis.

- The Institute of Nuclear Power Operations (INPO) forms a national academy – the National Academy for Nuclear Training – to accredit every nuclear power plant's training program.

- Lead-iron phosphate glass invented creating a more durable containment medium for storing nuclear wastes.

- A Soviet nuclear submarine suffered a reactor explosion while at a repair facility near Vladivostok, Russia. A cloud of radioactive material never reached the city, eight officers were killed in the explosion.

- Citizens Power, the first power marketer goes into business.

- The atomic force microscope is able to identify individual atoms, it is invented by Binning.

- The Super Phenix nuclear power plant in France with a power of 1200 MWe, is the largest power plant with a fast breeder reactor.

1986 – Chernobyl nuclear reactor melts down and fire occurs in the Ukraine. Massive quantities of radioactive material are released.

- U.S. FDA approves irradiation of fruits and vegetables and other foods up to doses of one kGy.

- The Perry nuclear power plant in Ohio becomes the 100th U.S. nuclear power plant in operation.

- Tests at EBR-ll demonstrate the inherent safety of the Integral Fast Reactor concept.

- A Soviet nuclear powered submarine suffered an explosion and sank with two reactors and 34 nuclear weapons on board….never recovered.

- The Father of the Nuclear Navy Admiral Hyman Rickover died.

- January 28, Space Shuttle Challenger disintegrates 73 seconds after liftoff, because of a failed O-ring on one of the booster rockets.

1987 – Nuclear Waste Policy Amendments Act designates Yucca Mountain, Nevada as candidate for nation's first geological repository for high level radioactive waste and spent nuclear fuel.

- The first electric steam generating unit to operate for 607 consecutive days establishing a world record took place at the Mountaineer Plant, West Virginia.

- The positron microscope is invented by van House and Rich.

- A massive storm destroys power lines putting southern England in a blackout.

- The Nuclear Utility Management and Resources Council (NUMARC) is created to address generic regulatory and technical issues. The Atomic Industrial Forum (AIF) was split to create NUMARC and give some responsibility to U.S. Council for Energy Awareness (USCEA).

- First ceramic conductors were discovered.

- Human population reaches 5 billion.

1988 – U.S. electricity demand is 50% higher than in 1973.

- Combustion Engineering (CE) was acquired by Asea Brown Boveri (ABB).

- ASEA and BBC Brown Boveri mergered to form Asea Brown Boveri (ABB).

- South Texas Project unit 1 reactor begins service, the largest reactors to be in service over 1,000 MWe or greater in the U.S.

1989 – 109 nuclear power plants provide 19% of the electricity used in the U.S., 46 units have entered service during this decade.

- The FDA approves the first positron radiopharmaceutical (rubidium-82) for myocardial perfusion imaging.

- South Texas Project 2 reactor begins service, one of the largest reactors to be in service over 1,200 MWe or greater in the U.S.

- The Galileo Mission is launched aboard the space shuttle Atlantis.

- The Department of Energy (DOE) shifts from nuclear materials production to one of environmental cleanup.

- The Nobel Prize in Physics is awarded to Norman Ramsey of Harvard University, Hans Dehmelt from the University of Washington and Wolfgang Paul from the University of Bonn for their work in the development of atomic clocks.

- A Soviet nuclear submarine Komsomolets caught fire and sank with two reactors…never recovered.

- A geomagnetic storm caused the Hydro-Qubec power failure which left 6 million people without power for numerous hours.

- The World Association of Nuclear Operators (WANO) was founded in 1989 after the Chernobyl accident to foster international cooperation and professional excellence within the nuclear industry. Every organization in the world that operates a nuclear electric generating station is a WANO member.

- Oil tanker Exxon Valdez ran aground and released oil onto the Alaskan coastline.

- The Berlin wall falls, as East Germany opens its borders with West Germany.

1990 – 110 nuclear power plants in the U.S. set a record for the amount of electricity generated, surpassing all fuel sources combined in 1956.

- The Ulysses mission is launched aboard the space shuttle Discovery.

- The FDA approves irradiation of packaged fresh or frozen uncooked poultry and supports it as an effective control of microorganisms responsible for a major portion of food borne illnesses, including Salmonella, Yersinia, and Campylobacter.

- The Clean Air Act amendments mandate additional pollution controls.

- The first combined cycle operation of a PFBC plant in North America occurred at the Tidd PFBC Demonstration Plant, Ohio.

1991 – 111 nuclear power plants operate commercially in the U.S. with a combined capacity of 99,673 megawatts. They produce almost 22 percent of the electricity generated commercially in the United States.

- Approximately 31 countries had nuclear power plants in commercial operation or under construction.

- The first conversion of a nearly completed nuclear plant to coal fired plant occurred at the Wm. H. Zimmer Generating Station, Ohio.

- Anders Olsson transmits solitary waves through an optical fiber with a data rate of 32 billion bits per second.

- A wind storm knocked out power to central North America affecting over 1 million people.

- A disgruntle employee at the Point Lepreau Nuclear Generating Station in Canada obtained a sample of heavy water from the primary heat transport loop and loaded it into the employees water cooler. Eight employees were affected with elevated radiation doses.

- The Cold War, which started after WWII ends with the collapse of the USSR.

- LEP experiments confirm the existence of 3 generations of elementary particles.

- The first commercial nuclear reactor in China, Qinshan 1, is connectd to the power grid.

- Operation Desert Storm, the beginning of the Persian Gulf War.

1992 – The FDA approves the first monoclonal antibody radiopharmaceutical for tumor imaging.

- The DOE signs cooperative agreement with the nuclear industry to co-fund the development of standard designs for advanced light water reactors.

- The fourth and final standardized nuclear power plant design is submitted to the Nuclear Regulatory Commission (NRC) for certification and approval. Getting the plant designs approved by the NRC is a step toward building uniform nuclear power plants in the U.S.

- President Bush signs into law the Energy Policy Act, which sets the U.S. on course for planning its energy needs, and reforms the licensing process for advanced, standardized nuclear power plants. The updated process affords the public more timely opportunities to participate in decisions to build new nuclear plants and is expected to create a more stable financial environment for investors.

- The National Energy Policy Act is established. It makes several important changes in the licensing process for nuclear power plants.

1993 – The nuclear energy industry positions itself for the future when 16 nuclear utilities sign the first of two contracts with U.S. nuclear plant manufacturers – each agreeing to develop first of a kind engineering on two advanced plant designs. General Electric signs in March and Westinghouse signs in June.

- Another nuclear power plant – the Comanche Peak unit 2 in Glen Rose, Texas – goes online to provide 1,150 MWs of electricity to U.S. consumers.

- Two decades after the first oil embargo, the 109 nuclear power plants operating in the United States generate 610 billion kilowatt hours of net electricity, providing about $1/5^{th}$ of the nation's electricity.

- The NBS-7 comes on line and eventually it achieves an uncertainty of 5×10^{-15}, or 20 times more accurate than NBS-6.

- The TFTR tokamak at Princeton (PPPL) experimented with 50% deuterium and 50% tritium eventually producing up to 10 Megawatts of power from a controlled fusion reaction.

- Fermilab in Chicago discovers the top quark, first predicted in 1984.

- The U.S. Department of Energy reveals that the United States conducted 204 secret underground nuclear tests over a 45-year period. These bring the total number of U.S. nuclear tests to 1051. The Energy Department also reveals that the U.S. deliberately exposed some Americans to dangerous levels of radiation in medical experiments without their consent.

1994 – The U.S. again leads the world in promoting the peaceful use of nuclear technology by signing a contract to buy uranium from Russia that could be blended down into nuclear power plant fuel ensuring that it will never again be used for warheads.

- The Nuclear Regulatory Commission issues final design approval for the first two of four advanced nuclear power plant designs – General Electric's Advanced Boiling Water Reactor (ABWR) and ABB Combustion Engineering's System 80+. The approval means that all major design and safety issues have been resolved to the satisfaction of the NRC staff and the Advisory Committee on Reactor safeguards. The two plants are the first to obtain final design approval under the NRC's new regulations for licensing standardized plant designs. The NRC will now prepare a rulemaking, which will include public comment, to codify the designs.

- Secret nuclear testing on humans is revealed by both the U.S. and C.I.S.

- The Experimental Breeder Reactor (EBR-ll) is officially shutdown.

- The U.K's first pressurized water reactor (PWR) was opened at Sizewell B in Suffolk.

- Peter Armbruster and other scientists created Ununnilium and Unununium.

- The Nuclear Energy Institute is created from the merger of NUMARC, USCEA, ANEC and AIF.

1995 – 178 nations renew the Non-proliferation Treaty.

- The U.S. announces a total ban on all U.S. nuclear weapons testing.

- Hurricane Opal killed 59 people, knocking out power to over 2 million customers across eastern and southern North America.

1996 – The NRC grants the Tennessee Valley Authority (TVA) a full power license for its Watts Bar 1 nuclear power plant, bringing the number of operating nuclear units in the U.S. to 112.

- The first of a kind engineering design is completed for the GE Advanced Boiling Water Reactor. Kashiwazaki-Kariwa 6, the world's first Advanced Boiling Water Reactor (1350 MWe), begins commercial service in Japan – ahead of schedule and under budget.

- The French President Chirac announces an end to French nuclear tests.

- The Treaty of Pelindaba creates a Nuclear Free Zone throughout Africa.

- The Fuel Conditioning Facility at Argonne-West began operations with the chopping of used fuel from EBR-ll.

- Peter Armbruster and other scientists created Ununbiium.

- The Western Intertie buckled under high summer heat causing a cascading power failure affecting nine western states and some parts of Mexico.

- A severe ice storm affected Washington state and Idaho causing power outages lasting up to two weeks.

- The Peninsular Malaysia electricity blackout crisis occurred.

1997 – The NRC issues design certification for the GE Advanced Boiling Water Reactor which is valid for the next 15 years.

- The NRC also issues design certification for the ABB Combustion Engineering System 80+ which is also valid for the next 15 years.

- The U.S. begins a round of underground nuclear related tests at the Nevada Test Site.

- Later that year the U.S. conducts a second underground explosive test on plutonium.

- New England ISO begins operation (first ISO). New England Electric sells power plants (first major plant divestiture).

- The JET tokamak in the UK produced 16 Megawatts of fusion power while generating four megawatts of alpha particle self-heating was achieved.

- Big Rock Point Nuclear plant was the longest running nuclear plant in the U.S. and retired after 35 years of service. Located in northern Michigan.

- First direct evidence of the tau neutrino is published.

1998 – President Clinton certifies that China supports international nuclear nonproliferation efforts, paving the way for the sale of U.S. nuclear technology to that country.

- Baltimore Gas and Electric Company (BGE) submits an application to the NRC to renew the license of its two units at Calvert Cliffs – the first U.S. company to apply for a 20 year extension of its 40 year license.

- Duke Power Company submits an application to the NRC to renew the license of its three units at Oconee nuclear power plant.

- In the month of May, India test five nuclear devices and Pakistan test six nuclear devices.

- California open market and ISO.

- Scottish Power (UK) to buy Pacificorp, first foreign takeover of US utility.

- National (UK) Grid then announces purchase of New England Electric System.

- An ice storm in North America knocked out power causing blackouts where transmission towers were destroyed and over 3.5 million customers were affected.

- PG & E placed a substation online while it was still grounded for maintenance causing a blackout for the San Francisco, California area.

1999 – Entergy Nuclear closes on it purchase of the Pilgrim Station from Boston Edison Company, the first completed nuclear plant sale in the U.S.

- NIST – F1 atomic clock begins operation with an uncertainty of 1.7 x 10-15, or accuracy to about one second in 20 million years, making it one of the most accurate clocks ever made.

- The electricity market enters the internet.

- The FERC issues Order 2000, promoting regional transmission.

- Numerous scientist from the Joint Institute of Nuclear Research in Dubna near Moscow, Russia and the Lawrence Livermore National Laboratory in California created Ununquadium.

- A blackout affecting about 70% of the Brazilian territory leaving more than 90 million people without power for hours. This caused panic and chaos throughout the event.

- In Taiwan, a transmission tower collapsed because of a landslide leaving 8.5 million customers without power.

- A severe accident at the uranium processing plant at Tokaimura, Japan exposes 55 workers to radiation.

2000 – The NRC issues the **first ever** license renewal to Constellation Energy's Calvert Cliff Nuclear Power Plant in southern Maryland, allowing an additional 20 years of operation.

- Y2K a widely publicized potential disaster caused by the failure of the world's computer systems to accommodate dates beyond 1999.

- A couple of months later the NRC approve a 20 year extension to the operating license of Duke Energy's three units at Oconee nuclear station.

- The world's first commercial wave power station on the Scottish island of Islay began to generate electricity. The station is called the Land Installed Marine Powered Energy Transformer (LIMPET) and can provide enough electricity for about 400 homes.

- George Chen of Cambridge University found a new way of extracting titanium through electrolysis....If produced on a larger scale could drop the cost of the metal.

- T.V. Oommen invented the environmentally friendly transformer fluid from vegetable oil.

- The 12 month California electricity crisis takes place due to regular power failures causing a major energy shortage.

- The World's Tallest Natural Draft Cooling Tower is built, northwest of Cologne, Germany standing over 200 meters tall.

- The last of the reactors at the Chernobyl nuclear power plant are shutdown.

- The Russian submarine Kursk sinks in the Barent Sea.

- Energy Providers Coalition for Education (EPCE) was created.

- July 25, Air France Concorde crashes 60 seconds after takeoff, suffering a tire blowout that ruptured a fuel tank. 113 people were killed in this event. The Concorde has experienced a tire blowout 57 times before this incident occurred.

2001 - Terrorists hijack planes and crash them into the World Trade Center and the Pentagon. The Twin Towers collapse and the Pentagon burns for days. President Bush declares a war on terrorism.

2002 – The DOE and the U.S. nuclear industry cooperate on site location for new nuclear power plant.

- House of Representatives approves Yucca Mountain as final disposal site for spent nuclear fuel.

- Davis Besse nuclear plant found reactor head partially eaten away by boric acid. A ¼ inch thick plate of stainless steel was the only thing holding the pressure in the reactor. This led to the entire nuclear industry reactor head inspection.

- There are 442 nuclear power plant units operating in 32 countries, with a total capacity of 357,000 MW. This is 17% of the total electricity production in the world. 41 further reactors are being built, with a total of 30,000 MW capacity

- North Korea claims it has a nuclear weapons program.

2003 – A power grid failure turns out the lights for nearly 50 million people in the Northeast. The blackout began in Ohio and spread through New England and parts of the Midwest and Canada. The outage took out 21 power plants in just 3 minutes.

- The Italy blackout resulted from a power failure that affected all of Italy cutting power to more than 56 million people.

- North Korea withdraws from the Nuclear Non-Proliferation Treaty while Libya ends its nuclear weapons program.

- February 1, Space Shuttle Columbia disintegrates over Texas after a piece of foam hits the shuttles wing damaging the insulating tile.

2004 – Finland ordered a 1,600 MWe European PWR (EPR).

- Two power plants in Greece shut down due to malfunctions within 12 hours of each other, during a period of high demand (heat wave) which led to the cascade failure causing the collapse of the entire southern half of Greece affecting several million people.

- Pakistani nuclear scientist Abdul Qadeer Khan is exposed as supplying a global black-market in nuclear technology.

- Ununtrium is discovered by Y.T. Oganessian, et. al.

2005 – Approximately 100 million people on Java Island, Indonesia, lost power for 7 hours. This was the largest blackout in history in terms of population affected.

- Hurricane Katrina knocked out power to Florida affecting 1.3 million people, then strengthening hurricane Katrina hits Louisiana, Mississippi and Alabama extensively destroying the power grid leaving millions of people without power.

- California experiences another blackout affecting millions.

- Palo Verde 2, west of Phoenix, AZ is the biggest capacity unit in the U.S. at 1,335 MWe.

2006 – Parts of Germany, France, Italy, Belgium, Spain and Portugal over 5 million people were without power after a major cascading breakdown. A German electric company E.ON switching off an electric line over the river Ems to allow a cruise ship to pass safely underneath.

- A major heat wave in New York (Queens) caused a major blackout due to excessive demand and dilapidated infrastructure leaving 50,000 people without power for numerous days.

- The NEI founded the Clean and Safe Energy Coalition (CASEnergy) to help build local support around the country for new nuclear building.

- Ununoctium discovered by Y.T. Oganessian, et. al.

2007 – A ceramic superconductor doped with thallium, mercury, copper, barium, calcium and oxygen currently holds the world record for superconductivity at 138^0K.

- A power failure in New York City affected 135,000 customers.

- The city of Barcelona suffered a near total blackout. Several areas remained without electricity for more than 3 days due to a massive electrical substation chain failure.

- A series of ice storms caused power to be cut to over 1 million homes and businesses from Oklahoma to Nebraska.

- A failed switch and fire at an electrical substation outside Miami triggered widespread blackouts in parts of Florida affecting over 4 million people. The failure knocked out power to customers in 35 southern Florida counties.

- At the end of 2007, there were 436 nuclear power plants operating worldwide, with 45 more under construction.

2008 – Calvert Cliffs achieves the highest capacity factor in the World at 101.37% for the year.

- On February 26, Turkey Point Generating Station both reactors were shut down due to the loss of off-site power during a widespread power outage in South Florida currently affecting 700,000 customers. At least 2.5 million people were without power. The blackout was initially caused by an overheated voltage switch that soon caught fire in a power substation in Miami, nowhere near the plant. The fire occurred at 1:08 pm which caused an automatic shutdown of the power plant. This led to a domino effect that caused outages as far north as Daytona Beach and Tampa. Power was restored by 4:30 pm.

2009 – Calvert Cliffs Nuclear Power Plant sets World Record for longest continuous running pressurized water reactor at 692 days before shutting down to refuel.

- November 10, two nuclear reactors at Brazil's Angra plant shut down when the country's electrical grid failed after bad weather caused three transmission lines to collapse. Some 60 million people in 800 towns were plunged into darkness.

2010 – Obama administration cancels Yucca Mountain site as a nuclear waste depository.

- Southern Nuclear Company is the first to receive government funding for two new nuclear plants (Vogtle 3 & 4).
- 8/21 Iran begins fueling its first nuclear reactor.

Future – The first commercial production of electricity from a nuclear fusion reactor is not expected to happen before the year 2040.

People in Nuclear History

Fourth Anniversary of the Nuclear Chain Reaction Reunion photo:

<u>Top</u> - **Norman Hiberry, Samual Allison, Thomas Brill, Robert Nobles, Warren Nyer and Marvin Wilkening**

<u>Middle</u> - **Harold Agnew, William Sturm, Harold Lichtenberger, Leona W. Marshall and Leo Szilard**

<u>Front</u> - **Enrico Fermi, Walter Zinn, Albert Wattenberg, and Herbert Anderson.**

A team led by Enrico Fermi in 1934 observed that bombarding uranium with neutrons produces the emission of beta rays (electrons or positrons; see beta particle). The fission products were at first mistaken for new elements of atomic numbers 93 and 94, which the Dean of the Faculty of Rome, Orso Mario Corbino, christened *ausonium* and *hesperium*, respectively. The experiments leading to the discovery of uranium's ability to fission (break apart) into lighter elements and release binding energy were conducted by Otto Hahn and Fritz Strassmann in Hahn's laboratory in Berlin. Lise Meitner and her nephew, physicist Otto Robert Frisch, published the physical explanation in February 1939 and named the process 'nuclear fission'. Soon after, Fermi hypothesized that the fission of uranium might release enough neutrons to sustain a fission reaction. Confirmation of this hypothesis came in 1939, and later work found that on average about 2 1/2 neutrons are released by each fission of the rare uranium isotope uranium-235. Further work found that the far more common uranium-238 isotope can be transmuted into plutonium, which, like uranium-235, is also fissionable by thermal neutrons.

On 2 December 1942, another team led by Enrico Fermi was able to initiate the first artificial nuclear chain reaction. Working in a lab below the stands of Stagg Field at the

University of Chicago, the team created the conditions needed for such a reaction by piling together 400 tons (360 tonnes) of graphite, 58 tons (53 tonnes) of uranium oxide, and six tons (five and a half tonnes) of uranium metal. Later researchers found that such a chain reaction could either be controlled to produce usable energy or could be allowed to go out of control to produce an explosion more violent than anything possible using chemical explosives.

Who was Hippolyte Pixii? Hippolyte Pixii, (1808-1835) a French inventor and instrument maker who created the first practical electric generator in 1832. This generator could produce both direct current and alternating current.   (Random House)

Who was Charles Curtis? Charles Curtis (1860-1953) born in Boston, MA. A civil engineer and inventor best known for his invention of the Curtis impulse steam turbine in 1896. He patented a multistage steam turbine that utilized both reaction and impulse turbine blading. This type of turbine started a revolution in power production because it was extremely efficient for its size and weight. In 1884, 12 years earlier, the reaction turbine had been patented by Sir Charles Parsons.

Who was Antoine Henri Becquerel? Antoine Henri Becquerel (1852-1908) was a French physicist who discovered radiation coming from uranium. Shared 1903 Noble prize for his discovery of radioactivity with Marie and Pierre Curie.   (Random House)

Who was Andre Marie Ampere? Andre Marie Ampere, (1775-1836) was a French physicist and mathematician. His name is given to the basic unit of electric current, (ampere, amp). He laid the foundation of science of electrodynamics through theoretical and experimental work after his discovery in 1820 of magnetic effects of electric currents. Ampere's law states that the magnetic force produced by two parallel current-carrying conductors to the product of their currents and the distance between the conductors.

Who was William Stanley? William Stanley (1858-1916) was an American electrical engineer and the inventor of the transformer (1885). His work also included the long-range transmission system for alternating current.

Who was Sebastian Ziani de Ferranti? Sebastian Ferranti (1864-1930) a British electrical engineer who established the principle of a national grid and an electricity-generating system based on alternating current. Ferranti also designed many other electrical and mechanical devices such as high-tension cables, circuit breakers, transformers, turbine, and other spinning machines.   (Random House)

Who was Gabriel Daniel Fahrenheit? Gabriel Daniel Fahrenheit (1686-1736) a Polish born Dutch physicist who developed the Fahrenheit scale. In 1724 he invented the first accurate thermometer. (Random House) This scale is commonly used in the United States yet most other countries use the Celsius scale.

Who was Albert Einstein? Albert Einstein (1879-1955) a German born U.S. physicist born in Ulm, Wurttemberg is the greatest scientist of modern times. He received the Nobel Prize in physics in 1921 for his discovery of the law of the photoelectric effect. In 1905 Einstein published four scientific papers, with each containing a great discovery; 1) was the presentation of the special theory of relativity, 2) the statement of the equivalence of mass and energy $E=mc^2$, 3) a theoretical explanation of Brownian motion, the incessant erratic movement of tiny particles suspended within a fluid, and 4) his application of Planck's quantum hypothesis to the investigation of the nature of light. (Random House)

Who was Thomas Alva Edison? Thomas Alva Edison (1847-1931) a U.S. scientist and inventor born in Milan, Ohio invented the electric light bulb by creating a filament and a vacuum bulb that enabled an incandescent light to burn steadily. Edison also invented the phonograph, movie camera (kinetograph), and many other devices. He also established the first electric power station at Pearl Street in New York City. Edison has over 1,000 patents to his name. (Random House) Edison conducted over 1,200 unsuccessful experiments before he successfully produced the light bulb in 1879. Whoever said being patient and persistent wouldn't pay off?

Who was John Dalton? John Dalton (1766-1844) an English chemist born near Cockermouth, in Cumbria who produced a theory about atoms. He has said that everything is made up of atoms, which cannot be created or destroyed. A Law of Partial Pressures was created by Dalton which states, that in a mixture of gases that are in equilibrium but do not react with each other, the total pressure equals the sum of those partial pressures that each gas would exert if it was by itself. A Dalton is a unit of measure used to express atomic mass, named after John Dalton. (Random House)

Who was Humphry Davy? Humphry Davy (1778-1829) an English chemist born in Penzance, Cornwall who used electrolytic methods to discover six elements: boron, barium, calcium, sodium, potassium, and magnesium. He has also invented the miner's safety lamp, better known as the Davy lamp.

Who was Prince Louis Victor Debroglie? Prince Louis Victor Debroglie (1892-1987) a French physicist who was awarded the 1929 Nobel Prize in physics for his discovery that the behavior of electrons, like that of light, could be explained in terms of wave motion. His theory led to the development of wave mechanics, which is an important part of quantum mechanics. (Random House)

Who was Arthur Jeffery Dempster? Arthur Jeffery Dempster (1886-1950) a U.S. physicist who developed the first mass spectrometer which is an instrument used to measure the mass of atomic nuclei and providing a way to analyze chemical compositions. Dempster had discovered many isotopes including uranium 235, which is used in atomic weapons and in nuclear reactors.

Who was Charles Augustin de Coulomb? Charles Augustin de Coulomb (1736-1806) a French physicist and inventor born in Angouleme who was noted for his research into the science of friction, electricity, magnetism, and torsion. He established Coulomb's Law of Electrostatic Forces that states, the principle that an electrostatic attraction or repulsion between electrically charged bodies is directly proportional to the product of the electric charges on each body, and inversely proportional to the square of the distance between the bodies. (Random House)

Who was Nicolas Leonard Sadi Carnot? Nicolas Leonard Sadi Carnot (1796-1832) a French physicist born in Paris who devised the Carnot cycle. The Carnot cycle demonstrates that the efficiency of a heat engine working at its maximum thermal efficiency does not depend on its mode of operation, but only on the temperatures at which it accepts and discards heat energy. Carnot founded the science of thermodynamics. (Random House)

Who was Anders Celsius? Anders Celsius (1701-1744) a Swedish astronomer, physicist, and mathematician born in Uppsala who devised the Celsius scale. (Random House) The Celsius scale is mostly used in scientific work and is also a common system of measurement in most parts of the world.

Who was Rudolf Julius Emmanuel Clausius? Rudolf Julius Emmanuel Clausius (1822-1888) a German theoretical physicist born in Pomerania (now Poland) first stated the second Law of Thermodynamics in 1850, that heat never flows from a cold surface to a hotter surface without work, and proposed the term entropy. He had also made contributions to the Kinetic theory and the theory of electrolysis. (Random House) He is also one of the founders of the science of thermodynamics. Clausius stated that there are two types of entropy; the conversion of heat into work, and the transfer of heat from high to a lower temperature.

Who was James Chadwick? James Chadwick (1891-1974) an English physicist born in Cheshire who was awarded the Nobel Prize in 1935 for his discovery of the neutron. (Random House)

Who was Ludwig Boltzmann? Ludwig Eduard Boltzmann (1844-1906) an Austrian physicist born in Vienna who made contributions in thermodynamics, classical statistical mechanics, and kinetic theory. (Random House) The Stefan-Boltzmann law states that the total energy radiated from a body is proportional to the fourth power of the temperature.

Who was Ernest Thomas Sinton Walton? Ernest Thomas Sinton Walton (1903-1995) an Irish nuclear physicist born in Country Waterford, shared a 1951 Nobel Prize with John D. Cockcroft in physics for developing the first particle accelerator where they initiated the first nuclear fission reaction applying a non-radioactive substance. Both of these men succeeded in splitting the atom in 1932. (Random House)

Who was George Westinghouse? George Westinghouse (1846-1914) a U.S. inventor and manufacturer who pioneered the use of high voltage AC electricity and founded the Westinghouse Electric Company in 1886 to develop AC induction motors and transmission equipment, he also founded 59 other companies. Westinghouse had more than 400 patents to his name. In 1869 he patented a powerful air brake designed for trains, allowing them to travel at higher speeds with greater loads while be able to slow-down faster. In 1895 the Westinghouse Electric Company used the water from Niagara Falls to generate electricity for the town of Buffalo. (Random House) In 1957 he was enshrined in the Hall of Fame for Great Americans at New York University.

Who was Charles Wheatstone? Charles Wheatstone (1802-1875) a British physicist and inventor born in Gloucester who popularized the "Wheatstone bridge" for measuring resistance. (Random House)

Who was Charles Alernon Parsons? Charles Alernon Parsons (1854-1931) an English engineer born in London who invented the Parsons steam turbine in 1884. His steam turbine was first used in marine applications and later used for electrical generation in driving a generator. (Random House)

Who was Alessandro Volta? Alessandro Giuseppe Antonio Anastasio Volta (1745-1827) an Italian physicist born in Como, who constructed the first battery, which provided science with the earliest source of continuous electric current. The volt is a unit of measure named after Volta is used to measure electricity. Volta invented a device called an electroscope, a device used for detecting an electric charge. (Random House)

Who was Joseph John Thomson? Joseph John Thomson (1856-1940) a British physicist born near Manchester won the Nobel Prize in physics in 1906 for his discovery of the electron in 1897. His discovery was made while he studied cathode rays that occurred in a glass tube under a vacuum when electric current was introduced. (Random House)

Who was Evangelista Torricelli? Evangelista Torricelli (1608-1647) an Italian physicist and mathematician born in Faenza, who invented the barometer in 1644. (Random House) Torricelli created a law that states that the velocity with which a liquid flows through an opening in a container equals the velocity of a body free falling from the surface of the liquid to the opening.

Who was Nikola Tesla? Nikola Tesla (1856-1943) a Croatian born American physicist and electrical engineer who invented the induction motor, which used alternating current. Tesla worked on many successful inventions including the Tesla coil, various generators, transformers, and an arc lighting system (fluorescent lighting). Tesla developed an alternating current (AC) electrical power system that is better for transmitting power over long distances than a direct current (DC) electrical power system. As strange as it may seem, Tesla neglected to patent many of his ideas and made little profit from them. One of Tesla's most ingenious ideas was to transmit electricity to anywhere in the world without using wires, just by using the Earth itself as an enormous oscillator. (Random House)

Who was Blaise Pascal? Blaise Pascal (1623-1662) a French philosopher and mathematician born in Clermont-Ferrand, who pioneered hydrodynamics and fluid mechanics, and created Pascal's Law, the basis of hydraulics.   (Random House)

Who was Hans Christian Oersted? Hans Christian Oersted (1777-1851) a Danish physicist born in Rudkobing, Langeland, who discovered that a magnetized needle could be manipulated by an electric current passing through a conductor (wire) gave birth to the science of electromagnetism. (Random House) An Oersted is a unit of measure used to measure the strength of a magnetic field. In 1820 Oersted discovered the magnetic field associated with an electric current.

Who was Georg Simon Ohm? Georg Simon Ohm (1789-1854) a Bavarian born German physicist born in Erlangen, Bavaria, who theorized Ohm's Law, from his work of electric current. (Random House) Ohm's law states that an electrical current in a circuit is directly proportional to the constant total electromotive force within the circuit. An Ohm is a unit of measure used to measure electrical resistance.

Who was James Clerk Maxwell? James Clerk Maxwell (1831-1879) a Scottish physicist born in Edinburgh who's most important work was in electricity, magnetism, and in the kinetic theory of gases. In creating the kinetic theory of gases, Maxwell had the proof that heat resides within the motion of molecules.   (Random House)

Who was Gustan Robert Kirchhoff? Gustan Robert Kirchhoff (1824-1887) a German physicist born and educated in Konigsberg (now Kaliningrad). (Random House) Known for his work on electrical conduction, showing that an electrical current passes through a conductor at the speed of light. There are two laws from Kirchhoff, the first law Kirchhoff's voltage law states that the sum of all voltages around any closed electrical circuit equals zero volts. And the second law, Kirchhoff's current law states the sum of all the currents at any point in an electrical circuit equal zero amps.

Who was James Prescott Joule? James Prescott Joule (1818-1889) an English physicist who determined the relationship between heat energy and mechanical energy, which in turn discovered the first Law of Thermodynamics. Joule created Joule's law, which states the relation between heat and electricity. A Joule is a unit of measure used to measure work.

Who was William Thomson Kelvin? William Thomson Kelvin (1824-1907) an Irish physicist born in Belfast formulated the second Law of Thermodynamics. His work on electromagnetism gave rise to the theory of an electromagnetic field. He also created the Kelvin scale for temperature. Kelvin was extremely instrumental in achieving the international adoption of many of our present day electrical units in 1881. (Random House)

Who was Otto Hahn? Otto Hahn (1879-1968) a German physical chemist born in Frankfurt-am-Main, who won the 1944 Nobel Prize for chemistry for splitting of a uranium atom and his discovery of the possibility of a chain reaction. (Random House) He had discovered nuclear fission.

Who was Heinrich Rudolph Hertz? Heinrich Rudolph Hertz (1857-1894) (Random House) a German physicist born in Hamburg, showed that an electrical current could produce electromagnetic waves when a high voltage current arced across the gap, from this radio broadcasting became a reality. The unit of frequency, called hertz, is named after him.

Who was Joseph Henry? Joseph Henry (1797-1878) a U.S. physicist born in Albany, New York the inventor of the electromagnetic motor and is best known for his studies in electromagnetism. The unit of inductance, the henry, is named after him. (Random House)

Who was Michael Faraday? Michael Faraday (1791-1867) was an English physicist and chemist born in Newington, Surrey. It was Faraday coined the terms anode, anion, cation, cathode, electrode, and electrolyte. Faraday also pointed out that the energy of a magnet is within the field around it and not within the magnet itself. (Random House) Faraday figured that if an electric current could produce a magnetic field, then a magnetic field should be able to produce an electric current. He also discovered that the water solutions of certain substances could conduct electric current. These substances he called electrolytes. The farad a unit of measure was named after Faraday to measure electrostatic capacitance.

Who was John Van Vleck? John Van Vleck (1899-1980) was an American physicist who contributed in the field of magnetism. In 1977 he was awarded the Nobel Prize for Physics. (Random House)

Who was William Gilbert? William Gilbert (1540-1603) was an English scientist who studied static electricity and magnetism. Gilbert had discovered many important facts about magnetism, like the laws of attraction and repulsion of a magnet. (Random House) The term Gilbert (Gb) is used as a unit of magnetism. He regarded the Earth as being like a giant magnet and explored other types of magnetic and electrostatic phenomena. He is considered the father of magnetism.

Who was Dmitri Ivanovich Mendeleyev? Dmitri Mendeleyev (1834-1907) was born in Tobolsk, Siberia and studied science at the University of St. Petersburg and in Germany in Heidelberg. (Random House, 328) In 1886, he devised the first logical arrangement of 63 known elements, which he called "The Periodic Table of the Elements". Because of the precise organization of this table, he was able to predict the existence of many unknown elements, including Gallium and Germanium. The element Mendelevium (101) was named in his honor.

Who was Jons Jacob Berzelius? Jons Berzelius (1779-1848) (Random House, 49) a Swedish scientist is regarded by many as the 'Father of Modern Chemistry'. Through his studies he discovered the elements Silicon, Cerium, Selenium and Thorium. He was the first to apply a standard symbol and name to the elements. In 1808, he wrote a chemistry book that was the standard text for 30 years. He also published a yearly review of chemical studies from 1821 to 1848. Berzelius was the recipient of many awards during his career for his contributions to the advancement of chemistry.

Who was Wilhelm Rontgen? Wilhelm Rontgen (1845-1923) taught at the University of Wurzburg. In 1895, while experimenting with a cathode ray generator called a "Crooke's Tube", he discovered that an image of the bones of his hand had formed on a photosensitive plate in his laboratory. He correctly deducted that some unknown form of radiation from the tube had caused this. He called this new form of radiation "X-rays". His discovery led to the study of radioactivity; a whole new area of Chemistry and Physics. He was awarded the Noble Prize for Physics in 1901.

Who was Joseph Priestley? Joseph Priestley (1733-1804) an English chemist and Unitarian minister with a strong interest in science in which he identified oxygen in 1774. (Random House, 393) This occurred when he was experimenting with the compound Mercuric oxide. Other discoveries made by Priestley were the gaseous forms of Carbon Dioxide, Sulfur Dioxide and Hydrogen Chloride. He was a close associate of the American scientist and statesman Benjamin Franklin and in 1794 moved to the U.S. where he lived the remainder of his life.

Who was Chien Shiung Wu? (1912 – 1997) Born in Shanghai, China she studied physics and continued to do so until 1936 when she moved to the U.S. to study at the University of California, Berkeley and received her PhD in 1940. She contributed to the Manhattan Project by developing a process to separate uranium isotopes by gaseous diffusion and by developing improved Geiger counters. She published a book "Beta Decay" in 1965 that is still used today by nuclear physicists.

Who was Katsuko Saruhashi? (1920 – 2007) Born in Tokyo, Japan she started studying CO2 levels in seawater. She earned her doctorate in chemistry in 1957 from the University of Tokyo. In 1954 after the Bikini Atoll nuclear test, the Japanese government began monitoring radioactivity in the seawater and in the rainfall. During these atomic tests a Japanese fishing trawler had been downwind from the tests when they occurred and its occupants became ill from the effects. Saruhashi found that it took a year and a half for the radioactivity to reach Japan in the seawater. In 1964 the radioactivity levels showed that the western and eastern North Pacific Ocean water had mixed completely and by 1969, traces of radioactivity had spread throughout the Pacific. This was some of the first research showing how the effects of fallout can spread across the entire world and not just affect the immediate area.

Who was Rolf Maximilian Sievert? (1896 – 1966) a medical physicist whose major contribution was in the study of biological effects of radiation. In 1964 he founded the International Radiation Protection Association and also chaired the United Nations Scientific Committee on the Effects of Atomic Radiation. In 1979, the unit for ionizing radiation dose equivalent was named after him and given the name sievert (Sv).

Who was Admiral Hyman Rickover? Admiral Hyman George Rickover, United States Navy, (January 27, 1900 – July 8, 1986) was known as the "**Father of the Nuclear Navy**", which as of July 2007 had produced 200 nuclear-powered submarines, and 23 nuclear-powered aircraft carriers and cruisers, though many of these U.S. vessels are now decommissioned and others under construction.

With his unique personality, political connections, responsibilities and depth of knowledge regarding naval nuclear propulsion, Rickover became the longest-serving active duty military officer in U.S. history with 63 years of continuous service.

Rickover's substantial legacy of technical achievements includes the United States Navy's continuing record of zero reactor accidents, as defined by the uncontrolled release of fission products subsequent to reactor core damage.

Who was Soren Sorenson? In 1909 Soren P.L. Sorenson suggested that the hydrogen ion concentration be used to categorize solutions in which he devised the pH scale. It is this scale that remains in use today.

Who was Enrico Fermi? Enrico Fermi (29 September 1901 – 28 November 1954) was an Italian physicist, particularly remembered for his work on the development of the first nuclear reactor, and for his contributions to the development of quantum theory, nuclear and particle physics, and statistical mechanics. Awarded the Nobel Prize in Physics in 1938 for his work on induced radioactivity, Fermi is widely regarded as one of the leading scientists of the 20th century, highly accomplished in both theory and experiment.[1] Fermium, a synthetic element created in 1952, the Fermi National Accelerator Lab, the Fermi Gamma-ray Space Telescope, and a type of particles called fermions are named after him.   Wikipedia

Who was Isaac Babbitt? Isaac Babbitt (b. July 26, 1799, Taunton, Massachusetts - d. May 26, 1862, Somerville, Massachusetts) was in 1839 the inventor of a low-friction tin-based metal alloy, Babbitt metal, that is used extensively in engine bearings today. Wikipedia

Who was Marie Curie? Marie Skłodowska Curie (7 November 1867 – 4 July 1934) was a physicist and chemist of Polish upbringing and, subsequently, French citizenship. She was a pioneer in the field of radioactivity, the first person honored with two Nobel Prizes, receiving one in physics and later, one in chemistry. She was the first woman to serve as professor at the University of Paris.

She was born Maria Skłodowska in Warsaw (then Vistula Land, Russian Empire; now Poland) and lived there until she was twenty-four years old. In 1891 she followed her elder sister, Bronisława, to study in Paris, where she obtained her higher degrees and conducted her subsequent scientific work. She founded the Curie Institutes in Paris and Warsaw. Her husband, Pierre Curie, was a Nobel co-laureate of hers, being awarded a Nobel Prize in physics at the same time. Her daughter, Irène Joliot-Curie, and son-in-law, Frédéric Joliot-Curie, also received Nobel prizes.

Her achievements include the creation of a theory of *radioactivity* (a term she coined), techniques for isolating radioactive isotopes, and the discovery of two new elements, polonium and radium. Under her direction, the world's first studies were conducted into the treatment of neoplasms (cancers), using radioactive isotopes.

While an actively loyal French citizen, she never lost her sense of Polish identity. She named the first new chemical element that she discovered (1898) polonium for her native country, and in 1932 she founded a Radium Institute (now the Maria Skłodowska–Curie Institute of Oncology) in her home town, Warsaw, which was headed by her sister, Bronisława, who was a physician.   Wikipedia

Who was Leo Szilard? Leó Szilárd (Hungarian: *Szilárd Leó*, February 11, 1898 – May 30, 1964) Hungarian-born American physicist who helped conduct the first sustained nuclear chain reaction and was instrumental in initiating the Manhattan Project for the development of the atomic bomb.

In 1922 Szilard received his Ph.D. from the University of Berlin and joined the staff of the Institute of Theoretical Physics there. When the Nazis came into power in 1933, he went to Vienna and, in 1934, to London, where he joined the physics staff of the medical college of St. Bartholomew's Hospital. There, with the British physicist T.A. Chalmers, Szilard developed the first method of separating isotopes (different nuclear forms of the same element) of artificial radioactive elements. In 1937 Szilard went to the United States and taught at Columbia University.

In 1939 Szilard, Edward Teller, and Eugene Wigner persuaded Albert Einstein to write the famous letter to President Franklin D. Roosevelt advocating the immediate development of an atomic bomb. From 1942 until the end of the war he conducted nuclear research at the University of Chicago, where he helped Enrico Fermi construct the first nuclear reactor. In 1946 he became professor of biophysics at Chicago.

After the atomic bomb was first used, Szilard became an ardent promoter of the peaceful uses of atomic energy and the international control of nuclear weapons, founding the

Council for a Livable World. In 1959 he received the Atoms for Peace Award. He published a collection of satirical sketches on the misuse of scientific knowledge entitled *The Voice of the Dolphins and Other Stories* (1961).

Who was Lise Meitner? Meitner, Lise (1878–1968), Austrian-Swedish physicist and mathematician. She was professor at the Univ. of Berlin (1926–33). A refugee from Germany after 1938, she became associated with the Univ. of Stockholm and with the Nobel Institute at Stockholm. In 1917, working with Otto Hahn, she isolated the most stable isotope of the element protactinium; she also investigated the disintegration products of radium, thorium, and actinium and the behavior of beta rays. In 1938 she participated in experimental research in bombarding the uranium nucleus with slow-speed neutrons. Meitner interpreted the results as a fission of the nucleus and calculated that vast amounts of energy were liberated. Her conclusion contributed to the development of the atomic bomb. In 1949, she became a Swedish citizen. The element with the atomic number 109 is named meitnerium in her honor. In 1966 United States Atomic Energy awarded the Enrico Fermi award to the entire team of Meitner, Hahn and Strassman. Meitner was the first woman to get this award. On 27th Oct. 1968 she passed away peacefully.

Denied the Nobel Prize and due recognition during her lifetime, Lise Meitner is now considered one of the most significant woman scientists of the century.

Who was Harvey Hubbell II? (1857 – 1927) Harvey Hubbell was a U.S. inventor who is best known for inventing the electrical plugs and the pull-chain light socket. Hubbell founded Harvey Hubbell, Incorporated in Bridgeport, Connecticut which is still in business today. Hubbell received over 45 patents on his inventions.

Nuclear Groups

What is the Energy Information Administration? The Energy Information Administration (EIA) was created in response to the need for additional Federal initiatives to collect and disseminate energy-related information, and to evaluate and analyze this information. These needs were revealed as the United States sought to respond to the energy crises of the 1970s. The first law to address these needs was the Federal Energy Administration Act of 1974 and, over the years, many subsequent laws have contributed to EIA's evolution and growth. Here is there contact information:

Energy Information Administration
1000 Independence Ave., SW
Washington, DC 20585
202-586-8800
InfoCtr@eia.doe.gov

What is the International Atomic Energy Agency? The IAEA was created in 1957 as an international organization that seeks to promote the peaceful use of nuclear energy and to inhibit its use for military purposes. Though established independently from the United Nations under its own international treaty, the IAEA reports to both the General Assembly and the Security Council. This organization is made up of 144 member states and is headquartered in Vienna, Austria. Their contact information is:

International Atomic Energy Agency
P.O. Box 100
Wagramer Strasse 5
A-1400 Vienna, Austria
Tel: (+431) 2600-0
Fax: (+431) 2600-7
http://www.iaea.org

What is the United Nations Atomic Energy Commission? The UNAEC was founded in 1946 by the first resolution of the United Nations General Assembly to deal with the problems raised by the discovery of atomic energy.

What is the NEI? The Nuclear Energy Institute (NEI) was founded in 1994 from the merger of several nuclear energy industry organizations, the oldest of which was created in 1953.

In 1953, the nuclear industry created the Atomic Industrial Forum (AIF) to focus on the beneficial uses of nuclear energy. This was two years before the international "Atoms for Peace" conference held in Geneva in 1955, marking the dawn of the nuclear age.

In 1987, the industry divided AIF to create NUMARC and USCEA. USCEA originated in 1979 as the U.S. Committee for Energy Awareness.

As the direct successor to AIF, the Nuclear Energy Institute represents more than 50 years' service to the nuclear technologies industry.

NEI has nearly 350 members in 19 countries. They include companies that operate nuclear power plants, design and engineering firms, fuel suppliers and service companies, companies involved in nuclear medicine and nuclear industrial applications, radionuclide and radiopharmaceutical companies, universities and research laboratories, and labor unions.

NEI members help set policy for the nuclear technologies industry. They also have access to all NEI technical reports, alerts on breaking news, our members-only Web site, reduced rates on conference registration, a weekly newsletter and more.

More than 6,000 industry professionals participate in NEI activities and programs, providing NEI broad industry representation and enabling NEI to focus industry expertise on crucial policy matters. Here is there contact information:

Nuclear Energy Institute
1776 I Street NW, Suite 400
Washington, DC 20006-3708
P: 202.739.8000
F: 202.785.4019
www.nei.org

What is the American Nuclear Society? The American Nuclear Society is a not-for-profit, international, scientific and educational organization. It was established by a group of individuals who recognized the need to unify the professional activities within the diverse fields of nuclear science and technology. December 11, 1954, marks the Society's historic beginning at the National Academy of Sciences in Washington, D.C. ANS has since developed a multifarious membership composed of approximately 11,000 engineers, scientists, administrators, and educators representing 1,600 plus corporations, educational institutions, and government agencies. It is governed by four officers and a board of directors elected by the membership. Here is their contact information:

American Nuclear Society
555 North Kensington Avenue
La Grange Park, Illinois 60526
www.ans.org

What is IUPAC? The International Union of Pure and Applied Chemistry (IUPAC) serves to advance the worldwide aspects of the chemical sciences and to contribute to the application of chemistry in the service of Mankind. IUPAC was formed in 1919 by chemists from the industry and academia. As a scientific international non-governmental and objective body, IUPAC can address many global issues involving the chemical sciences. Here is their contact information:

International Union of Pure and Applied Chemistry
IUPAC Secretariat
P.O. Box 13757
Research Triangle Park, NC 27709-3757, USA
www.iupac.org

What is the Advisory Committee on Reactor Safeguards? The Advisory Committee on Reactor Safeguards (ACRS) is statutorily mandated by the Atomic Energy Act of 1954, as amended. The Committee has three primary purposes:

- to review and report on safety studies and reactor facility license and license renewal applications;
- to advise the Commission on the hazards of proposed and existing production and utilization facilities and the adequacy of proposed safety standards;
- to initiate reviews of specific generic matters or nuclear facility safety-related items; and
- to provide advice in the areas of health physics and radiation protection.

At the request of the Commission, the ACRS also reviews the NRC's Research Activities and provides a biannual report (NUREG-1635) to the Commission. Upon request from the Department of Energy, the ACRS reviews and provides reports on U.S. Naval reactor designs under a reimbursable agreement. Upon request, and with the Commission's consent, the ACRS is required to provide advice to the Defense Nuclear Facilities Safety Board in accordance with Public Law 100-456.

The ACRS is independent of the NRC staff and reports directly to the Commission, which appoints its members. The operational practices of the ACRS are governed by the provisions of the Federal Advisory Committee Act (FACA). Advisory committees are structured to provide a forum where experts representing many technical perspectives can provide independent advice that is factored into the Commission's decision-making process. Most Committee meetings are open to the public and any member of the public may request an opportunity to make an oral statement during the committee meeting. Those who wish to do so should contact the ACRS contact indicated on the Public Meeting Schedule page. 							Courtesy of the NRC

What is INPO? The Institute of Nuclear Power Operations which was established by the nuclear power industry in December 1979, the Institute of Nuclear Power Operations is a not-for-profit organization headquartered in Atlanta.

The Kemeny Commission – set up by President Jimmy Carter to investigate the March 1979 accident at the Three Mile Island nuclear power plant – had recommended that:

The (nuclear power) industry should establish a program that specifies appropriate safety standards including those for management, quality assurance, and operating procedures and practices, and that conducts independent evaluations.

There must be a systematic gathering, review, and analysis of operating experience at all nuclear power plants coupled with an industry-wide international communications network to facilitate the speedy flow of this information to affected parties.

In addressing those recommendations, the nuclear power industry:

- Established INPO – the Institute of Nuclear Power Operations
- Charged INPO with a mission that remains the same today:

To promote the highest levels of safety and reliability – to promote excellence – in the operation of nuclear electric generating plants. Here is their contact information:

Institute of Nuclear Power Operations
700 Galleria Parkway, SE, Suite 100
Atlanta, GA 30339-5943
www.inpo.info

What is EPCE? The Energy Providers Coalition for Education (EPCE) is a group of industry representatives that develops, sponsors, and promotes industry-driven, standardized, quality online learning programs to meet the workforce needs of the energy industry. EPCE is a collaboration of energy employers, associations, contractors, labor organizations and education providers that work together to develop and implement quality online learning solutions. Industry stakeholders join EPCE as members to lead the development and use of online learning for energy workforce development. Members also join EPCE to network with peers, receive discounts on tuition and fees, and direct the future of EPCE. Here is there contact information:

Energy Providers Coalition for Education
c/o CAEL
6021 S. Syracuse Way, Suite 213
Greenwood Village, CO 80111
Phone: 303-804-4672
Fax: 303-773-0026
info@epceonline.org

What is WANO? The World Association of Nuclear Operators (WANO) was founded in 1989 after the Chernobyl accident to foster international cooperation and professional excellence within the nuclear industry. Every organization in the world that operates a nuclear electric generating station is a WANO member. WANO operates four regional centers in Tokyo, Paris, Moscow, and Atlanta, with a coordinating center in London.

WANO-Atlanta Centre
700 Galleria Parkway SE
Suite 100
Atlanta, GA 30339, USA
www.wano.org.uk/

What is NANTeL? NANTeL -- National Academy for Nuclear Training electronic-Learning -- is a national Web-based system that provides standardized, generic training for the supplemental workforce. Training topics will initially include plant access, radiation worker, and human performance tools. INPO and the National Academy for Nuclear Training manage and operate the system for the U.S. nuclear industry. The courses are required before you can perform most jobs at a nuclear power plant. All U.S. nuclear power plants have agreed to accept training delivered through NANTeL. That means you can avoid repeating training courses if you move from one nuclear site to another. It means that nuclear power plants agree to share your training records through NANTeL, and you get credit for training that you've already completed on NANTeL. Here is their contact information:

NANTeL
770-644-8900
nanteladmin@inpo.org
www.nantel.org

What is the National Board of Boiler and Pressure Vessel Inspectors? The NB was created in 1919 to promote greater safety to life and property through uniformity in the construction, installation, repair, maintenance and inspection of boilers and pressure vessels. This membership oversees adherence to codes involving the construction and repair of boilers and pressure vessels. Here is their address:

National Board of Boilers And Pressure Vessel Inspectors
1055 Crupper Ave
Columbus, OH 43229
614-888-8320
www.nationalboard.com

What is the N.E.C.? The N.E.C. is the National Electrical Code. This code sets a standard that provides minimum safety requirements for wiring installations in residential and commercial applications. In which is a book that is updated and published every three years by the National Fire Protection Agency. It contains rules and regulations for the proper installation of electrical wiring and devices in order to prevent fires and ensure electrical safety.

What is the U.S. Electric Power Research Institute? EPRI was created from a crisis. In the great northeastern United States in 1965 revealed for the first time the serious vulnerabilities in the nation's electrical supply system. Threatened by Senate proposals in 1971 to create a special federal agency to conduct electricity related research and development projects, in response, America's public and private utilities banded together to develop an industry-organized alternative, now known as EPRI. With almost 30 years of success, the company serves about 1,000 energy related organizations in 40 countries. EPRI's program is designed to help clients cut costs, improve efficiency, meet the needs of energy end-users, and build for the future. Here is their address:

What are the Reliability councils of the NERC? There are nine Reliability councils of the NERC, they are: ERCOT- Electrical Reliability Council of Texas, WSEC- Western Systems Coordinating Council, the Eastern interconnection is made up of the other seven councils which are: ECAR- East Central Area Reliability Coordination Agreement, MAAC- Mid-Atlantic Area Council, MAIN- Mid-America Interconnected Network, MAPP- Mid-Continent Area Power Pool, NPCC- Northeast Power Coordinating Council (includes Hydro-Quebec), SERC- Southeastern Electric Reliability Council, and SPP- Southwest Power Pool. Here is their contact information:

What is the Federal Energy Regulatory Council? The FERC is a federal government agency that was created in 1977 and is responsible for setting rates on transportation and the sale of electricity, the transportation of oil through a pipeline, and other regulations that apply to the utility industry. Here is their address:

What is the I.E.E.E.? The I.E.E.E. stands for Institute of Electrical and Electronic Engineers which was formed through a merger in 1963 between the AIEE (American Institute of Electrical Engineers) created in 1884 and the IRE (Institute of Radio Engineers) created in 1884. The IEEE is a non-profit organization that has 350,000 members in 150 countries. Because of its members, the IEEE is the leading authority in technical areas ranging from computer engineering, biomedical technology, telecommunications, electric power production, aerospace, consumer electronics and several other areas. From its beginning, the IEEE has advanced in the theory and application of electro-technology and other sciences, served as a catalyst for technological innovation and supported the needs of its members through a wide variety of programs and services. Here is their address:

IEEE
445 Hoes Lane
P.O.Box 1331
Piscataway, New Jersey
08855-1331
732-981-0060
www.iccc.org

What does A.S.M.E. stand for? A.S.M.E. stands for The American Society of Mechanical Engineers. It was founded in 1880 by 3 engineers, Alexander Holley (1832-1882), Henry Worthingon (1817-1880) and John Sweet (1832-1882). ASME is a worldwide engineering society that focuses on technical, educational and research issues. It conducts one of the largest technical publishing operations in the world with 125,000 members and sets many industrial and manufacturing standards. Here is their address:

ASME
3 Park Ave
New York, NY 10016-5990
www.asme.org

What is A.S.T.M.? A.S.T.M. is the American Society for Testing and Materials. A.S.T.M. is the foremost developer and provider of voluntary consensus standards, related technical information and services having internationally recognized quality and applicability that improves the overall quality of life, contributes to the reliability of materials, products, systems and services. Here is their address:

ASTM
100 Barr Harbor Drive
West Conshohocken, PA 19428-2959
610-832-9585
www.astm.org

What is OSHA? The Occupational Safety and Health Administration (OSHA) was established in 1970 by the U.S. Department of Labor. This agency is to regulate health and safety standards in the work force by conducting on-site inspections and promotes public awareness of on the job hazards. Here is their address:

U.S. Department of Labor
Occupational Safety and Health Administration
200 Constitution Avenue
Washington, D.C. 20210
1-800-321-6742
www.osha.com

What is the AIChE? The AIChE is the American Institute of Chemical Engineers, founded in 1908 with a membership approximately 57,000. It is a non-profit organization that provides leadership to the chemical engineering profession. Their address is:

AIChE
3 Park Ave
New York, NY
10016-5991
1-800-242-4363
www.aiche.org

What is the AIME? The AIME is the American Institute of Mining, Metallurgical and Petroleum Engineers. The AIME was founded in 1871 by 22 mining engineers and today has nearly 90,000 members. The goal of AIME is to advance the knowledge of engineering and the arts and sciences involved in the production and use of minerals, metals, materials and energy resources, while disseminating significant developments in these areas of technology. Here is their contact information:

AIME
Three Park Avenue
New York, NY
10016-5998
212-419-7676
www.idis.com/aime/welcome.htm

What is the Idaho National Laboratory? In operation since 1949, INL is a science-based, applied engineering national laboratory dedicated to supporting the U.S. Department of Energy's missions in nuclear and energy research, science, and national defense.

Mission

Ensure the nation's energy security with safe, competitive, and sustainable energy systems and unique national and homeland security capabilities.

Vision

By 2015, INL will be the pre-eminent nuclear energy laboratory with synergistic, world-class, multi-program capabilities and partnerships.

Safety and Environmental Stewardship

The health and safety of every employee, both on and off-the-job, is critical to our mission, and we demonstrate world-leading safety behavior, safety performance and environmental stewardship.

BEA and Partners

INL is operated for the Department of Energy (DOE) by Battelle Energy Alliance (BEA) and partners, each providing unique educational, management, research and scientific assets into a world-class national laboratory.

Organization

Leading a senior management team of eighteen is Laboratory Director John J. Grossenbacher, a Vice Admiral, U.S. Navy (ret.), a Naval Academy graduate and one of the nation's most respected leaders. At the time of his retirement from the Navy, Mr. Grossenbacher commanded all United States submarine forces.

INL History

The world's first usable amount of electricity from nuclear energy was generated in Idaho in 1951. Here is their contact information:

Idaho National Laboratory
2025 Fremont Avenue
Idaho Falls, ID 83415
866-495-7440
www.inl.gov

What is AREVA? The AREVA group is born, lead by CEA Industrie, through the merger of all Framatome ANP, COGEMA, CEA and FCI operations. Framatome's contribution makes AREVA the sole supplier of next-generation EPR reactors and number one worldwide for fuel fabrication and supply, with a strong presence in Europe and the United States. AREVA is the only group in the world with expertise and active involvement in every sector of the nuclear power industry, including the nuclear fuel cycle, reactors, instrumentation, nuclear measurement systems and engineering. Here is their contact information:

AREVA NP INC.
3315 Old Forest Road
Lynchburg, VA 24501
www.us.areva-np.com

What is Électricité de France? Electricité de France (EDF) is the world's largest utility company with €64.28 billion in revenues 2008, operating a diverse portfolio of 120,000+ megawatts of generation capacity in Europe, Latin America, Asia, the Middle-East and Africa. EDF was founded on April 8, 1946, as a result of the nationalisation of a number of electricity producers, transporters and distributors by the minister of industrial production Marcel Paul and has become the main electricity generation and distribution company in France. Among these prior electricity producers was Société Toulousaine d'Electricité du Bazacle at that time probably the oldest limited company in the world. [2]

Until November 19, 2004, it was a government corporation, but it is now a limited-liability corporation under private law (*société anonyme*). The French government partially floated shares of the company on the Paris Stock Exchange in November 2005,[3] although it retains almost 85% ownership as of the end of 2008.[4]

EDF held a monopoly in the distribution, but not the production, of electricity in France until 1999, when the first European Commission directive to harmonize regulation of electricity markets was implemented.

EDF is one of the world's largest producers of electricity. In 2003, it produced 22% of the European Union's electricity, primarily from nuclear power:

- nuclear: 74.5%
- hydro-electric: 16.2%
- thermal: 9.2%
- wind power and other renewable sources: 0.1%

Its 58 active nuclear reactors (in 2004) are spread out over 20 sites (nuclear power plants). They comprise 34 reactors of 900 MWe, 20 reactors of 1.3 GWe, and 4 reactors of 1450 MWe, all PWRs.

On September 24, 2008 EDF announced that its wholly owned subsidiary, Lake Acquisitions, had made a recommended offer for the shares of British Energy Group PLC, which generates about 20 percent of British electricity, mainly from 8 nuclear stations.

What is the Atomic Energy of Canada Limited? The AECLis a leading-edge nuclear technology and services company providing services to utilities worldwide. Over 5,000 highly skilled employees enthusiastically deliver a range of nuclear services: from R&D support, construction management, design and engineering to specialized technology, waste management and decommissioning in support of CANDU reactor products.

AECL is committed to supporting its Canadian and international customers in all aspects of nuclear power technology management. We provide on-site expertise, closely supported by our nuclear science laboratories, testing capability and engineering facilities. CANDU reactors supply about 16% of Canada's electricity and are an important component of clean-air energy programs on four continents.

For over 50 years AECL has been safely and passionately developing peaceful and productive applications of nuclear energy. Here is their contact information:

AECL
1-866-513-2325 Public Request for Information
librarycr@aecl.ca or librarysp@aecl.ca
www.aecl.ca

Nuclear Accidents

What happened at Chalk River? On December 12, 1952 a combination of mechanical failure and human error led to a now-famous power excursion and fuel failure in the NRX (Nuclear Thermal Reactor) reactor at AECL Chalk River Laboratories. At the time NRX was one of the most significant research reactors in the world (rated at that time for 30 MW operation), in its sixth year of operation.

During preparations for a reactor-physics experiment at low power, a defect in the NRX shut-off rod mechanism combined with a number of operator errors to cause a temporary loss of control over reactor power. Power surged ultimately to somewhere between 60 and 90 MW over a period of about a minute (the total energy surge is estimated to be approximately 4000 MW-seconds). This energy load would normally not have been a problem, but several experimental fuel rods that were at that moment receiving inadequate cooling for high power operation ruptured and melted. About 10,000 Curies of fission products were carried by about a million gallons of cooling water into the basement of the reactor building. This water was subsequently pumped to Chalk River Laboratories' waste management facility, where the long-term ground water outflow was monitored thereafter to ensure adherence to the drinking water standard. The core of the reactor was left severely damaged.

This accident is historically important, not only because it was the first of its type and magnitude, but also because of its legacy to Canadian and international practice in reactor safety and design. Nobody was killed or hurt in the incident, but a massive clean-up operation was required that involved hundreds of AECL staff, as well as Canadian and American military personnel, and employees of an external construction company working at the site. In addition the reactor core itself was rendered unusable for an extended period. Environmental effects outside the plant were negligible, as was radiation exposure to members of the public. The health record of AECL and Canadian military personnel involved in the clean-up was scientifically reviewed in the 1980s (no significant health effects were observed).

Several of today's fundamental safety principles of reactor design and operation stem from the lessons learned at this formative stage of Canada's nuclear program, making Canada an early leader in this field. Among these were:

- the need for an independent, reliable, fast-acting shutdown system, separate from routine reactor control;
- the need for shutdown capability even in a reactor that is already shutdown (i.e., the safest reactor configuration may not be one with all neutron absorbers in the core);
- the need for a reactor trip on rate of change in power, in addition to a high power threshold;
- the importance of written and thoroughly reviewed procedures for every operational and experimental activity;

- the importance of an efficient human-machine interface in the control room;
- the need to balance thorough safety coverage with simplicity that does not interfere unduly with operations.

The accident also demonstrated that, due to a combination of redundant safety features, emergency procedures, and a level of inherent "forgiveness" (or robustness) in the technology, a major fuel-melt accident in a nuclear reactor can occur without significant environmental effects and radiation exposure to the surrounding population.

Courtesy of www.nuclearfaq.ca/cnf_sectionD.htm

What happened at Windscale? After the Second World War, in 1946, despite the participation of many British scientists in the Manhattan Project, and formal agreement of a joint technology-exchange program, the United States government passed legislation that closed its nuclear weapons program to all other countries.

The British government, not wanting to be left behind as a world power in an emerging arms race, then embarked on a program to build its own atomic bomb as quickly as possible.

The reactors were built near the tiny village of Seascale, Cumberland, and were known as Windscale Pile 1 and Windscale Pile 2, housed in large, concrete buildings a few hundred feet from one another. The reactors were graphite-moderated and air-cooled. Because nuclear fission produces large amounts of heat, it was necessary to cool the reactor cores by blowing cold air through channels in the graphite. Hot air was then exhausted out of the back of the core and up the chimney. Filters were added late into construction at the insistence of Sir John Cockcroft and these were housed in galleries at the very top of the discharge stacks. They were deemed unnecessary, a waste of money and time and presented something of an engineering headache, being added very late in construction in large concrete houses at the top of the 400-ft chimneys. Due to this, they were known as "Cockcroft's Folly" by workers and engineers. As it was, "Cockcroft's Folly" probably prevented a disaster from becoming a catastrophe.

The reactors themselves were built of a solid graphite core, with horizontal channels through which cans of uranium and isotope cartridges could be passed, to expose the isotope cartridges to neutron radiation from the uranium and produce plutonium and radioisotopes, respectively. Fuel and isotopes were fed into the channels in the front of the reactor, the "charge face", and spent fuel was then pushed all the way through the core and out of the back—the "discharge face"—into a water duct for initial cooling prior to retrieval and processing to extract the plutonium.

Unenriched uranium metal in aluminum cans with fins to improve cooling was used for the production of plutonium. As this plutonium was intended for weapons purposes, the burn-up of the fuel would have been kept low to reduce production of the heavier plutonium isotopes.

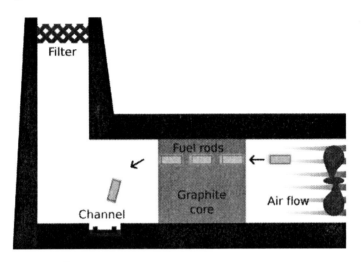

On 7 October, 1957, operators began an annealing cycle for Windscale Pile №1 by switching the cooling fans to low power and stabilizing the reactor at low power. The next day, to carry out the annealing, the operators increased the power to the reactor. When it appeared that the annealing process was taking place, control rods were lowered back into the core to shut down the reactor, but it soon became apparent that the Wigner energy release was not spreading through the core, but dwindling prematurely. The operators withdrew the control rods again to apply a second nuclear heating and complete the annealing process. Because some thermocouples were not in the hottest parts of the core, the operators were not aware that some areas were considerably hotter than others. This, and the second heating, are suspected of having been the deciding factors behind the fire, although the precise cause remains unknown. The official report suggests that a can of uranium ruptured and oxidized causing further overheating and the fire, but a more recent report suggests that it may actually have been a magnesium/lithium isotope cartridge. All that was visible on the instruments was a gentle increase in temperature, which was to be expected during the Wigner release.

Early in the morning on 10 October, it was suspected that something unusual was going on. The temperature in the core was supposed to gradually fall as Wigner release ended, but the monitoring equipment showed something more ambiguous was going on and one thermocouple indicated that core temperature was instead rising. In an effort to help cool the pile, more air was pumped through the core. This lifted radioactive materials up the chimney and into the filter galleries. It was then that workers in the control room realized that the radiation monitoring devices which measured activity at the top of the discharge stack were at full scale reading. In accordance with written guidelines, the foreman declared a site emergency. No one at Windscale was now in any doubt that Pile Number 1 was in serious trouble.

Operators tried to examine the pile with a remote scanner but it had jammed. Tom Hughes, second in command to the Reactor Manager, suggested examining the reactor personally and so he and another operator went to the charge face of the reactor, clad in protective gear. A fuel channel inspection plug was taken out close to a thermocouple registering high temperatures and it was then that the operators saw that the fuel was red hot.

"An inspection plug was taken out," said Tom Hughes in a later interview, "and we saw, to our complete horror, four channels of fuel glowing bright cherry red."

There was no doubt that the reactor was now on fire, and had been for almost 48 hours. Reactor Manager Tom Tuohy donned full protective equipment and breathing apparatus and scaled the 80 feet to the top of the reactor building, where he stood atop the reactor lid to examine the rear of the reactor, the discharge face. Here he reported a dull red luminescence visible, lighting up the void between the back of the reactor and the rear containment. Red hot fuel cartridges were glowing in the fuel channels on the discharge face. He returned to the reactor upper containment several times throughout the incident, at the height of which a fierce conflagration was raging from the discharge face and playing on the back of the reinforced concrete containment—concrete whose specifications insisted that it must be kept below a certain temperature to prevent its disintegration and collapse.

Operators were unsure what to do about the fire. First, they tried to blow the flames out by putting the blowers onto full power and increasing the cooling, but predictably this simply fanned the flames. Tom Hughes and his colleague had already created a fire break by ejecting some undamaged fuel cartridges from around the blaze and Tom Tuohy suggested trying to eject some from the heart of the fire, by bludgeoning them through the reactor and into the cooling pond behind it with scaffolding poles. This proved impossible and the fuel rods refused to budge, no matter how much force was applied. The poles were withdrawn with their ends red hot and, once, a pole was returned red hot and dripping with molten metal. Hughes knew this had to be molten irradiated uranium and this caused serious radiation problems on the charge hoist itself.

Next, the operators tried to extinguish the fire using carbon dioxide. The new gas-cooled Calder Hall reactors next door had just received a delivery of 25 tons of liquid carbon dioxide and this was rigged up to the charge face of Windscale Pile 1, but there were problems getting it to the fire in useful quantities. The fire was so hot that it stripped the oxygen from what carbon dioxide could be applied.

On the morning of Friday 11 October and at its peak, eleven tons of uranium was ablaze. Temperatures were becoming extreme (one thermocouple registered 1,300 degrees

Celsius) and the biological containment around the stricken reactor was now in severe danger of collapse. Faced with this crisis, the operators decided to use water. This was incredibly risky: molten metal oxidizes in contact with water, stripping oxygen from the water molecules and leaving free hydrogen, which could mix with incoming air and explode, tearing open the weakened containment. But there was no other choice. About a dozen hoses were hauled to the charge face of the reactor; their nozzles were cut off and the lines themselves connected to scaffolding poles and fed into fuel channels about a meter above the heart of the fire.

Tom Tuohy then ordered everyone out of the reactor building except himself and the Fire Chief. All cooling and ventilating air entering the reactor was shut off. Tuohy once again hauled himself atop the reactor shielding and ordered the water to be turned on, listening carefully at the inspection holes for any sign of a hydrogen reaction as the pressure was increased. Tuohy climbed up several times and reported watching the flames leaping from the discharge face slowly dying away. During one of the inspections, Tuohy found that the inspection plates—which were removed with a metal hook to facilitate viewing of the discharge face of the core—were stuck fast. This, Tuohy reported, was the fire trying to suck air in from wherever it could.

Finally he managed to pull the inspection plate away and was greeted with the unfathomable sight of the fire dying away.

"First the flames went, then the flames reduced and the glow began to die down," he described, "I went up to check several times until I was satisfied that the fire was out. I did stand to one side, sort of hopefully," he went on to say, "but if you're staring straight at the core of a shutdown reactor you're going to get quite a bit of radiation."

Water was kept flowing through the pile for a further 24 hours until it was completely cold.

No one was evacuated from the surrounding area, but there was concern that milk might be dangerously contaminated. Milk from about 500 km^2 of nearby countryside was destroyed by (dumping it in the Irish Sea) for about a month.

What happened at SL-1? The SL-1, or Stationary Low-Power Reactor Number One, was a United States Army experimental nuclear power reactor which underwent a steam explosion and meltdown on January 3, 1961, killing its three operators. The direct cause was the improper withdrawal of the main control rod, responsible for 80% of neutron moderation in the poorly designed reactor core. The event is the only fatal reactor accident in the United States.

The facility, located at the National Reactor Testing Station approximately forty miles (60 km) west of Idaho Falls, Idaho, was part of the Army Nuclear Power Program and was known as the Argonne Low Power Reactor (ALPR) during its design and build phase. It was intended to provide electrical power and heat for small, remote military facilities, such as radar sites near the Arctic Circle, and those in the DEW Line. The design power was 3 MW (thermal). Operating power was 200 kW electrical and 400 kW thermal for space heating. NASA system failure studies have cited that the core power level reached nearly 20 GW in just four milliseconds, precipitating the reactor accident and steam explosion.

On December 21, 1960, the reactor was shut down for maintenance, calibration of the instruments, installation of auxiliary instruments, and installation of 44 flux wires to monitor the neutron flux levels in the reactor core. The wires were made of aluminum, and contained slugs of aluminum-cobalt alloy.

On January 3, 1961 the reactor was to be restarted after a shutdown of eleven days over the holidays. Maintenance procedures commenced, which required the main central control rod to be withdrawn a few inches; at 9:01 p.m. this rod was suddenly withdrawn too far, causing SL-1 to go prompt critical instantly. In four milliseconds, the heat generated by the resulting enormous power surge caused water surrounding the core to begin to explosively vaporize. The water vapor caused a pressure wave to strike the top of the reactor vessel. This propelled the control rod and the entire reactor vessel upwards, which killed the operator who had been standing on top of the vessel, leaving him impaled to the ceiling by the control rod. The other two military personnel, a supervisor and a trainee, were also killed. The victims were Army Specialists John A. Byrnes (age 25) and Richard L. McKinley (age 22), and Navy Electrician's Mate Richard C. Legg (age 25).

Courtesy of Radiationworks.com

What happened at Fermi 1? The Fermi I reactor was a breeder located at Lagoona Beach, 30 miles from Detroit. On October 5, 1966, high temperatures were measured (700 compared to normal 580F) and radiation alarms sounded involving two fuel rod subassemblies. The reactor scrammed and there was indication of fuel melting. After a month of sweating, they tested out enough subassemblies to limit the damage to 6 subassemblies. By January 67 they had learned that 4 subassemblies were damaged with two stuck together, but it took until May to remove the assemblies.

When they had checked the sodium flow earlier, they had detected a clapping noise. In August 67 they were able to lower a periscope device into the meltdown pan and found that a piece of zirconium cladding had come loose and was blocking the sodium coolant nozzles. The zirconium cladding was part of the lining of the meltdown cone designed to direct the distribution of fuel material should a meltdown of the fuel occur. Such structures are necessary in a breeder reactor because of the possibility of molten fuel reassembling itself in a critical configuration. This is not a possibility in an ordinary light water reactor because of the low level of enrichment of the uranium, but a fast breeder reactor is operated with a much higher level of enrichment. The phrase "China syndrome" was coined in regard to this accident as they were contemplating the possibilities should a meltdown of fuel with critical reassembly take place. The uncontrolled fission reaction could create enough heat to melt its way into the earth, and some engineer remarked "it could go all the way to China".

With ingenious tools designed and built for the purpose, the piece of zirconium was fished out in April of 1968. In May of 1970, the reactor was ready to resume operation, but a sodium explosion delayed it until July of 1970. In October it finally reached a level of 200 megawatts. The total cost of the repair was about $132 million. In August of 1972 upon denial of the extension of its operating license, the shutdown process for the plant was initiated.   Courtesy of Hyperphysics.phy-astr.gsu.edu

What happened at Browns Ferry? At noon on March 22, 1975, both Units 1 and 2 at the Brown's Ferry plant in Alabama were operating at full power, delivering 2200 megawatts of electricity to the Tennessee Valley Authority.

Just below the plant's control room, two electricians were trying to seal air leaks in the cable spreading room, where the electrical cables that control the two reactors are separated and routed through different tunnels to the reactor buildings. They were using strips of spongy foam rubber to seal the leaks. They were also using candles to determine whether or not the leaks had been successfully plugged -- by observing how the flame was affected by escaping air.

The electrical engineer put the candle too close to the foam rubber, and it burst into flame.

The resulting fire, which disabled a large number of engineered safety systems at the plant, including the entire emergency core cooling system (ECCS) on Unit 1, and almost resulted in a boil off/meltdown accident, demonstrates the vulnerability of nuclear plants to "single failure" events and human fallibility.

Approximately 15 minutes passed between the time the fire started (12:20 pm) and the time at which a fire alarm was turned in. It was not until one of the electricians told a plant guard inside the turbine building that a fire had broken out that an alarm was sounded. However, confusion over the correct telephone number for the fire alarm delayed its being sounded.

Despite the fire alarm, the reactor operators in the plant control room did not shut down the two reactors, but continued to let them run. At 12:40, five minutes after the fire alarm sounded, the Unit 1 reactor operator noticed that all of the pumps in the emergency core cooling system (ECCS) had started.

Beginning at 12:55, the electrical supply was lost both to control and power the emergency core cooling system and other reactor shutdown equipment on Unit 1. The normal feedwater system was lost; the reactor core spray system was lost; the low-pressure ECCS was lost; the reactor core isolation cooling system was lost; and most of the instrumentation which tells the control room what is going on in the reactor was lost.

Meanwhile, a few feet away on the Unit 2 side of the control room, warning lights had also been going off for some time. At 1:00 pm the Unit 2 operator observed decreasing reactor power, many scram alarms, and the loss of some indicating lights. The operator put the reactor in shutdown mode. Some of the shutdown equipment began failing on Unit 2, and the high-pressure ECCS was lost at 1:45 pm. Control over the reactor relief valves was lost at 1:20 pm and not restored until 2:15 pm, at which time the reactor was depressurized by using the relief valves and brought under control.

The electrical cables continued to burn for another six hours, because the fire fighting was carried out by plant employees, despite the fact that professional firemen from the Athens, Alabama, fire department had been on the scene since about 1:30 pm. As the Athens fire chief pointed out, "I was aware that my effort was in support of, and under the direction of, Browns Ferry plant personnel, but I did recommend, after I saw the fire in the cable spreading room, to put water on it. The Plant Superintendent was not receptive to my ideas.

"I informed him that this was not an electrical fire and that water could and should be used because CO_2 and dry chemical were not proper for this type of fire. The problem was to cool the hot wires to prevent recurring combustion. CO_2 and dry chemical were not capable of providing the required cooling. Throughout the afternoon, I continued to recommend the use of water to the Plant Superintendent. He consulted with people over the phone, but apparently was told to continue to use CO_2 and dry chemical. Around 6:00 pm, I again suggested the use of water . . . . The Plant Superintendent finally agreed and his men put out the fire in about 20 minutes.

"They were using type B and C extinguishers on a type A fire; the use of water would have immediately put the fire out."

Even when the decision to put the fire out with water had been taken, further difficulties developed. The fire hose had not been completely removed from the hose rack, so that full water pressure did not reach the nozzle. The fire-fighters did not know this, however, and decided that the nozzle was defective. They borrowed a nozzle from the Athens fire department, "but it had incorrect type threads and would not stay on the hose."

Lessons learned from this accident was to separate redundant equipment so that a single point failure would be less likely to take place. The ability to not only prevent fires but to detect them and put them out quicker. Improving the use and control of combustible materials. www.ccnr.org/browns_ferry.html

How did Three Mile Island get its name? Many years before the plant was built, the local farmers owned the land. It was approximately 3 miles from the island to the nearest town...therefore it was called Three Mile Island.

What happened at Three Mile Island? The accident at the Three Mile Island Unit 2 (TMI-2) nuclear power plant near Middletown, Pa., on March 28, 1979, was the most serious in U.S. commercial nuclear power plant operating history, even though it led to no deaths or injuries to plant workers or members of the nearby community. But it brought about sweeping changes involving emergency response planning, reactor operator training, human factors engineering, radiation protection, and many other areas of nuclear power plant operations. It also caused the U.S. Nuclear Regulatory

Commission to tighten and heighten its regulatory oversight. Resultant changes in the nuclear power industry and at the NRC had the effect of enhancing safety.

The sequence of certain events – equipment malfunctions, design-related problems and worker errors – led to a partial meltdown of the TMI-2 reactor core but only very small off-site releases of radioactivity.

The accident began about 4:00 a.m. on March 28, 1979, when the plant experienced a failure in the secondary, non-nuclear section of the plant. The main feedwater pumps stopped running caused by either a mechanical or electrical failure, which prevented the steam generators from removing heat. First the turbine, then the reactor automatically shut down. Immediately, the pressure in the primary system (the nuclear portion of the plant) began to increase. In order to prevent that pressure from becoming excessive, the pilot-operated relief valve (a valve located at the top of the pressurizer) opened. The valve should have closed when the pressure decreased by a certain amount, but it did not. Signals available to the operator failed to show that the valve was still open. As a result, cooling water poured out of the stuck-open valve and caused the core of the reactor to overheat.

As coolant flowed from the core through the pressurizer, the instruments available to reactor operators provided confusing information. There was no instrument that showed the level of coolant in the core. Instead, the operators judged the level of water in the core by the level in the pressurizer, and since it was high, they assumed that the core was properly covered with coolant. In addition, there was no clear signal that the pilot-operated relief valve was open. As a result, as alarms rang and warning lights flashed, the operators did not realize that the plant was experiencing a loss-of-coolant accident. They took a series of actions that made conditions worse by simply reducing the flow of coolant through the core.

Because adequate cooling was not available, the nuclear fuel overheated to the point at which the zirconium cladding (the long metal tubes which hold the nuclear fuel pellets) ruptured and the fuel pellets began to melt. It was later found that about one-half of the core melted during the early stages of the accident. Although the TMI-2 plant suffered a severe core meltdown, the most dangerous kind of nuclear power accident, it did not produce the worst-case consequences that reactor experts had long feared. In a worst-case accident, the melting of nuclear fuel would lead to a breach of the walls of the containment building and release massive quantities of radiation to the environment. But this did not occur as a result of the three Mile Island accident.

The accident caught federal and state authorities off-guard. They were concerned about the small releases of radioactive gases that were measured off-site by the late morning of March 28 and even more concerned about the potential threat that the reactor posed to the surrounding population. They did not know that the core had melted, but they immediately took steps to try to gain control of the reactor and ensure adequate cooling to the core. The NRC's regional office in King of Prussia, Pa., was notified at 7:45 a.m. on March 28. By 8:00, NRC Headquarters in Washington, D.C., was alerted and the NRC

Operations Center in Bethesda, Md., was activated. The regional office promptly dispatched the first team of inspectors to the site and other agencies, such as the Department of Energy and the Environmental Protection Agency, also mobilized their response teams. Helicopters hired by TMI's owner, General Public Utilities Nuclear, and the Department of Energy were sampling radioactivity in the atmosphere above the plant by midday. A team from the Brookhaven National Laboratory was also sent to assist in radiation monitoring. At 9:15 a.m., the White House was notified and at 11:00 a.m., all non-essential personnel were ordered off the plant's premises.

By the evening of March 28, the core appeared to be adequately cooled and the reactor appeared to be stable. But new concerns arose by the morning of Friday, March 30. A significant release of radiation from the plants auxiliary building, performed to relieve pressure on the primary system and avoid curtailing the flow of coolant to the core, caused a great deal of confusion and consternation. In an atmosphere of growing uncertainty about the condition of the plant, the governor of Pa., Richard L. Thornburgh, consulted with the NRC about evacuating the population near the plant. Eventually, he and NRC Chairman Joseph Hendrie agreed that it would be prudent for those members of society most vulnerable to radiation to evacuate the area. Thornburgh announced that he was advising pregnant women and pre-school-age children within a 5-mile radius of the plant to leave the area.

Within a short time, the presence of a large hydrogen bubble in the dome of the pressure vessel, the container that holds the reactor core, stirred new worries. The concern was that the hydrogen bubble might burn or even explode and rupture the pressure vessel. In that event, the core would fall into the containment building and perhaps cause a breach of containment. The hydrogen bubble was a source of intense scrutiny and great anxiety, both among government authorities and the population, throughout the day on Saturday, March 31. The crisis ended when experts determined on Sunday, April 1, that the bubble could not burn or explode because of the absence of oxygen in the pressure vessel. Further, by that time, the utility had succeeded in greatly reducing the size of the bubble.

Detailed studies of the radiological consequences of the accident have been conducted by the NRC, the Environmental Protection Agency, the Department of Health, Education and Welfare (now Health and Human Services), the Department of Energy, and the State of Pa.. Several independent studies have also been conducted. Estimates are that the average dose to about 2 million people in the area was only about 1 millirem. To put this into context, exposure from a chest x-ray is about 6 millirem. Compared to the natural radioactive background dose of about 100-125 millirem per year for the area, the collective dose to the community from the accident was very small. The maximum dose to a person at the site boundary would have been less than 100 millirem.

In the months following the accident, although questions were raised about possible adverse effects from radiation on human, animal, and plant life in the TMI area, none could be directly correlated to the accident. Thousands of environmental samples of air, water, milk, vegetation, soil, and foodstuffs were collected by various groups monitoring the area. Very low levels of radionuclides could be attributed to releases from the accident. However, comprehensive investigations and assessments by several well respected organizations have concluded that in spite of serious damage to the reactor, most of the radiation was contained and that the actual release had negligible effects on the physical health of individuals or the environment.

The accident was caused by a combination of personnel error, design deficiencies, and component failures. There is no doubt that the accident at Three Mile Island permanently changed both the nuclear industry and the NRC. Public fear and distrust increased, NRC's regulations and oversight became broader and more robust, and management of the plants was scrutinized more carefully. The problems identified from careful analysis of the events during those days have led to permanent and sweeping changes in how NRC regulates its licensees – which, in turn, has reduced the risk to public health and safety.

Here are some of the major changes which have occurred since the accident:

- Upgrading and strengthening of plant design and equipment requirements. This includes fire protection, piping systems, auxiliary feedwater systems, containment building isolation, reliability of individual components (pressure relief valves and electrical circuit breakers), and the ability of plants to shut down automatically;
- Identifying human performance as a critical part of plant safety, revamping operator training and staffing requirements, followed by improved instrumentation and controls for operating the plant, and establishment of fitness-for-duty programs for plant workers to guard against alcohol or drug abuse;
- Improved instruction to avoid the confusing signals that plagued operations during the accident;
- Enhancement of emergency preparedness to include immediate NRC notification requirements for plant events and an NRC operations center that is staffed 24 hours a day. Drills and response plans are now tested by licensees several times a year, and state and local agencies participate in drills with the Federal Emergency Management Agency and NRC;
- Establishment of a program to integrate NRC observations, findings, and conclusions about licensee performance and management effectiveness into a periodic, public report;
- Regular analysis of plant performance by senior NRC managers who identify those plants needing additional regulatory attention;
- Expansion of NRC's resident inspector program – first authorized in 1977 – whereby at least two inspectors live nearby and work exclusively at each plant in the U.S. to provide daily surveillance of licensee adherence to NRC regulations;
- Expansion of performance-oriented as well as safety-oriented inspections, and the use of risk assessment to identify vulnerabilities of any plant to severe accidents;
- Strengthening and reorganization of enforcement as a separate office within the NRC;
- The establishment of the Institute of Nuclear Power Operations (INPO), the industry's own "policing" group, and formation of what is now the Nuclear

Energy Institute to provide a unified industry approach to generic nuclear regulatory issues, and interaction with NRC and other government agencies;

- The installing of additional equipment by licensees to mitigate accident conditions, and monitor radiation levels and plant status;
- Employment of major initiatives by licensees in early identification of important safety-related problems, and in collecting and assessing relevant data so lessons of experience can be shared and quickly acted upon; and
- Expansion of NRC's international activities to share enhanced knowledge of nuclear safety with other countries in a number of important technical areas.

Today, the TMI-2 reactor is permanently shut down and defueled, with the reactor coolant system drained, the radioactive water decontaminated and evaporated, radioactive waste shipped off-site to an appropriate disposal site, reactor fuel and core debris shipped off-site to a Department of Energy facility, and the remainder of the site being monitored. In 2001, FirstEnergy acquired TMI-2 from GPU. FirstEnergy has contracted the monitoring of TMI-2 to Exelon, the current owner and operator of TMI-1. The companies plan to keep the TMI-2 facility in long-term, monitored storage until the operating license for the TMI-1 plant expires, at which time both plants will be decommissioned.

Below is a chronology of highlights of the TMI-2 cleanup from 1980 through 1993.

Date	Event
July 1980	Approximately 43,000 curies of krypton were vented from the reactor building.
July 1980	The first manned entry into the reactor building took place.
Nov. 1980	An Advisory Panel for the Decontamination of TMI-2, composed of citizens, scientists, and State and local officials, held its first meeting in Harrisburg, PA.
July 1984	The reactor vessel head (top) was removed.
Oct. 1985	Defueling began.
July 1986	The off-site shipment of reactor core debris began.
Aug. 1988	GPU submitted a request for a proposal to amend the TMI-2 license to a "possession-only" license and to allow the facility to enter long-term monitoring storage.
Jan. 1990	Defueling was completed.
July 1990	GPU submitted its funding plan for placing $229 million in escrow for radiological decommissioning of the plant.
Jan. 1991	The evaporation of accident-generated water began.
April 1991	NRC published a notice of opportunity for a hearing on GPU's request for a license amendment.
Feb. 1992	NRC issued a safety evaluation report and granted the license amendment.

Aug. 1993	The processing of 2.23 million gallons accident-generated water was completed.
Sept. 1993	NRC issued a possession-only license.
Sept. 1993	The Advisory Panel for Decontamination of TMI-2 held its last meeting.
Dec. 1993	Post-Defueling Monitoring Storage began.

Courtesy of NRC

What happened at Chernobyl? At the time of the Chernobyl accident, on 26 April 1986, the Soviet Nuclear Power Program was based mainly upon two types of reactors, the WWER, a pressurized light-water reactor, and the RBMK, a graphite moderated light-water reactor. While the WWER type of reactor was exported to other countries, the RBMK design was restricted to republics within the Soviet Union.

The Chernobyl Power Complex, lying about 130 km north of Kiev, Ukraine, and about 20 km south of the border with Belarus (Figure 1), consisted of four nuclear reactors of the RBMK-1000 design, Units 1 and 2 being constructed between 1970 and 1977, while Units 3 and 4 of the same design were completed in 1983 (IA86). Two more RBMK reactors were under construction at the site at the time of the accident.

To the South-east of the plant, an artificial lake of some 22 km2 , situated beside the river Pripyat, a tributary of the Dniepr, was constructed to provide cooling water for the reactors.

This area of Ukraine is described as Belarussian-type woodland with a low population density. About 3 km away from the reactor, in the new city, Pripyat, there were 49 000 inhabitants. The old town of Chernobyl, which had a population of 12 500, is about 15 km to the South-east of the complex. Within a 30-km radius of the power plant, the total population was between 115 000 and 135 000.

The RBMK-1000 reactor

The RBMK-1000 (Figure 2) is a Soviet designed and built graphite moderated pressure tube type reactor, using slightly enriched (2% 235U) uranium dioxide fuel. It is a boiling light water reactor, with direct steam feed to the turbines, without an intervening heat-exchanger. Water pumped to the bottom of the fuel channels boils as it progresses up the pressure tubes, producing steam which feeds two 500 MWe [megawatt electrical] turbines. The water acts as a coolant and also provides the steam used to drive the turbines. The vertical pressure tubes contain the zirconium-alloy clad uranium-dioxide fuel around which the cooling water flows. A specially designed refueling machine allows fuel bundles to be changed without shutting down the reactor.

The moderator, whose function is to slow down neutrons to make them more efficient in producing fission in the fuel, is constructed of graphite. A mixture of nitrogen and helium is circulated between the graphite blocks largely to prevent oxidation of the graphite and to improve the transmission of the heat produced by neutron interactions in the graphite, from the moderator to the fuel channel. The core itself is about 7 m high and about 12 m in diameter. There are four main coolant circulating pumps, one of which is always on standby. The reactivity or power of the reactor is controlled by raising or lowering 211 control rods, which, when lowered, absorb neutrons and reduce the fission rate. The power output of this reactor is 3 200 MWt (megawatt thermal) or 1000 MWe, although there is a larger version producing 1 500 MWe. Various safety systems, such as an emergency core cooling system and the requirement for an absolute minimal insertion of 30 control rods, were incorporated into the reactor design and operation.

The most important characteristic of the RBMK reactor is that it possesses a "positive void coefficient". This means that if the power increases or the flow of water decreases, there is increased steam production in the fuel channels, so that the neutrons that would have been absorbed by the denser water will now produce increased fission in the fuel. However, as the power increases, so does the temperature of the fuel, and this has the effect of reducing the neutron flux (negative fuel coefficient). The net effect of these two opposing characteristics varies with the power level. At the high power level of normal operation, the temperature effect predominates, so that power excursions leading to excessive overheating of the fuel do not occur. However, at a lower power output of less than 20% the maximum, the positive void coefficient effect is dominant and the reactor becomes unstable and prone to sudden power surges. This was a major factor in the development of the accident.

Events leading to the accident

The Unit 4 reactor was to be shut-down for routine maintenance on 25 April 1986. It was decided to take advantage of this shutdown to determine whether, in the event of a loss of station power, the slowing turbine could provide enough electrical power to operate the emergency equipment and the core cooling water circulating pumps, until the diesel emergency power supply became operative. The aim of this test was to determine whether cooling of the core could continue to be ensured in the event of a loss of power.

This type of test had been run during a previous shut-down period, but the results had been inconclusive, so it was decided to repeat it. Unfortunately, this test, which was considered essentially to concern the non-nuclear part of the power plant, was carried out without a proper exchange of information and co-ordination between the team in charge of the test and the personnel in charge of the operation and safety of the nuclear reactor. Therefore, inadequate safety precautions were included in the test program and the operating personnel were not alerted to the nuclear safety implications of the electrical test and its potential danger.

The planned program called for shutting off the reactor's emergency core cooling system (ECCS), which provides water for cooling the core in an emergency. Although subsequent events were not greatly affected by this, the exclusion of this system for the whole duration of the test reflected a lax attitude towards the implementation of safety procedures.

As the shutdown proceeded, the reactor was operating at about half power when the electrical load dispatcher refused to allow further shutdown, as the power was needed for the grid. In accordance with the planned test program, about an hour later the ECCS was switched off while the reactor continued to operate at half power. It was not until about 23:00 hr on 25 April that the grid controller agreed to a further reduction in power.

For this test, the reactor should have been stabilized at about 1 000 MWt prior to shut down, but due to operational error the power fell to about 30 MWt, where the positive void coefficient became dominant. The operators then tried to raise the power to 700 - 1000 MWt by switching off the automatic regulators and freeing all the control rods manually. It was only at about 01:00 hr on 26 April that the reactor was stabilized at about 200 MWt.

Although there was a standard operating order that a minimum of 30 control rods was necessary to retain reactor control, in the test only 6-8 control rods were actually used. Many of the control rods were withdrawn to compensate for the buildup of xenon which acted as an absorber of neutrons and reduced power. This meant that if there were a power surge, about 20 seconds would be required to lower the control rods and shut the reactor down. In spite of this, it was decided to continue the test program.

There was an increase in coolant flow and a resulting drop in steam pressure. The automatic trip which would have shut down the reactor when the steam pressure was low had been circumvented. In order to maintain power the operators had to withdraw nearly all the remaining control rods. The reactor became very unstable and the operators had to make adjustments every few seconds trying to maintain constant power.

At about this time, the operators reduced the flow of feedwater, presumably to maintain the steam pressure. Simultaneously, the pumps that were powered by the slowing turbine were providing less cooling water to the reactor. The loss of cooling water exaggerated the unstable condition of the reactor by increasing steam production in the cooling channels (positive void coefficient), and the operators could not prevent an overwhelming power surge, estimated to be 100 times the nominal power output.

The sudden increase in heat production ruptured part of the fuel and small hot fuel particles, reacting with water, caused a steam explosion, which destroyed the reactor core. A second explosion added to the destruction two to three seconds later. While it is not known for certain what caused the explosions, it is postulated that the first was a steam/hot fuel explosion, and that hydrogen may have played a role in the second.

The accident

The accident occurred at 01:23 hr on Saturday, 26 April 1986, when the two explosions destroyed the core of Unit 4 and the roof of the reactor building.

In the IAEA Post-Accident Assessment Meeting in August 1986, much was made of the operators' responsibility for the accident, and not much emphasis was placed on the design faults of the reactor. Later assessments suggest that the event was due to a combination of the two, with a little more emphasis on the design deficiencies and a little less on the operator actions.

The two explosions sent fuel, core components and structural items and produced a shower of hot and highly radioactive debris, including fuel, core components, structural items and graphite into the air and exposed the destroyed core to the atmosphere. The plume of smoke, radioactive fission products and debris from the core and the building rose up to about 1 km into the air. The heavier debris in the plume was deposited close to the site, but lighter components, including fission products and virtually the entire noble gas inventory were blown by the prevailing wind to the North-west of the plant.

Fires started in what remained of the Unit 4 building, giving rise to clouds of steam and dust, and fires also broke out on the adjacent turbine hall roof and in various stores of diesel fuel and inflammable materials. Over 100 fire-fighters from the site and called in from Pripyat were needed, and it was this group that received the highest radiation exposures and suffered the greatest losses in personnel. A first group of 14 firemen arrived on the scene of the accident at 1.28 a.m. Reinforcements were brought in until about 4 a.m., when 250 firemen were available and 69 firemen participated in fire control activities. By 2.10 a.m., the largest fires on the roof of the machine hall had been put out, while by 2.30 a.m., the largest fires on the roof of the reactor hall were under control. These fires were put out by 05:00 hr of the same day, but by then the graphite fire had started. Many firemen added to their considerable doses by staying on call on site. The intense graphite fire was responsible for the dispersion of radionuclides and fission fragments high into the atmosphere. The emissions continued for about twenty days, but were much lower after the tenth day when the graphite fire was finally extinguished.

The graphite fire

While the conventional fires at the site posed no special firefighting problems, very high radiation doses were incurred by the firemen, resulting in 31 deaths. However, the graphite moderator fire was a special problem. Very little national or international expertise on fighting graphite fires existed, and there was a very real fear that any attempt to put it out might well result in further dispersion of radionuclides, perhaps by steam production, or it might even provoke a criticality excursion in the nuclear fuel.

A decision was made to layer the graphite fire with large amounts of different materials, each one designed to combat a different feature of the fire and the radioactive release. The first measures taken to control fire and the radionuclides releases consisted of dumping neutron-absorbing compounds and fire-control material into the crater that resulted from the destruction of the reactor. The total amount of materials dumped on the reactor was about 5 000 t including about 40 t of boron compounds, 2 400 t of lead, 1 800 t of sand and clay, and 600 t of dolomite, as well as sodium phosphate and polymer liquids (Bu93). About 150 t of material were dumped on 27 April, followed by 300 t on 28 April, 750 t on 29 April, 1 500 t on 30 April, 1 900 t on 1 May and 400 t on 2 May. About 1 800 helicopter flights were carried out to dump materials onto the reactor; during the first flights, the helicopter remained stationary over the reactor while dumping materials. As the dose rates received by the helicopter pilots during this procedure were too high, it was decide that the materials should be dumped while the helicopters travelled over the reactor. This procedure caused additional destruction of the standing structures and spread the contamination. Boron carbide was dumped in large quantities from helicopters to act as a neutron absorber and prevent any renewed chain reaction. Dolomite was also added to act as heat sink and a source of carbon dioxide to smother the fire. Lead was included as a radiation absorber, as well as sand and clay which it was hoped would prevent the release of particulates. While it was later discovered that many of these compounds were not actually dropped on the target, they may have acted as thermal insulators and precipitated an increase in the temperature of the damaged core leading to a further release of radionuclides a week later.

The further sequence of events is still speculative, although elucidated with the observation of residual damage to the reactor. It is suggested that the melted core materials settled to the bottom of the core shaft, with the fuel forming a metallic layer below the graphite. The graphite layer had a filtering effect on the release of volatile compounds. But after burning without the filtering effect of an upper graphite layer, the release of volatile fissions products from the fuel may have increased, except for non-volatile fission products and actinides, because of reduced particulate emission. On day 8 after the accident, the corium melted through the lower biological shield and flowed onto the floor. This redistribution of corium would have enhanced the radionuclide releases, and on contact with water corium produced steam, causing an increase of radio nuclides at the last stage of the active period.

By May 9, the graphite fire had been extinguished, and work began on a massive reinforced concrete slab with a built-in cooling system beneath the reactor. This involved digging a tunnel from underneath Unit 3. About four hundred people worked on this tunnel which was completed in 15 days, allowing the installation of the concrete slab. This slab would not only be of use to cool the core if necessary, it would also act as a barrier to prevent penetration of melted radioactive material into the groundwater.

In summary

The Chernobyl accident was the product of a lack of "safety culture". The reactor design was poor from the point of view of safety and unforgiving for the operators, both of which provoked a dangerous operating state. The operators were not informed of this and were not aware that the test performed could have brought the reactor into an explosive condition. In addition, they did not comply with established operational procedures. The combination of these factors provoked a nuclear accident of maximum severity in which the reactor was totally destroyed within a few seconds.

The left photo is an aerial view of the remaining section of the unit. The right photo is called the elephants foot… Congealed core material from Chernobyl's meltdown. Dose rates are over 1000 R/Hr.   www.nea.fr/html/rp/chernobyl/c01.html

What happened to Davis Besse? As with Chernobyl, the Davis Besse incident will show that management and all workers must guard against complacency brought on by good performance metrics and past successes. This incident was deemed so egregious that all management and supervisory personnel are required to review the event in their initial and continuing training programs to ensure that this never happens again. Here is the timeline of the incident.

March 21, 1997: The NRC regional administrator tells FirstEnergy during a public meeting, "You obviously are doing very well in identifying problems. This is certainly one of the better, if not the best, performers in the region."

April 25, 1998: A worker at Davis-Besse following a videotaped inspection of the reactor vessel head the previous day "indicated several 'fist' size clumps of Boric Acid.… Where clumps were not present, a light dusting of Boric Acid was found covering the surface area of the vessel head."

July 30, 1999: results from laboratory analysis of materials deposited on radiation monitors inside containment. The results indicated the material was primarily iron oxide with an indication that corrosion caused the iron oxide granules.

April 6, 2000: Workers performing the initial reactor vessel head inspection the previous day reported "Boron deposits were 'lava-like' and originating from the 'mouse hole' and CRDM flanges." A copy of the condition report, with seven color photographs attached including the now infamous "Red Photo," was handed to an NRC inspector evaluating in-service inspection program.

April 23, 2001: the filters on the radiation monitors inside containment were being changed out more frequently. Filters that used to last several months were being replaced every 14 days. According to the CR, "All filters contained Boron crystals."

July 23, 2001: the filters on the radiation monitors inside containment were being changed out more frequently. The replacement frequency increased to every two to seven days. The CR stated, "Currently we still have a small RCS [reactor coolant system] leak in CTMT [containment]. This is indicated by boron deposits on the clogged filters."

August 3, 2001: The NRC issued Bulletin 2001-01 to the owners of pressurized water reactors, including FirstEnergy. The owners were required to submit information to the NRC regarding the condition of CRDM nozzles. Owners of plants susceptible to CRDM nozzle cracking that were not planning to conduct nozzle inspections on or before December 31, 2001, were requested to provide the NRC with justification for continued operation.

September 28, 2001: The NRC staff called the owners of Davis-Besse, to discuss the CRDM nozzle issue. This was a reactor identified by the NRC has having high susceptibility to CRDM nozzle cracking but no plans to inspect the nozzles by December 31, 2001.

October 11, 2001: FirstEnergy representatives briefed the NRC commissioners' technical assistants about Davis-Besse's position relative to Bulletin 2001-01. According to the NRC summary of this meeting, First Energy stated, "All CRDM penetrations were verified to be free from the characteristic boron deposits using video recordings from the previous 2 refueling outages. These videos were made before and after cleaning the head."

October 15, 2001: The NRC staff distributed a draft order requiring the shutdown of reactors by December 31, 2001, for CRDM nozzle inspections.

October 24, 2001: The NRC staff met with FirstEnergy representatives about Davis-Besse and the CRDM nozzle issue. FirstEnergy informed the NRC that 24 CMDR nozzle penetrations were not examined during the April 2000 refueling outage.

November 28, 2001: NRC staff met with FirstEnergy. The company committed to a number of measures if Davis-Besse could operate until February 16, 2002, instead of shutting down by the end of 2001.

November 29, 2001: The NRC staff briefed the EDO about the decision NOT to issue the order requiring Davis-Besse to be shut down by December 31, 2001, but to allow the reactor to operate until February 16, 2002. In the final summation slide, the NRC staff outlined the five criteria governing the decision and how Davis-Besse basically did not meet them:

- Current regulations are met. "It is likely that current regulations are not met with respect to TS [technical specifications; part of the operating license issued by the NRC] requirements and GDC [general design criteria in federal regulations]"
- Defense-in-depth philosophy maintained. "It is likely that one of 3 barriers is lost."
- Sufficient safety margins are maintained. "It is likely that safety margins are reduced."
- Only a small increase in CDF [core damage frequency] results. "Incremental ΔCDF (no comp measures) is 1.1E-06/ry to 1.3E-04/ry."
- The basis of risk measurement is monitored during performance. "Will not occur until inspection is performed."

December 4, 2001: The NRC informed FirstEnergy that Davis-Besse could operate until February 16, 2002, without conducting the CRDM nozzle inspections requested in Bulletin 2001-01.

January 9, 2002: The NRC staff briefed the staff of the House Energy & Commerce Committee about the CRDM nozzle issue.

February 16, 2002: Operators shut down Davis-Besse for a scheduled refueling outage.

February 21, 2002: workers discovered boric acid on the reactor vessel head: "The large boron accumulation is in the same region as seen in 12RFO [the 12th refueling outage in April 2000], but not as deep."

February 26, 2002: workers noted "more boron than expected was found on the top of the head." Davis-Besse (Outage dates: February 16, 2002 to March 16, 2004)

February 27, 2002: "Ultrasonic testing (UT) performed on the #3 Control Rod Drive Mechanism (CRDM) nozzle (location G9) revealed indications of through wall axial flaws in the weld region.... These indications represent potential leakage paths." The CRDM nozzle crack indication was reported to the NRC that same day.

February 28, 2002: workers completed the ultrasonic examination of all 69 CRDM nozzles. Five CRDM nozzles were determined to have axial crack indications with three of them exhibiting through-wall leakage.

March 4, 2002: The NRC sent FirstEnergy its annual assessment of performance at Davis-Besse. The NRC stated:

" *Davis-Besse operated in a manner that preserved public health and safety and fully met all cornerstone objectives. Plant performance for the most recent quarter, as well as the first two quarters of the assessment cycle, was within the Licensee Response Column of the NRC's Action Matrix, based on all inspection findings being classified as having very low safety significance (Green) and all PIs indicating performance at a level requiring no additional NRC oversight (Green)."*

March 8, 2002: "Evaluation of bottom up ultrasonic test data in the area of reactor pressure vessel head nozzle number 3 shows significant degradation of the reactor vessel head pressure boundary." The NRC publicly reported significant "metal loss" in the reactor vessel head.

March 11, 2002: The NRC wrote to the Nuclear Energy Institute requesting answers by March 13, 2002, to questions about whether CRDM nozzle inspections conducted at other plants could have identified the reactor vessel head damage discovered at Davis-Besse.

March 12, 2002: The NRC chartered an Augmented Inspection Team to travel to Davis-Besse and determine (1) the history of reactor coolant system operational leakage, (2) the history of reactor vessel head material condition issues, (3) the history of reactor vessel head inspections, and (4) the characterization of all reactor vessel head corrosion damage.

March 12, 2002: The NRC issued information to the owners of all other operating pressurized water reactors about the reactor vessel head damage discovered at Davis-Besse.

May 2004: The NRC gives Davis-Besse permission after all work has been completed, to restart the unit.

Erosion of the 6" thick carbon steel reactor head, caused by a persistent leak of borated water.

The major contributor to the event was a shift in the focus at all levels of the organization from implementing high standards to justifying minimum standards. This resulted from excessive focus on meeting short-term production goals, a lack of management oversight, symptom based problem solving, justification of plant problems, isolationism, ineffective use of operating experience and a lack of sensitivity to nuclear safety.

Courtesy of www.ucsusa.org/assets/documents/nuclear_power/davis-besse-ii.pdf

What was the first reactor to have a core meltdown? The EBR-1 experimental nuclear reactor near Arco, Idaho was the first to experience a partial meltdown.

What is the Demon core? The Demon core was the nickname given to a 6.2 kg spherical subcritical mass of plutonium that accidentally went critical on two separate instances at the Los Alamos laboratory, in 1945 and 1946. Each incident resulted in the acute radiation poisoning and subsequent death of a scientist. After these incidents, the core was referred to as the Demon core.

On August 21, 1945, the plutonium core produced a burst of ionizing radiation that irradiated Harry Daghlian, a physicist who made a mistake while working alone doing neutron reflection experiments on the core. The core was placed within a stack of neutron-reflective bricks, moving the assembly closer to criticality. Harry Daghlian, while attempting to stack another brick around the assembly, accidentally dropped one of the bricks onto the core causing the assembly to go supercritical. Despite moving the brick off the assembly quickly, Daghlian received a fatal dose of radiation.

Then on May 21, 1946, physicist Louis Slotin and other scientists were in a Los Alamos laboratory conducting an experiment that involved creating a fission reaction by placing two half-spheres of beryllium (a neutron reflector) around the same plutonium core. Slotin's hand holding a screwdriver separating the hemispheres slipped, the beryllium neutron reflector hemispheres closed, and the core went supercritical, releasing a very high dose of radiation. He quickly pulled the two halves apart, stopping the chain reaction and hence saving the lives of the others in the laboratory. Louis Slotin died 9 days later from acute radiation poisoning.

The Demon core was most likely intended to be used in another nuclear weapon against Japan. However, The core was instead used in the ABLE test of the Crossroads series, demonstrating that the criticality experiments of Daghlian and Slotin increased the efficiency of the weapon.

What is a criticality accident? A criticality accident, sometimes referred to as an excursion or a power excursion, occurs when a nuclear chain reaction accidentally occurs in fissile material, such as enriched uranium or plutonium. This releases neutron radiation which is highly dangerous to surrounding personnel and causes induced radioactivity in the surroundings.

Nuclear fission normally is supposed to occur inside reactor cores and inside some test facilities. However, if fission occurs due to an accidental cause, such as a criticality accident, the radiation emitted poses a high risk of serious injury or even death to workers up to at least 20 metres (66 feet) away. Although dangerous, the low densities of fissile material and the long insertion time involved in these events limit the fission yield and peak power, preventing them from becoming a large scale nuclear explosion.

Have there been any deaths from criticality accidents? Unfortunately yes. Since 1945 there have been at least 21 deaths from criticality accidents; 7 in the United States, 10 in the Soviet Union, 2 in Japan, 1 in Argentina, and 1 in Yugoslavia. 9 have been due to process accidents, with the remaining from research reactor accidents.

Criticality accidents have occurred both in the context of <u>nuclear weapons</u> and <u>nuclear reactors</u>.

On **June 4, 1945,** Los Alamos an experiment to determine the critical mass of enriched uranium became critical when water leaked into the polyethylene box holding the metal. The radiation gave three people non-fatal doses.

On **August 21, 1945,** Los Alamos scientist Harry K. Daghlian, Jr. suffered fatal radiation poisoning after dropping a tungsten carbide brick onto a sphere of plutonium. The brick acted as a neutron reflector, bringing the mass to criticality. This was the first known criticality accident causing a fatality.

On **May 21, 1946,** another Los Alamos scientist, Louis Slotin, accidentally irradiated himself during a similar incident, when a critical mass experiment with the very same sphere of plutonium (see demon core) took a wrong turn. Immediately realizing what had happened he quickly disassembled the device, likely saving the lives of seven fellow scientists nearby. Slotin succumbed to radiation poisoning nine days later.

On **October 15, 1958**, a criticality excursion in the heavy water RB reactor at the Boris Kidrič Institute of Nuclear Sciences in Vinča, Yugoslavia killed one and injured five.

On **July 23, 1964** – Wood River Junction facility in Charlestown, Rhode Island. A criticality accident occurred at the plant, designed to recover uranium from scrap material left over from fuel element production. An operator accidentally added a concentrated uranium solution to an agitated tank containing sodium carbonate, resulting in a critical nuclear reaction. This criticality exposed the operator to a fatal radiation dose of 10,000 rad (100 Gy). Ninety minutes later a second excursion happened, exposing two cleanup crews to doses of up to 100 rad (1 Gy) without ill effect.

On **December 10, 1968** Russia, Mayak, a nuclear fuel processing center in central Russia was experimenting with plutonium purification techniques. Two operators were using an "unfavorable geometry vessel in an improvised and unapproved operation as a temporary vessel for storing plutonium organic solution." In other words, the operators were decanting plutonium solutions into the wrong type of vessel. After most of the solution had been poured out, there was a flash of light, and heat. "Startled, the operator dropped the bottle, ran down the stairs, and from the room." After the complex had been evacuated, the shift supervisor and radiation control supervisor re-entered the building. The shift supervisor then deceived the radiation control supervisor and entered the room of the incident and possibly attempted to pour the solution down a floor drain, causing a large nuclear reaction that irradiated the shift supervisor with a fatal dose of radiation. The shift supervisor's actions are the subject of a Darwin Award.

On **September 23, 1983**, an operator at the RA-2 research reactor in Constituyentes, Argentina received a fatal radiation dose of 3700 rads (37 Gy) while changing the fuel rod configuration with moderating water in the reactor. Two others were injured.

Between **June 24, 1990 and July 1, 1990**, about four years after the Chernobyl accident, signs of a sub-critical neutron multiplication event occurred inside room 304/3 at the damaged reactor (see Russian Research Centre Kurchatov Institute report). The neutron increase was by a factor of about 60, much less than the increase that would result from a criticality. A gadolinium solution was injected to absorb neutrons and the neutron level returned to the original level.

In **1999** at a Japanese uranium reprocessing facility in Tokai, Ibaraki, workers put a mixture of uranyl nitrate solution into a precipitation tank which was not designed to dissolve this type of solution and caused an eventual critical mass to be formed, and resulted in the death of two workers from radiation poisoning.

Emergency Preparedness

What does an emergency evacuation route map look like?

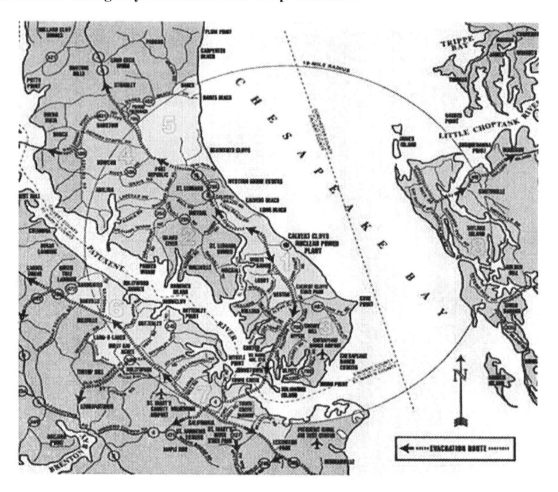

What is the purpose of the International Nuclear Event Scale? The INES was introduced by the International Atomic Energy Agency (IAEA) in order to enable prompt communication of safety significance information in case of a nuclear accident. A number of criteria and indicators are defined to assure coherent reporting of nuclear events by different official authorities. There are 8 levels on the INES scale:

Level 0 – this is a below scale event of no safety significance.

Level 1 – this is an anomaly beyond the authorized operating regime.

Level 2 – this is an incident with no off-site impact, a significant spread of contamination on-site may have occurred or over-exposure of a worker or incidents with significant failures in safety provisions. E.g. Catalonia, Spain April 2008, The Forsmark Nuclear Plant, Sweden 2006.

Level 3 – a very small off-site impact, public exposure at levels below the prescribed limits or severe spread of contamination on-site and/or acute health effects to one or more workers or it is a "near accident" event, when no safety layers are remaining. E.g. THORP plant in Sellafield, United Kingdom 2005, Paks Nuclear Plant, Hungary 2003

Level 4 – minor off-site impact resulting in public exposure of the order of the prescribed limits or significant damage to a reactor core/radiological barriers or the fatal exposure of a worker. E.g. Sellafield, United Kingdom 1955 & 1979, Saint-Laurent Nuclear Plant 1980, Buenos Aires, Argentina 1983, Tokaimura nuclear accident, Japan 1999

Level 5 – limited off-site release, likely to require partial implementation of planned countermeasures or severe damage to reactor core/radiological barriers. E.g. Windscale fire, United Kingdom 1957, Three Mile Island 1979

Level 6 – significant off-site release, likely to require full implementation of planned countermeasures. E.g. Mayak accident, Soviet Union 1957

Level 7 – a large off-site impact with widespread health and environmental effects. So far, the Chernobyl disaster is the only accident to reach level 7. An example of a non- nuclear accident of roughly the same magnitude would be the Bhopal disaster which resulted in thousands of off-site deaths.

How does the United States scale a nuclear accident? The U.S. through the NRC has created an emergency classification list to determine plant status. There are four categories for plant status and are listed in order of increasing severity.

1. **Notification of Unusual Event** – events are in process or have occurred which indicate potential degradation in the level of safety of the plant. No release of radioactive material requiring offsite response or monitoring is expected unless further degradation occurs.

2. **Alert** – events are in process or have occurred which involve an actual or potential substantial degradation in the level of safety of the plant. Any release of radioactive material from the plant is expected to be limited to a small fraction of the Environmental Protection Agency (EPA) protective action guides.

3. **Site Area Emergency** – events in process or which have occurred that result in actual or likely major failures of plant functions needed for protection of the public. Any releases of radioactive material are not expected to exceed the EPA PAGs except near the site boundary.

4. **General Emergency** – involves actual or imminent substantial core damage or melting of reactor fuel with the potential for loss of containment integrity. Radioactive releases during a general emergency can reasonably be expected to exceed the EPA PAGs for more than the immediate site area.

The International Nuclear Event Scale (INES) consists of a 7-level event classification system. Events of greater safety significance (levels 4-7) are termed "accidents", events of lesser safety significance (levels 1-3) are termed "incidents", and events of no safety significance (level 0 or below scale) are termed "out-of-scale deviations".

How does the United States military scale their nuclear incidents? The United States Department of Defense has created a series of terms to represent the criticality of a nuclear event. There are nine of these terms that are used only for the U.S. and are neither NATO nor global standards.

Pinnacle – denotes an incident of interest to the Major Commands, Department of Defense and the National Command Authority, in that it: Generates a higher level of military action, causes a national reaction, affects international relationships, causes immediate widespread coverage in news media, is clearly against the national interest and affects current national policy.

Bent Spear – refers to incidents involving nuclear weapons, warheads, components or vehicles transporting nuclear material that are of significant interest but are not categorized as Pinnacle – Nucflash or Pinnacle – Broken Arrow. Bent Spear incidents include violations or breaches of handling and security regulations.

Broken Arrow – refers to an accidental event that involves nuclear weapons, warheads or components, but which does not create the risk of nuclear war. These include: accidental or unexplained nuclear detonation, non-nuclear detonation or burning of a nuclear weapon, radioactive contamination, loss in transit of nuclear asset with or without its carrying vehicle, jettisoning of a nuclear weapon or nuclear component, public hazard, actual or implied.

NUCFLASH – refers to detonation or possible detonation of a nuclear weapon which creates a risk of an outbreak of nuclear war. Events which may be classified Nucflash may include:

- Accidental, unauthorized, or unexplained or possible nuclear detonation.

- Accidental, unauthorized launch of a nuclear armed or nuclear capable missile in the direction of, or having the capability to reach another nuclear capable country.

- Unauthorized flight of or deviation from an approved flight plan by a nuclear armed or nuclear capable aircraft with the capability to penetrate the airspace of another nuclear capable country.

- Detection of unidentified objects by a missile warning system or interference that appears threatening and could create a risk of nuclear war.

The term Pinnacle Nucflash is a report that has the highest precedence in the reporting structure. All other reporting terms are secondary to this report.

Emergency Disablement – Pinnacle emergency disablement refers to operations involving the emergency destruction of nuclear weapons.

Emergency Evacuation – Pinnacle emergency evacuation refers to operations involving the emergency evacuation of nuclear weapons.

Empty Quiver – Pinnacle empty quiver refers to the seizure, theft, or loss of a functioning nuclear weapon.

Faded Giant – refers to an event involving a nuclear reactor or other radiological accident not involving nuclear weapons.

Dull Sword – is an Air Force term that marks reports of minor incidents involving nuclear weapons, components or systems, or which could impair their deployment. This could involve actions involving vehicles capable of carrying nuclear weapons but with no nuclear weapons on board at the time of the accident.

What is a design basis accident? The Design Basis Accident (DBA) for a nuclear power plant is the most severe possible single accident that the designers of the plant and the regulatory authorities could imagine. It is, also, by definition, the accident the safety systems of the reactor are designed to respond to successfully, even if it occurs when the reactor is in its most vulnerable state. Courtesy of NRC

What is meant by Defense in Depth? Defense in Depth is a term used to define three barriers between the nuclear fuel and the environment; 1) the fuel cladding, 2) the reactor coolant system and 3) the reactor containment building, which is typically a pre-stressed concrete structure with an inner steel liner.

What are tech specs? Technical specifications are the licensed limits for operation. These specifications are based on the analyses and evaluations of the final safety analysis report which include the following: Safety limits, Limiting conditions for operations, Surveillance requirements, design features and administrative controls.

Does the nuclear industry share information with each other? YES, the nuclear industry does share information about their individual site events. When the nuclear industry first started, sharing information was never thought of…until Three Mile Island. TMI opened the eyes of the nuclear industry and determined in the benefit of mankind that information must be shared between the nuclear sites in order not to repeat the lessons learned from other sites.

How are nuclear reactors defended against adversaries? Commercial nuclear power plants are heavily fortified with well-trained and armed guards. They also have layered physical security measures, such as access controls, water barriers, intrusion detection and strategically placed guard towers. Together, these make up the plants' response to the Design Basis Threat – usually called the DBT. The DBT is developed from real-world intelligence information and describes the adversary force – coming from both ground and water – the plants must defend against. DBT specifics are not public in order to protect sensitive information that could aid terrorists. The NRC regularly reviews the DBT and adds new requirements when necessary.

Category I Fuel-Cycle Facilities

There are two NRC-licensed Category I Fuel-Cycle Facilities in the U.S. that make reactor fuel for nuclear plants. Since these plants handle nuclear material that could be targeted by adversaries, they also must defend against a DBT similar to that for nuclear power plants.

How are nuclear plants prepared to respond to emergencies? No matter how small the risk, the NRC requires all nuclear power plants to have and periodically test emergency plans that are coordinated with federal, state and local responders. The goal of preparedness is to reduce the risk to the public during an emergency.

In an emergency, the NRC and the licensee would activate their Incident Response Programs. Licensee specialists would evaluate the situation and identify ways to end the emergency, while the NRC would monitor the event closely, keeping government offices informed. If a radiation release occurred, the plant would make protective action recommendations to state and local officials, such as evacuating areas around the plant.

• Emergency Planning Zones (EPZs)

Each nuclear power plant has two EPZs. Each EPZ considers the specific conditions and geography at the site, and the community. The first is the Plume Exposure Pathway EPZ, which has a radius of about 10 miles from the reactor. People living there may be asked to evacuate or "shelter in place" during an emergency, to avoid or reduce their radiation dose. The second is the Ingestion Exposure Pathway EPZ. This has a radius of about 50 miles from the reactor. Protective action plans for this area aim to avoid or reduce the radiation dose from consuming contaminated food and water.

How safe are nuclear reactors from aircraft attack? Since 9/11, the issue of an airborne attack on this nation's infrastructure, including both operating and potential new nuclear power plants, has been widely discussed. The NRC has comprehensively studied the effect of an airborne attack on nuclear power plants. Shortly after 9/11, the NRC began a security and engineering review of operating nuclear power plants. Assisting the NRC were national experts from Department of Energy laboratories, who used state-of-the-art experiments, and structural and fire analyses.

These classified studies confirm that there is a low likelihood that an airplane attack on a nuclear power plant would affect public health and safety, thanks in part to the inherent robustness of the structures. A second study identified new methods plants could use to minimize damage and risk to the public in the event of any kind of large fire or explosion. Nuclear power plants subsequently implemented many of these methods.

The NRC is now considering new regulations for future reactors' security. The goal is to include inherent safety and security features to minimize potential damage from an airborne attack.

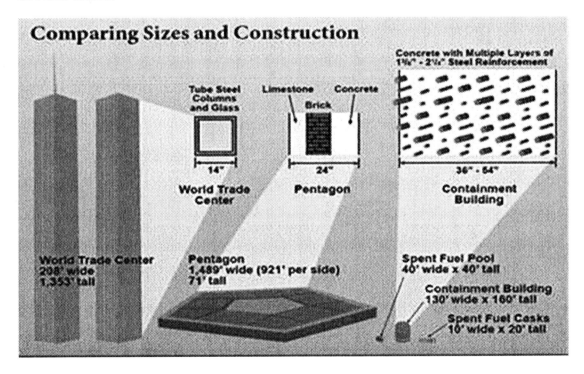

How does a local nuclear power plant prepare their local community in an emergency?
Every nuclear power plant in the United States has been tasked with preparing numerous

emergency contingencies in order to protect the surrounding community. Each nuclear power plant location must perform drills, have local and federal agencies involved with these drills. This is designed not only to prepare for any emergency that may develop but to get the community involved. There are numerous redundancies built into each drill to ensure that there is plenty of personnel to take care of any emergency that arises.

What is Potassium Iodine? Potassium iodine (KI) is an ingredient that is added to table salt making it iodized salt. Potassium iodine contains approx. 76% iodine. The thyroid gland is very vulnerable to atomic energy since a radioactive isotope of iodine is one of the major components of fallout. There is no medicine that can be taken to prevent nuclear radiation from damaging the cells in the human body. The thyroid gland can absorb both non-radioactive and radioactive iodine, and will retain much of this element in either or both forms. The purpose of potassium iodine is to flood the thyroid gland with non-radioactive iodine so that it can only absorb about 1% more additional iodine (radioactive) if contact should occur. Any excess iodine in the blood is eliminated through the kidneys. If potassium iodine is taken in time it can block the absorption of radioactive iodine into the thyroid gland, which could help prevent thyroid cancers and other diseases caused from a nuclear accident. Here is a sample of just how strong a potassium iodine pill is: to achieve an intake of 130 milligrams of potassium iodine would require ingesting 250 teaspoons of salt per day (5 cups)(Blood pressure what). Salt water from the sea is not any better; you would have to 33 kilograms of seawater a day (nope, not me).

If the risk of a nuclear plant emergency is so small, why prepare? Everything people do every day carries some risk. Although the risk of an emergency at a nuclear plant is remote, the community can take comfort in knowing that preparation, not panic, is the order of the day. The nuclear industry plans and drills for every possible situation. With such extensive planning, the industry knows how to respond effectively and immediately to protect the community, families, co-workers, pets and livestock. It is important that you know what to do too. It is also important that you fully cooperate with government authorities responding to the emergency.

How will you know if there is a nuclear plant emergency? Nuclear plants are required to have emergency plans to handle any incident. These plans are approved by local, state and federal government agencies. These plans are tested regularly in emergency drills that include the nuclear plant, local, state and federal agencies.

What is an alert notification system? In a nuclear emergency, the plant will inform local, county and state officials immediately. County officials will activate the siren alert system within a 10-mile radius of the plant. The sirens are installed to alert the public to tune into your local radio stations for information and instructions on any emergency situation and are not a signal to evacuate. The sirens may be sounded for testing or weather warnings.

What might you need in an emergency? In the event of an emergency you should have at least the following: safety items, first aid kit, cell phone, tool box, portable radio / tv, flashlights, extra batteries, medical supplies, money, personal hygiene items, clothing for several days, baby supplies, blankets, pillows, sleeping bags, road map, non-perishable foods, bottled water (at least 1 gallon per person for 2 to 3 days), and family contact information.

Uranium

What are fission by-products of uranium?

$_{52}Te^{135}$ β $_{53}I^{135}$ β $_{54}Xe^{135}$ β $_{55}Cs^{135}$ β $_{56}Ba^{135}$ β (stable)

n $_{54}Xe^{136}$ (stable)

$n \rightarrow$ U-235 n n n

$_{40}Zr^{98}$ β $_{41}Nb^{98}$ β $_{42}Mo^{98}$ β (stable)

Why is the nuclear fuel enriched? The concentration of the fissionable isotope, U-235 (0.71% in natural uranium) is less than that required to sustain a nuclear chain reaction in light water reactor cores. Natural UF_6 thus must be enriched in the fissionable isotope for it to be used as nuclear fuel. The different levels of enrichment required for a particular nuclear fuel application are specified by the customer: light-water reactor fuel normally is enriched to 3 to 5% U-235, but uranium enriched to lower concentrations also is required. Enrichment is accomplished using some one or more methods of isotope separation. Gaseous diffusion and gas centrifuge are the commonly used uranium enrichment technologies, but new enrichment technologies are currently being developed.

The bulk (96%) of the byproduct from enrichment is depleted uranium (DU), which can be used for armor, kinetic energy penetrators, radiation shielding and ballast. Still, there are vast quantities of depleted uranium in storage. The United States Department of Energy alone has 470,000 tons. About 95% of depleted uranium is stored as uranium hexafluoride (UF_6).

How big is a nuclear fuel pellet? As you can see from the photo below, a single fuel pellet is extremely small for the amount of energy it contains within itself.

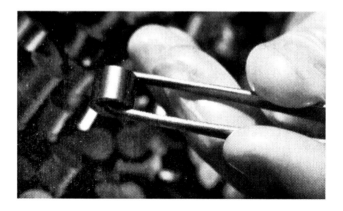

What is meant by spent fuel? Spent fuel is nuclear fuel that has been irradiated to the point where it is no longer useful in sustaining a nuclear reaction. . Fuel assemblies that are discharged from reactors during refueling are called spent fuel. During the course of the irradiation, much of the U-235 is depleted by fission and transmutation to U-236. A small fraction of the U-238 changes to plutonium isotopes. Even smaller amounts of other transuranic isotopes, such as Np, Am, and Cm, are also produced by neutron transmutation. Among the Pu isotopes that are produced, Pu-239 and Pu-241 are fissile and contribute to energy production. The generation of decay heat continues long after the fuel assemblies have been removed from the core, so the spent fuel assemblies are stored in spent fuel pools immersed in water to remove the decay heat. The water in the spent fuel pools is circulated through coolers to remove the decay heat. After allowing several years for the decay of the radioactive isotopes in the spent fuel, the spent fuel assemblies can be removed from the pools and stored in dry casks.

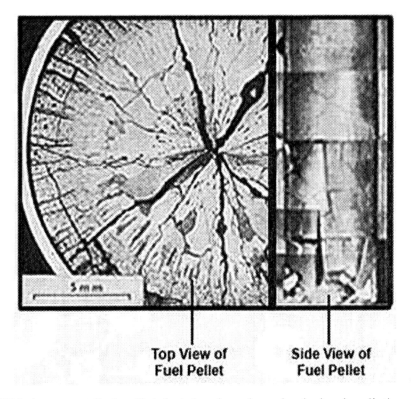

Top View of
Fuel Pellet

Side View of
Fuel Pellet

This is a spent fuel pellet that developed cracks during irradiation.

How is average fuel burn-up figured? The term burn-up is used to represent the amount of energy produced per unit mass of a material. The units of burn-up are megawatt days per metric ton (MWd/tm) of initial heavy metal such as uranium. One megawatt day represents about 2.6×10^{21} fissions. A nuclear fuel assembly is typically discharged from the reactor when it has achieved a burn-up of 50,000 MWd/tm. In commercial power reactors that are fueled with uranium comprising mainly the isotope U-238, with 4% or less U-235, most of the energy produced comes from the fissioning of U-235. At burn-ups of 50,000MWd/tm only about 5% of the initial uranium content of the fuel assembly has fissioned. Even when the other nuclear reactions that cause loss of the original uranium atoms are considered, there is still more than 90% of the initial uranium remaining in the fuel assembly when it is discharged. The amount of energy extracted from nuclear fuel is called its burn-up, which is expressed in terms of the heat energy produced per initial unit of fuel weight. Burn up is commonly expressed as megawatt days thermal per metric ton of initial heavy metal.   Courtesy of Babcock and Wilcox

1 Uranium ore - the principal raw material of nuclear fuel

2 Yellowcake - the form in which uranium is transported to a conversion plant

3 UF$_6$ - used in enrichment

4 Nuclear fuel - a compact, inert, insoluble solid

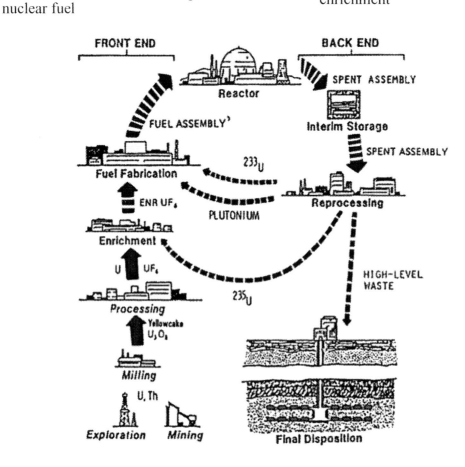

Courtesy of the NRC

What are stages of the nuclear fuel cycle? The nuclear fuel cycle uses uranium in different chemical and physical forms. Stages of this cycle typically include the following:

- mining (extracting from ore) and milling
- conversion to uranium hexafluoride
- enrichment
- fuel fabrication into uranium oxide
- use of the fuel in nuclear power, research, or naval propulsion reactors
- Interim Storage
- Reprocessing (currently not done in the U.S.)
- High-Level Waste

Where are the uranium milling facilities located? The majority of the uranium milling facilities is located in Wyoming. There are several facilities in New Mexico, Utah, and Nebraska.   Courtesy of the NRC

Why is uranium used in nuclear reactors more than any other radioactive elements? Natural uranium is a mixture of three isotopes: Uranium-234 (0.01%), Uranium-235 (0.71%) and Uranium-238 (99.28%). U-234 is present in small amounts and is not significant. U-235 is a fissionable isotope and U-238 is known as a fertile isotope which can be transmuted to a fissionable material. These isotopes are radioactive and emit alpha particles. U-235 is fissionable because its nucleus can absorb a neutron, become unstable and split into multiple pieces. By itself, U-238 cannot sustain a nuclear chain reaction, but it can be fissioned by high energy neutrons. Sufficient U-235 must be present to initiate the fission process and drive the chain reaction. U-238 when exposed to neutrons, its nucleus will capture a neutron and transform into plutonium-239 (Pu-239) a fissile isotope. Pu-239 is capable of sustaining a chain reaction and when fissioned, produces a similar number of free neutrons as U-235. U-238 is therefore considered to be fertile because of this transmutable property. U-238 is 140 times more abundant in nature than U-235 and is the main component of nuclear fuel.   Courtesy of Babcock & Wilcox

What do you know about Uranium?

Uranium is a white/black metallic chemical element in the actinide series of the periodic table that has the symbol U and atomic number 92. It has 92 protons and electrons, 6 of them valence electrons. It can have between 141 and 146 neutrons, with 143 and 146 in its most common isotopes. Uranium has the highest atomic weight of the naturally occurring elements. Uranium is approximately 70% more dense than lead and is weakly radioactive. It occurs naturally in low concentrations (a few parts per million) in soil, rock and water, and is commercially extracted from uranium-bearing minerals such as uraninite. Its atomic weight 238.03. The melting point is 1132°C (2070°F) and the boiling point is 3818°C (6904°F). Uranium is one of the actinide series. It is known as one of the heavy metals.

In nature, uranium atoms exist as uranium-238 (99.275%), uranium-235 (0.711%), and a very small amount of U-234 (0.0058%). Uranium decays slowly by emitting an alpha particle. The half-life of U-238 is about 4.47 billion years and that of U-235 is 704 million years,[1] making them useful in dating the age of the Earth (see uranium-thorium dating, uranium-lead dating and uranium-uranium dating). Along with thorium and plutonium, uranium is one of the three fissile elements, meaning it can easily break apart to become lighter elements. While U-238 has a small probability to fission spontaneously or when bombarded with fast neutrons, the much higher probability of U-235 and to a lesser degree U-233 to fission when bombarded with slow neutrons generates the heat in nuclear reactors used as a source of power, and provides the fissile material for nuclear weapons. Both uses rely on the ability of uranium to produce a sustained nuclear chain reaction. Depleted U-238 is used in kinetic energy penetrators and armor plating.

Uranium is used as a colorant in uranium glass, producing orange-red to lemon yellow hues. It was also used for tinting and shading in early photography. The 1789 discovery of uranium in the mineral pitchblende is credited to Martin Heinrich Klaproth, who named the new element after the planet Uranus. Eugène-Melchior Péligot was the first person to isolate the metal, and its radioactive properties were uncovered in 1896 by Antoine Becquerel. Research by Enrico Fermi and others starting in 1934 led to its use as a fuel in the nuclear power industry and in *Little Boy*, the first nuclear weapon used in war. An ensuing arms race during the Cold War between the United States and the Soviet Union produced tens of thousands of nuclear weapons that used enriched uranium and uranium-derived plutonium. The security of those weapons and their fissile material following the breakup of the Soviet Union in 1991 along with the legacy of nuclear testing and nuclear accidents is a concern for public health and safety.

Characteristics

An induced nuclear fission event involving U-235

When refined, uranium is a silvery white, weakly radioactive metal, which is slightly softer than steel, strongly electropositive and a poor electrical conductor. It is malleable, ductile, and slightly paramagnetic. Uranium metal has very high density, being approximately 70% more dense than lead, but slightly less dense than gold.

Uranium metal reacts with nearly all nonmetallic elements and their compounds, with reactivity increasing with temperature. Hydrochloric and nitric acids dissolve uranium, but non-oxidizing acids attack the element very slowly. When finely divided, it can react with cold water; in air, uranium metal becomes coated with a dark layer of uranium oxide. Uranium in ores is extracted chemically and converted into uranium dioxide or other chemical forms usable in industry.

Uranium reacts reversibly with hydrogen to form UH_3 at 250°C (482°F). Correspondingly, the hydrogen isotopes form uranium deuteride, UD_3, and uranium tritide, UT_3. The uranium-oxygen system is extremely complicated. Uranium monoxide, UO, is a gaseous species which is not stable below 1800°C (3270°F). In the range UO_2 to UO_3, a large number of phases exist. The uranium halides constitute an important group of compounds. Uranium tetrafluoride is an intermediate in the preparation of the metal and the hexafluoride. Uranium hexafluoride, which is the most volatile uranium compound, is used in the isotope separation of U-235 and U-238. The halides react with oxygen at elevated temperatures to form uranyl compounds and ultimately U_3O_8.

Uranium was the first element that was found to be fissile. Upon bombardment with slow neutrons, its U-235 isotope becomes a very short-lived U-236 isotope, which immediately divides into two smaller nuclei, releasing nuclear binding energy and more neutrons. If these neutrons are absorbed by other U-235 nuclei, a nuclear chain reaction occurs and, if there is nothing to absorb some neutrons and slow the reaction, the reaction is explosive. As little as 15 lb (7 kg) of U-235 can be used to make an atomic bomb. The first atomic bomb worked by this principle (nuclear fission).

Uranium atoms are unstable; that is, their nuclei tend spontaneously to fission or break down into smaller nuclei, fast particles (including neutrons), and high-energy photons. The fission of an isolated uranium nucleus is a randomly timed event; however, collision with a neutron may trigger a uranium nucleus to fission immediately. Crowding large numbers of uranium atoms together can enable the neutrons emitted by a few nuclei undergoing fission to cause other nuclei to fission, whose released neutrons in turn trigger still other nuclei, and so on. If this chain reaction proceeds at a constant rate, it may be used to generate electricity; if it proceeds at an exponentially increasing rate, a nuclear explosion results.

History

Pre-discovery use

The use of uranium in its natural oxide form dates back to at least the year 79, when it was used to add a yellow color to ceramic glazes. Yellow glass with 1% uranium oxide was found in a Roman villa on Cape Posillipo in the Bay of Naples, Italy by R. T. Gunther of the University of Oxford in 1912. Starting in the late Middle Ages, pitchblende was extracted from the Habsburg silver mines in Joachimsthal, Bohemia (now Jáchymov in the Czech Republic) and was used as a coloring agent in the local glassmaking industry. In the early 19th century, the world's only known source of uranium ores were these old mines.

Discovery

Antoine Becquerel discovered the phenomenon of radioactivity by exposing a photographic plate to uranium (1896).

The discovery of the element is credited to the German chemist Martin Heinrich Klaproth. While he was working in his experimental laboratory in Berlin in 1789, Klaproth was able to precipitate a yellow compound (likely sodium diuranate) by dissolving pitchblende in nitric acid and neutralizing the solution with sodium hydroxide. Klaproth mistakenly assumed the yellow substance was the oxide of a yet-undiscovered element and heated it with charcoal to obtain a black powder, which he thought was the newly discovered metal itself (in fact, that powder was an oxide of uranium). He named the newly discovered element after the planet Uranus, which had been discovered eight years earlier by William Herschel.

In 1841, Eugène-Melchior Péligot, who was Professor of Analytical Chemistry at the Conservatoire National des Arts et Métiers (Central School of Arts and Manufactures) in Paris, isolated the first sample of uranium metal by heating uranium tetrachloride with potassium. Uranium was not seen as being particularly dangerous during much of the 19th century, leading to the development of various uses for the element. One such use for the oxide was the aforementioned but no longer secret coloring of pottery and glass.

Antoine Becquerel discovered radioactivity by using uranium in 1896. Becquerel made the discovery in Paris by leaving a sample of uranium on top of an unexposed photographic plate in a drawer and noting that the plate had become 'fogged'. He determined that a form of invisible light or rays emitted by uranium had exposed the plate.

Fission research

Production and mining

Yellowcake is a concentrated mixture of uranium oxides that is further refined to extract pure uranium.

Uranium ore is mined in several ways: by open pit, underground, in-situ leaching, and borehole mining (see uranium mining). Low-grade uranium ore typically contains 0.1 to 0.25% of actual uranium oxides, so extensive measures must be employed to extract the metal from its ore. High-grade ores found in Athabasca Basin deposits in Saskatchewan, Canada can contain up to 70% uranium oxides, and therefore must be diluted with waste rock prior to milling. Uranium ore is crushed and rendered into a fine powder and then leached with either an acid or alkali. The leachate is then subjected to one of several sequences of precipitation, solvent extraction, and ion exchange. The resulting mixture, called yellowcake, contains at least 75% uranium oxides. Yellowcake is then calcined to remove impurities from the milling process prior to refining and conversion.

Commercial-grade uranium can be produced through the reduction of uranium halides with alkali or alkaline earth metals. Uranium metal can also be made through electrolysis of KUF$_5$ or UF$_4$, dissolved in a molten calcium chloride (CaCl$_2$) and sodium chloride (NaCl) solution. Very pure uranium can be produced through the thermal decomposition of uranium halides on a hot filament.

Resources and reserves

It is estimated that there is 4.7 million tonnes of uranium ore reserves (economically mineable) known to exist, while 35 million tonnes are classed as mineral resources (reasonable prospects for eventual economic extraction). An additional 4.6 billion tonnes of uranium are estimated to be in sea water (Japanese scientists in the 1980s proved that extraction of uranium from sea water using ion exchangers was feasible).

Exploration for uranium is continuing to increase with US$200 million being spent worldwide in 2005, a 54% increase on the previous year.

Australia has 40% of the world's uranium ore resources—the most of any country. In fact, the world's largest single uranium deposit is located at the Olympic Dam Mine in South Australia. Almost all the uranium is exported under strict International Atomic Energy Agency safeguards to satisfy the Australian people and government that none of the uranium is used in nuclear weapons. As of 2006, the Australian government was advocating an expansion of uranium mining, although issues with state governments and indigenous interests complicate the issue.

The largest single source of uranium ore in the United States was the Colorado Plateau located in Colorado, Utah, New Mexico, and Arizona. The U.S. federal government paid discovery bonuses and guaranteed purchase prices to anyone who found and delivered uranium ore, and was the sole legal purchaser of the uranium. The economic incentives resulted in a frenzy of exploration and mining activity throughout the Colorado Plateau from 1947 through 1959 that left thousands of miles of crudely graded roads spider-webbing the remote deserts of the Colorado Plateau, and thousands of abandoned uranium mines, exploratory shafts, and tailings piles. The frenzy ended as suddenly as it had begun, when the U.S. government stopped purchasing the uranium.

Enrichment

Cascades of gas centrifuges are used to enrich uranium ore to concentrate its fissionable isotopes.

Enrichment of uranium ore through isotope separation to concentrate the fissionable U-235 is needed for use in nuclear power plants and nuclear weapons. A majority of neutrons released by a fissioning atom of U-235 must impact other U-235 atoms to sustain the nuclear chain reaction needed for these applications. The concentration and amount of U-235 needed to achieve this is called a 'critical mass.'

To be considered 'enriched', the U-235 fraction has to be increased to significantly greater than its concentration in naturally-occurring uranium. Enriched uranium typically has a U-235 concentration of between 3 and 5%. The process produces huge quantities of uranium that is depleted of U-235 and with a correspondingly increased fraction of U-238, called depleted uranium or 'DU'. To be considered 'depleted', the U-235 isotope concentration has to have been decreased to significantly less than its natural concentration. Typically the amount of U-235 left in depleted uranium is 0.2% to 0.3%. As the price of uranium has risen since 2001, some enrichment tailings containing more than 0.35% U-235 are being considered for re-enrichment, driving the price of these depleted uranium hexafluoride stores above $130 per kilogram in July, 2007 from just $5 in 2001.

The gas centrifuge process, where gaseous uranium hexafluoride (UF_6) is separated by weight using high-speed centrifuges, has become the cheapest and leading enrichment process (lighter UF_6 concentrates in the center of the centrifuge). The gaseous diffusion process was the previous leading method for enrichment and the one used in the Manhattan Project. In this process, uranium hexafluoride is repeatedly diffused through a silver-zinc membrane, and the different isotopes of uranium are separated by diffusion rate (U-238 is heavier and thus diffuses slightly slower than U-235). The molecular laser isotope separation method employs a laser beam of precise energy to sever the bond between U-235 and fluorine. This leaves U-238 bonded to fluorine and allows U-235 metal to precipitate from the solution. Another method is called liquid thermal diffusion.

Precautions

Exposure

A person can be exposed to uranium (or its radioactive daughters such as radon) by inhaling dust in air or by ingesting contaminated water and food. The amount of uranium in air is usually very small; however, people who work in factories that process phosphate fertilizers, live near government facilities that made or tested nuclear weapons, live or work near a modern battlefield where depleted uranium weapons have been used, or live or work near a <u>coal-fired</u> power plant, facilities that mine or process uranium ore, or enrich uranium for reactor fuel, may have increased exposure to uranium. Houses or structures that are over uranium deposits (either natural or man-made slag deposits) may have an increased incidence of exposure to radon gas.

Almost all uranium that is ingested is excreted during digestion, but up to 5% is absorbed by the body when the soluble uranyl ion is ingested while only 0.5% is absorbed when insoluble forms of uranium, such as its oxide, are ingested. However, soluble uranium compounds tend to quickly pass through the body whereas insoluble uranium compounds, especially when ingested via dust into the lungs, pose a more serious exposure hazard. After entering the bloodstream, the absorbed uranium tends to bioaccumulate and stay for many years in bone tissue because of uranium's affinity for phosphates. Uranium does not absorb through the skin, and alpha particles released by uranium cannot penetrate the skin.

Effects

The greatest health risk from large intakes of uranium is toxic damage to the kidneys, because, in addition to being weakly radioactive, uranium is a toxic metal. Radiological effects are generally local because this is the nature of alpha radiation, the primary form from U-238 decay. No human cancer has been seen as a result of exposure to natural or depleted uranium, but exposure to some of its decay products, especially radon, does pose a significant health threat. Exposure to strontium-90, iodine-131, and other fission products is unrelated to uranium exposure, but may result from medical procedures or exposure to spent reactor fuel or fallout from nuclear weapons. Although accidental inhalation exposure to a high concentration of uranium hexafluoride has resulted in human fatalities, those deaths were not associated with uranium itself. Finely-divided uranium metal presents a fire hazard because uranium is pyrophoric, so small grains will ignite spontaneously in air at room temperature.

Isotopes of Uranium

Isotope	Half Life	Isotope	Half Life
U-230	20.8 DAYS	U-236	2.34E7 YEARS
U-231	4.2 DAYS	U-237	6.75 DAYS
U-232	70 YEARS	U-238	4.47E9 YEARS
U-233	159,000 YEARS	U-239	23.5 MINUTES
U-234	247,000 YEARS	U-240	14.1 HOURS
U-235	7.004E8 YEARS		

Atomic Structure U-238

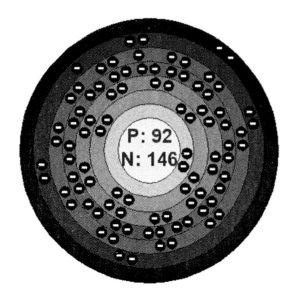

Number of Energy Levels: 7

First Energy Level: 2
Second Energy Level: 8
Third Energy Level: 18
Fourth Energy Level: 32
Fifth Energy Level: 21
Sixth Energy Level: 9
Seventh Energy Level: 2

What are uranium mill tailings? Uranium mill tailings are primarily the sandy process waste material from a conventional mill. This ore residue contains the radioactive decay products from the uranium chains (mainly the U-238 chain) and heavy metals. By definition in 10 CFR Part 40, the tailings or wastes produced by the extraction or concentration of uranium or thorium from any ore processed primarily for its source material content is byproduct material. This includes discrete surface waste resulting from uranium solution extraction processes, such as in situ leach, heap leach, and ion-exchange. Byproduct material does not include underground ore bodies depleted by solution extraction. The wastes from these solution extraction facilities are transported to a mill tailings impoundment for disposal.

Most of the NRC regulations for this type of byproduct material are in Appendix A to 10 CFR Part 40 - Criteria Relating to the Operation of Uranium Mills and the Disposition of Tailings or Wastes Produced by the Extraction or Concentration of Source Material from Ores Processed Primarily for Their Source Material Content. The criteria in Appendix A cover the sighting and design of tailings impoundments, disposal of tailings or wastes, decommissioning of land and structures, groundwater protection standards, testing of the radon emission rate from the impoundment cover, monitoring programs, airborne effluent and offsite exposure limits, inspection of retention systems, financial surety requirements for decommissioning and long-term surveillance and control of the tailings impoundment, and eventual government ownership of the tailings site under a NRC general license. Courtesy of NRC

How are the fuel rods constructed? Fuel rod manufacture begins with the Zircaloy-4 tubing/cladding. The tubing is 100% ultrasonically tested for defects before it is released into production.

The cladding is then prepared for end cap welding. A press fit weld joint configuration is machined on each end and a cleaning plug is passed through the cladding to remove resultant chips. A plenum spring is then loaded and one end cap is welded in place in a chamber pressurized with inert gas. The rod end being welded is inserted into the chamber through a sealing hole and the rod is held stationary while the welding head rotates a tungsten electrode about the weld joint.

After welding, a unique serial number is stamped on the end cap. Quality data for all components of the fuel rods are traceable by this number. Each end cap weld is automatically tested with ultrasonic equipment within minutes of completing the weld, providing rapid feedback to the welder operator.

The tubes are transferred to the laser drilling station. Here, the capped end of the cladding is inserted in a small chamber, the chamber is evacuated and a focused laser beam is directed at the center of the end cap. Discharge of the laser generates a small hole in the cap. This hole serves several purposes: it removes back-pressure during fuel pellet loading, it provides a path to remove moisture from the fuel pellets and rod internals and it permits access to pressurize the fuel rod. The cladding with one laser drilled end cap in place is then ready for fuel loading.

The fuel pellets containing the proper per rod U-235 content are weighed; this weight is computer documented. The stack of pellets is visually examined for pellet quality and length and the fuel column, with appropriate nonfuel internals, is inserted into the cladding. The plenum at the open rod end is then measured to verify seating and column length and the lower fuel rod spring is inserted. The second cap is installed in the open end of the rod, causing the plenum spring to preload the fuel column. This end cap is welded and inspected as was the first cap.

Caped fuel rods are accumulated and loaded into a vacuum furnace in lots. Sample rods are included for subsequent hydrogen analysis. The loaded furnace is evacuated, backfilled with inert gas, heated and re-evacuated. This cycle removes remaining moisture from the fuel pellets and internal components. At completion of the cycle, the chamber and rods are refilled with inert gas at atmospheric pressure and the sample rod is evaluated by quality control personnel.

Once the correct hydrogen content of the sample pellets is confirmed, the fuel rods are removed from the drying retort. Within an established time limit, the drilled end cap of each rod is inserted into a small chamber for pressurizing and sealing. Once in the chamber, the laser drilled hole is aligned with another laser. The rod is secured and the chamber is sealed and evacuated. The fuel rod is pressurized with dry helium and the laser is discharged to fuse the hole in the end cap. An additional automatic ultrasonic inspection follows the seal weld.

The sealed rods are then transferred to a neutron scanner. Here, the rods are exposed to the neutron beam from a californium source. The resulting activity is measured and analyzed to nondestructively verify the fuel column integrity and length and the enrichment of the pellets. The rods are then transported from the scanner to automatic cleaning equipment. The entire fuel rod surface is scrubbed, rinsed with demineralized water and dried in preparation for final inspection.

The final inspection, the end of the rod through which the pellets were loaded is subjected to alpha counting to ensure that there is no uranium contamination on the outside surface. Following dimensional inspection and a complete visual inspection, helium leak testing is conducted. This is a functional test of the cladding, end caps and welds, verifying total rod sealing.

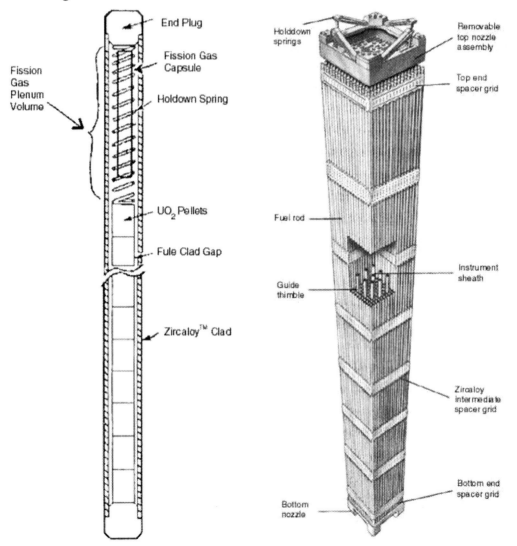

Courtesy of the University of Florida

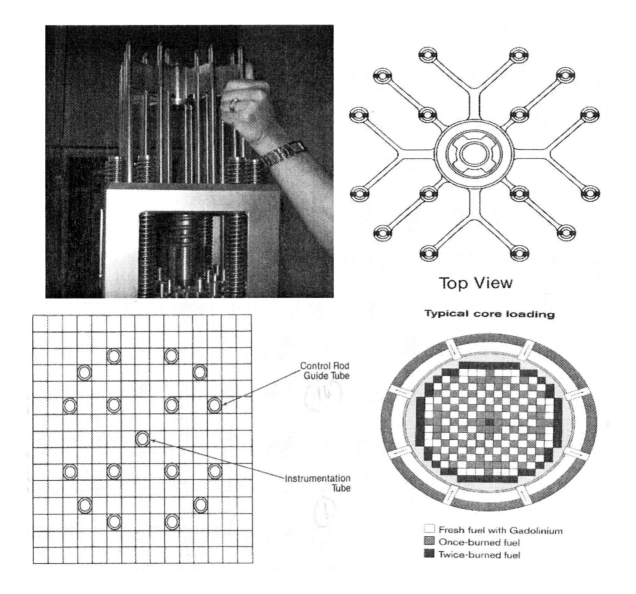

Top View

Typical core loading

Control Rod
Guide Tube

Instrumentation
Tube

☐ Fresh fuel with Gadolinium
▨ Once-burned fuel
▧ Twice-burned fuel

How are the fuel rods affected by radiation? The study of the nuclear fuel cycle includes the study of the behavior of nuclear materials both under normal conditions and under accident conditions. For example, there has been much work on how uranium dioxide based fuel interacts with the zirconium alloy tubing used to cover it. During use, the fuel swells due to thermal expansion and then starts to react with the surface of the zirconium alloy, forming a new layer which contains both fuel and zirconium (from the cladding). Then, on the fuel side of this mixed layer, there is a layer of fuel which has a higher caesium to uranium ratio than most of the fuel. This is because xenon isotopes are formed as fission products that diffuse out of the lattice of the fuel into voids such as the narrow gap between the fuel and the cladding. After diffusing into these voids, it decays to caesium isotopes. Because of the thermal gradient which exists in the fuel during use, the volatile fission products tend to be driven from the center of the pellet to the rim area.

Below is a graph of the temperature of uranium metal, uranium nitride and uranium dioxide as a function of distance from the center of a 20 mm diameter pellet with a rim temperature of 200°C. The uranium dioxide (because of its poor thermal conductivity) will overheat at the center of the pellet, while the more thermally conductive other forms of uranium remain below their melting points.

What is a fuel pin? A cylindrical hollow rod that will hold the fuel pellets. The pin is sealed at both ends and is charged with helium and held in place with a spring. The fuel pin can also be called a fuel rod.

What is the purpose of the fuel assembly? The fuel assembly is designed to align all the fuel pins in a symmetric pattern to ensure the most efficient conversion of energy of fission to thermal energy (heat).

How many fuel pins (rods) are there in one fuel assembly? It will depend on the design of the reactor core as well as the design of the type of reactor being used. Example: a PWR designed to hold 220 fuel assemblies with each assembly designed in a 14x14 matrix and each assembly holding 176 fuel pins (rods). Not all of the fuel assemblies will be holding fuel, there must be some room for in-core instrumentation as well as the control rods.

What are burnable poison pins? Use to reduce moderator temperature coefficient at the beginning of a fuel cycle.

What does the reactor core fuel rod tube temperature profile look like?

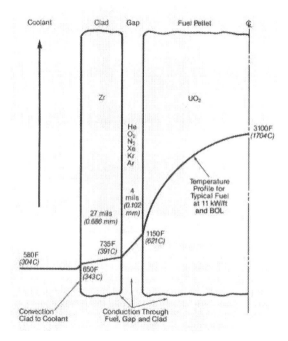

What is the purpose of pressurizing the fuel rods with helium? Helium is used to pressurize the fuel rods in order to help minimize pellet cladding interaction (PCI) which can lead to

fuel rod failure over long periods.   The helium gas also helps to improve the
conduction of heat from the fuel to the cladding.   Courtesy of Areva

How can reactor power be changed in a PWR and a BWR?

BWR Core Changes – Vs – Reactor Power

Moderator Temp.	↑	PWR	↓		Voids	↑	PWR	↓
Moderator Temp.	↓	PWR	↑		Voids	↓	PWR	↑
Fuel Temp.	↑	PWR	↓		Reactor Press.	↑	PWR	↑
Fuel Temp.	↓	PWR	↑		Reactor Press.	↓	PWR	↓
Recirc. Flow	↑	PWR	↑		Control Rods	↓	PWR	↓
Recirc. Flow	↓	PWR	↓		Control Rods	↑	PWR	↑

PWR Core Changes – Vs – Reactor Power

Moderator Temp.	↑	PWR	↓		Voids	↑	PWR	↓
Moderator Temp.	↓	PWR	↑		Voids	↓	PWR	↑
Fuel Temp.	↑	PWR	↓		Reactor Press.	↑	PWR	↑
Fuel Temp.	↓	PWR	↑		Reactor Press.	↓	PWR	↓
Sec. Wtr Temp.	↑	PWR	↓		Boron Concent.	↑	PWR	↓
Sec. Wtr Temp.	↓	PWR	↑		Boron Concent.	↓	PWR	↑
Sec. Wtr Flow	↑	PWR	↑		Control Rods	↓	PWR	↓
Sec. Wtr Flow	↓	PWR	↓		Control Rods	↑	PWR	↑

What might cause fuel failure? Fuel failures can be caused by foreign material exclusion (FME), CRUD induced localized high fuel rod temperatures, vibration within fuel assembly, reactor coolant temperature transients, and fuel hydriding.

What is meant by fuel failure? Fuel failure is a term used to indicate that the radioactive material that was contained within the fuel rod has somehow become damaged and has released some of its content into the reactor coolant.

What is fuel hydriding? Fuel hydriding is when the inside of the fuel rod becomes corrosive and begins to corrode from the inside of the fuel rod to the outside.

How can you determine if you have a fuel leak in the reactor core? The first noticeable effect of the release will generally show up as an increase in the Xenon and krypton activity in the plant's off-gas system (BWR). In a PWR, fuel damage would be indicated in increases in Xe and I in the RCS which is monitored closely.

What is a bio-shield? To safeguard workers inside the plant, a concrete biological shield surrounds the reactor vessel.

Radiation

What is radiation? When an atom has too much energy, it becomes unstable and emits packets of energy or particles from the electron cloud or nucleus. It may require several emissions before becoming stable. These emissions are called radiation. Often, the reactor creates unstable atoms by forcing a stable atom to accept a neutron or energy particle.

Radiation refers to the propagation of waves and particles through space and includes both electromagnetic radiation and atomic and subatomic particle radiation. Electromagnetic radiation has a broad, continuous spectrum of energy (see The Radiation Spectrum) that includes visible light, radio waves, microwaves, x-rays, gamma rays, and infrared and ultraviolet radiation. All electromagnetic radiation travels at the speed of light. Particle radiation includes alpha and beta particles, neutrons, protons and heavy ions. The speed and energy of particle radiation depends upon the source of the radiation and any subsequent interaction of the particle with other matter.

While there are many different sources of radiation, it generally arises from or is produced by radioactive decay, energy change of an atomic electron or nucleus, motion of atoms or molecules, or the interaction between particles or electromagnetic radiation and atoms or nuclei. There are many sources and types of naturally occurring radiation such as the sun, radioactive materials, visible light, solar and cosmic radiation, and thermal radiation (what is typically referred to as heat). Radiation can also be generated to diagnose and treat illness, eliminate or reduce harmful microorganisms to enhance the safety of medical equipment and the food supply, cook food, transmit information (radio, television, cellular phones, etc.), and many other applications addressed on this website.

Radioactivity refers to the property of spontaneous emission of particles or electromagnetic radiation exhibited by certain materials. This radiation is emitted by unstable atoms as they undergo a transition to a more stable state; the transition is called radioactive decay. Unstable atoms that exist in nature are said to be naturally radioactive. Examples of radioactive atoms found in nature are carbon-14, potassium-42, radon-222, uranium-235, uranium-238 and thorium-232. In addition to naturally occurring radioactive materials, radioactive atoms can be produced when the nucleus of an atom is made to interact with a particle or electromagnetic radiation to form an unstable nucleus; this is typically done in nuclear reactors and particle accelerators.

As radioactive atoms go through the transition to a stable state they emit radiation in several forms as follows:
- charged particles (alpha particles, beta particles and positrons)
- uncharged particles (neutrons)
- electromagnetic radiation (gamma rays and x-rays)

Courtesy of the American Nuclear Society

What are sources of radiation? Most radiation comes to us from the sun and from cosmic radiation -- so that people at higher elevations like Colorado and adjacent Rocky

Mountain States receive more than those who live at sea level. However, a lot of radiation also comes from the soil and rocks around us. Granite and marble have background levels of radioactivity. A relatively small additional amount comes from our man-made technology.

Courtesy of the American Nuclear Society

What is electromagnetic radiation? Electromagnetic radiation consists of electric and magnetic field components which oscillate in phase perpendicular to each other and perpendicular to the direction of energy propagation. It can be classified in several different types depending on its frequency: radio waves, microwaves, infrared radiation, visible light, ultraviolet radiation, X-rays, and gamma rays.

What are the four basic types of radiation and how are they different?
    Alpha – a helium atom, is the least penetrating particle. The source is reactor fuel and radon gas. Can be stopped by a piece of paper or a dead layer of skin. Can be shielded by clothing and is primarily an internal hazard.
    Beta – a particle consisting of an electron, carrying a negative charge. Beta particles are produced in certain types of radioactive decay. A stream of them is called beta radiation. Has more penetrating power than an alpha particle. Source is a fluid within a pipe. Can be shielded by a thin layer of aluminum or plastic. A hazard to the skin and eye lens.
    Gamma – no charge, no mass, travel at speed of light, massive radiation. Has high penetrating ability. Sources are fluids and corrosion products within the primary system, including steam in a BWR. Shielded by dense material such as lead and steel. Gives a whole body dose and is the primary concern in the plant.
    Neutron – a particle carrying no charge. Has high penetrating ability. Source is the reactor core at power. Shielded by water or concrete. Gives a whole body dose.

What is meant by ionization? Radiation may interact with other atoms, causing a loss of electrons. This loss is called ionization and it can change the properties of an atom.

What is contamination? The deposition of unwanted radioactive material on the surfaces of structures, areas, objects, or people where it may be external or internal.

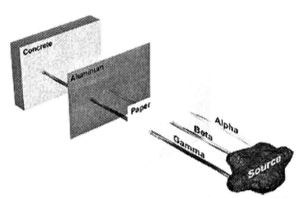

What is an absorber? Any material that stops ionizing radiation. Lead, concrete, and steel attenuate gamma rays. A thin sheet of paper or metal will stop or absorb alpha particles and most beta particles.

What is the average exposure for an everyday person? The average person in the U.S. is exposed to about 360 millirem of radiation annually. Natural (background) radiation accounts for more than three quarters of the yearly exposure.

What is ALARA? ALARA is an acronym meaning As Low As REASONABLY Achievable. 10CFR20.1101 mandates that nuclear power plants keep radiation doses minimized by using the ALARA process. All radiation levels are monitored not only throughout the plant and site but as well as all of the employees who work in those areas of possible exposure. ALARA was developed during the cleanup efforts of **SL-1** to help reduce radiation dose for any one individual.

Why use ALARA? By using the ALARA process, each millirem of radiation exposure we decrease, decreases the risk of health problems in the future.

What are four major components of ALARA? The four major components of ALARA are: Time, Distance, Shielding and Minimization. The amount of time workers spend in a radioactive area determines how much dose they will receive. The distance between the workers and the sources of radiation they are working near will determine how much dose they will receive…the closer they are to the source the more like they are to receive a higher dose of radiation. The use of shielding greatly reduces the dose rate should a worker need to work close to a radiation source. And finally, the fourth major component of ALARA is to minimize the creation of radioactive sources. Whenever a material such as iron is exposed to radiation, it becomes a source of radiation and has to be dealt with in a similar method that the actual radioactive source would go through. By following the time, distance, shielding and minimization of radiation sources, helps to keep the workers healthier by reducing their chance of lethal exposure.

What is the heaviest of the four types of radiation? The heaviest of the four types of radiation is the alpha type radiation which is basically a helium atom composed of two protons and two neutrons.

How is radiation measured? Radiation can be measured in different ways using different units of measure. Those commonly used are: curies, roentgens, rads, rems and mrems.

What is infrared electromagnetic radiation? Infrared electromagnetic radiation that has a wavelength range of 0.75 nanometer to about 1 millimeter. This energy is heat radiation that occupies a space on the electromagnetic spectrum between microwaves and visible light.

What is the radiation spectrum? Light is a mixture of colors. White light can be dispersed into its component colors: violet, indigo, blue, green, yellow, orange, and red. This ordered arrangement of colors is called the visible spectrum. A particular color of light can be specified by either its frequency or its wavelength with the property that the frequency multiplied by the wavelength is equal to the speed of light, about 300 million meters per second (about 186,000 miles per second).

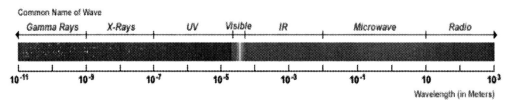

The rainbow is a natural spectrum, produced by meteorological phenomena. A similar effect can be produced by sunlight passing through a glass prism. The English mathematician and physicist Sir Isaac Newton advanced the first correct explanation of the phenomenon in 1666. When a ray of light passes from one transparent medium, such as air, into another, such as glass or water, it is bent; upon reemerging into the air, it is bent again. This bending is called refraction; the amount of refraction depends on the wavelength of the light. Violet light, for example, is bent more than red light in passing from air to glass or from glass to air. A mixture of red and violet light is thus dispersed into the two colors when it passes through a wedge-shaped glass prism.

A device for producing and observing a spectrum visually is called a spectroscope. A device for observing and recording a spectrum photographically is called a spectrograph. A device for measuring the brightness of the various portions of spectra is called a spectrophotometer; and the science of using the above devices to study spectra is called spectroscopy.

For accurate spectroscopic measurements, an interferometer is used. During the 19th century, scientists discovered that beyond the violet end of the spectrum, radiation could be detected that was invisible to the human eye but that had marked photochemical action; this radiation was termed ultraviolet. Similarly, beyond the red end of the spectrum, infrared radiation was detected that, although invisible, transmitted energy, as shown by their ability to raise the temperature of a thermometer. The definition of spectrum was then revised to include this invisible radiation, and has since been extended to include radio waves beyond the infrared, and X-rays and gamma rays beyond the ultraviolet.   Courtesy of the American Nuclear Society

What is Alpha Decay?

Alpha Decay

Alpha decay is one process that unstable atoms can use to become more stable. During alpha decay, an atom's nucleus sheds two protons and two neutrons in a packet that scientists call an alpha particle.

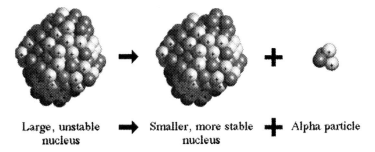

Large, unstable   ➡   Smaller, more stable   ➕   Alpha particle
nucleus   nucleus

Since an atom loses two protons during alpha decay, it changes from one element to another. For example, after undergoing alpha decay, an atom of uranium (with 92 protons) becomes an atom of thorium (with 90 protons). Courtesy of Jefferson Lab

What is Gamma Decay?

Gamma Decay

In gamma decay, a nucleus changes from a higher energy state to a lower energy state through the emission of electromagnetic radiation (photons). The number of protons (and neutrons) in the nucleus does not change in this process, so the parent and daughter atoms are the same chemical element. In the gamma decay of a nucleus, the emitted photon and recoiling nucleus each have a well-defined energy after the decay. The characteristic energy is divided between only two particles.

What is Beta Decay?

Beta Decay

Beta decay is one process that unstable atoms can use to become more stable. There are two types of beta decay, beta-minus and beta-plus.

During beta-minus decay, a neutron in an atom's nucleus turns into a proton, an electron and an antineutrino. The electron and antineutrino fly away from the nucleus, which now has one more proton than it started with. Since an atom gains a proton during beta-minus decay, it changes from one element to another. For example, after undergoing beta-minus decay, an atom of carbon (with 6 protons) becomes an atom of nitrogen (with 7 protons).

During beta-plus decay, a proton in an atom's nucleus turns into a neutron, a positron and a neutrino. The positron and neutrino fly away from the nucleus, which now has one less proton than it started with. Since an atom loses a proton during beta-plus decay, it changes from one element to another. For example, after undergoing beta-plus decay, an atom of carbon (with 6 protons) becomes an atom of boron (with 5 protons).

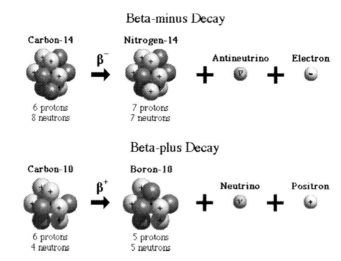

Although the numbers of protons and neutrons in an atom's nucleus change during beta decay, the total number of particles (protons + neutrons) remains the same.

Courtesy of Jefferson Lab

What is meant by radioactive decay? The spontaneous emission of radiation by a nuclide in order to achieve a greater stability. Will lower the atom's excitation state and result in closer proximity to the ground state.

Does an ion exchanger remove radiation from contaminated water? Yes and No, the ion exchanger does remove radiation from the primary system contaminated water but only at the ion level…it is an ion exchanger and not a filter to trap actual particles.

What is the purpose of the radiation monitoring system? The radiation monitoring system is designed to warn plant personnel of radiation levels in and around the plant area. This system allows the plant to monitor for increased/decreased levels of radiation continuously. It also provides an early warning of any plant malfunctions which may result in a radiation hazard and/or plant damage. These limits are set by Part 20 of Title 10 of the Code of Federal Regulations.

What is the purpose of the atmosphere monitors? The atmosphere monitors is a system of monitors used to determine if any radioactive particles are suspended in the air around the plant. It consists of an off-line Geiger-Mueller tube which has a range from 10^1 to 10^6 counts per minute (cpm). These monitors provide indications to the control room and/or chart recorders that are placed throughout the plant area.

What is Cherenkov radiation? Cherenkov radiation is electromagnetic radiation emitted when a charged particle passes through an insulator at a constant speed greater than the speed of light in that medium. The characteristic blue glow of nuclear reactors is due to Cherenkov radiation. It was named after Russian scientist Pavel Alekseyevich Cherenkov. Cherenkov observed the emission of blue light from a bottle of water subjected to radioactive bombardment. This phenomenon, associated with charged atomic particles moving at velocities greater than the speed of light in the local medium.

Reactor Core

What is alpha radiation? Radiated particle consisting of 2 protons and 2 neutrons. Emitted by heavy nuclides with a low neutron/proton ratio (alpha decay will increase the neutron/proton ratio). Interact with matter by direct collision or charged particle interaction. Large size (4 AMU) and charge (+2) result in rapid energy transfer to other atoms.

What is beta radiation? A beta particle is a high energy electron emitted from the nucleus. The process of beta decay involves electron ejection and the transformation of a neutron into a proton. Beta particles are emitted by nuclides with a high neutron/proton ratio (below the line of stability). The majority of the product nuclides produced during fission are beta emitters.

What is positron radiation? A positron is a high energy particle (positively charged electron) emitted form the nucleus. The process of positron decay involves positron ejection and the transformation of a proton into a neutron. Positrons are emitted by nuclides with a low neutron/proton ratio (above the line of stability).

What is gamma radiation? Gamma radiation can also be called photons. Has no charge, no rest mass, and travels at light speed. Interact with matter via three types of photo-interactions (Photoelectric effect, Compton Scattering and Pair Production). Highly energetic electromagnetic waves emitted from the nucleus of an atom as it attempts to achieve a more stable energy state. Energies and wave lengths correspond to discrete quantum (metastable) energy levels in the nucleus.

What is ionizing radiation? Any radiation which causes the formation of ion pairs, typically by interaction with the orbital electrons in an atom.

What kind of symptoms would be caused by different radiation exposure levels?

0.05 – 0.2 Sv (5 – 20 REM)

No symptoms. Potential for cancer and mutation of genetic material.

0.2 – 0.5 Sv (20 – 50 REM)

No noticeable symptoms. Red blood count decreases temporarily.

0.5 – 1.0 Sv (50 – 100 REM)

Mild radiation sickness with headache and increased risk of infection due to disruption of immunity cells. Temporary male sterility is possible.

1.0 – 2.0 Sv (100 – 200 REM)

Light radiation poisoning, 10% fatality after 30 days (LD 10/30). Typical symptoms include mild to moderate nausea (50% probability at 2 Sv), with occasional vomiting, beginning 3 to 6 hours after irradiation and lasting for up to one day. This is followed by a 10 to 14 day latent phase, after which light symptoms like general illness and fatigue appear (50% probability at 2 Sv). The immune system is depressed, with convalescence extended and increased risk of infection. Temporary male sterility is common. Spontaneous abortion or stillbirth will occur in pregnant women.

2.0 – 3.0 Sv (200 – 300 REM)

Moderate radiation poisoning, 35% fatality after 30 days (LD 35/30). Nausea is common (100% Sv), with 50% risk of vomiting at 2.8 Sv. Symptoms onset at 1 to 6 hours after irradiation and last for 1 to 2 days. After that, there is a 7 to 14 day latent phase, after which the following symptoms appear: loss of hair all over the body (50% probability at 3 Sv), fatigue and general illness. There is a massive loss of leukocytes (white blood cells), greatly increasing the risk of infection. Permanent female sterility is possible. Convalescence takes one to several months.

3.0 – 4.0 Sv (300 – 400 REM)

Severe radiation poisoning, 50% fatality after 30 days (LD 50/30). Other symptoms are similar to the 2 – 3 Sv dose, with uncontrollable bleeding in the mouth, under the skin and in the kidneys (50% probability at 4 Sv) after the latent phase.

4.0 – 6.0 Sv (400 – 600 REM)

Acute radiation poisoning, 60% fatality after 30 days (LD 60/30). Fatality increases from 60% at 4.5 Sv to 90% at 6 Sv (unless there is intense medical care). Symptoms start half an hour to two hours after irradiation and last for up to 2 days. After that, there is a 7 to 14 day latent phase, after which generally the same symptoms appear as with 3 – 4 Sv irradiation, with increased intensity. Female sterility is common at this point. Convalescence takes several months to a year. The primary cause of death (in general 2 to 12 weeks after irradiation) are infections and internal bleeding.

6.0 – 10 Sv (600 – 1,000 REM)

Acute radiation poisoning, near 100% fatality after 14 days (LD 100/14). Survival depends on intense medical care. Bone marrow is nearly or completely destroyed, so a bone marrow transplant is required. Gastric and intestinal tissue is severely damaged. Symptoms start 15 to 30 minutes after irradiation and last for up to 2 days. Subsequently, there is a 5 to 10 day latent phase, after which the person dies from infection or internal bleeding. Recovery would take several years and probably would never be complete.

10 – 50 Sv (1,000 – 5,000 REM)

Acute radiation poisoning, 100% fatality after 7 days (LD 100/7). An exposure this high leads to spontaneous symptoms after 5 to 30 minutes. After powerful fatigue and immediate nausea caused by direct activation of chemical receptors in the brain by the irradiation, there is a period of several days of comparative well-being, called the latent (walking ghost) phase. After that, cell death in the gastric and intestinal tissue, causing massive diarrhea, intestinal bleeding and loss of water, leads to water electrolyte imbalance. Death sets in with delirium and coma due to breakdown of circulation. Death is currently inevitable; the only treatment that can be offered is pain therapy.

More than 50 Sv (> 5,000 REM)

A worker receiving 100 Sv (10,000 REM) in an accident at Wood River, Rhode Island in July 1964 survived for 49 hours after exposure.

How is metal affected by radiation? Metals suffer radiation damage primarily when neutrons bounce off atoms in the material. If the neutron transfers more energy than that holding the atom in the crystal lattice, typically around 40 electron volts, then the atom is displaced. This displacement damage consists of two localized flaws, or point defects: a hole in the lattice, known as a vacancy, and an interstitial, where the atom comes to rest between other atoms in the lattice.

Nuclear fission generates fast-moving neutrons with energies from 1 to 10 million electron volts. Only a tiny fraction reach the pressure vessel without being slowed down significantly. But a neutron that does reach the vessel with so much energy has a great effect: its collision with a single atom causes many displacements within a small volume as that atom collides with other atoms, forcing them out of position and starting a displacement cascade.

Calculations of the molecular dynamics of cascades show that vacancies and interstitials can cluster to form larger defects or combine to effectively reduce the damage. These calculations also show that very energetic collisions are less efficient at producing damage than previously thought.

Many more slow-moving neutrons with energies less than about 0.1 million electron volts reach the vessel. But while these lower-energy neutrons will always produce fewer displacements, calculations show that the smaller cascades created provide less opportunity for recombination to occur.

In the past, the accumulated dose of fast-moving neutrons was used as a measure of radiation damage. The dose was measured with a dosimeter consisting of materials that capture neutrons to form radioactive isotopes. The isotopes formed, which are identified by their rates of decay, are a measure of the neutron dose. But this procedure takes no account of the effect of low-energy neutrons that have been slowed down, or moderated, in the reactor to improve efficiency of the fission process.

A better measure of radiation damage is the displacement dose, which is defined as the number of times each atom in a material has been knocked from a crystal lattice position. To record this, engineers must know the energy spectrum of the neutrons and be able to calculate how the displacements are produced. This has led to the development of a device that relies on sapphire crystals, which darken progressively the more displacements they sustain.

Typical displacement doses accumulated by pressure vessels in the oldest operating reactors range from 0.01 to 0.1 displacements per atom (dpa). Structures closer to the core and structural materials in the fuel elements may accumulate doses of several dpa. Changes of mechanical properties start to be observed at doses of around 0.01 dpa. At doses of 1 dpa dimensional changes start to be seen.

The fast breeder reactors and fusion reactors of the future will have displacement doses of more than 100 dpa by the end of their working lives. This means that every atom will have been knocked off a lattice position more than a hundred times. Fusion generates neutrons with such high energies, typically around 14 mega-electron volts, that helium gas is produced in the bombarded structure as well as displacement damage. As a result any fusion reactor would not only distort, or 'swell', but form surface blisters too.

Distortion, or swelling, of steel at the Prototype Fast Reactor at Dounreay in Scotland, which has been operating for 18 years, has been reduced by developing new materials for the plant, such as alloys containing as much as 40 per cent nickel. However, the PFR is due to close in two years' time, following the government's confirmation last month that public funding of the plant will end in 1994.

What are some benefits of radiation? Radiation and radioactive materials can be used by humans in a number of ways. Here is a short list:

Agriculture – the increase in the volume and quality of grains and cereals has been vastly improved by selectively growing superior strains labeled by radioactive isotopes. These improvements are helping to alleviate famine in third world countries.

Environmental Measurements – the movement of pollutants through the environment – its ground waters and rivers – can be accurately measured by the use of radioactive tracers.

Eradication of pests – a number of pest flies are no longer the problem that they once were in California since their numbers have been cut drastically following the release of sterile male flies in the region.

Food – such as chicken, beef and other meats that have been sterilized by irradiation has a longer shelf life and is free of E. coli (a bacterium that has killed several children as a result of eating poorly cooked fast food hamburgers). An extension of food irradiation could save the lives of many children and would be particularly useful in developing countries where refrigeration is not available.

Generation of electricity – over 400 nuclear plants around the world contributes some 16% of the world's electrical energy needs. Over a hundred nuclear plants in the U.S. contribute 22% of the U.S.'s consumption of electricity in 2005.

Medical Diagnostics – the use of radiation in the medical world extends from X-rays, through magnetic resonance imaging (MRI), to the use of radioactive tracers to diagnose such varied conditions as faulty thyroid glands or bone problems. The use of radioactive tracers often takes the place of invasive surgical diagnosis.

Oil Drilling – isotopes are used to measure the quality of steam before it is injected into almost defunct oil wells to force out residual supplies.

Polymerization of Plastics – plastics can be polymerized by radiation instead of damaging heat treatments. The polymerized plastics are used in such applications as car dashboards, which would, otherwise, crack badly under the heat in the summer.

Quality Control of Metal Parts – the integrity of metal parts such as aircraft engine turbine blades can be verified by radiophotography on a conveyor belt instead of having to destroy a sampling of blades to ensure they are intact.

Research in Biology – the use of radioactive tracers allows the non-invasive tracking of elements and drugs through the body for both metabolic studies and medicine.

Space Power – when small amounts of power are needed in space in regions in which solar power is inefficient, plutonium batteries are ideal producers of compact energy.

Treatment of Cancers – cancerous cells can be selectively killed by the use of radioactivity, either in the form of directed beams, as for breast cancer, or as radioactive bullets that are designed to migrate directly to the cancerous cells that need killing. The only alternative, chemotherapy, which involves the use of invasive drugs, is a very difficult alternative for the patient.
Courtesy of American Nuclear Society

What does most of the natural radiation come from? Most radiation comes to us from the sun and from cosmic radiation – so that people at higher elevations receive more radiation than those living at sea level. A lot of radiation also comes from the soil and rocks around us. Granite and marble have background levels of radioactivity. A relatively small additional amount comes from our man-made technology. Courtesy of ANS

What is Cutaneous Radiation Syndrome (CRS)? The complex syndrome resulting from radiation exposure of more than 200 rads to the skin. The immediate effects can be reddening and swelling of the exposed area, blisters, ulcers on the skin, hair loss, and severe pain. Very large doses can result in permanent hair loss, scarring, altered skin color, deterioration of the affected body part, and death of the affected tissue (requiring surgery). Courtesy of the CDC.

What are three ways to minimize radiation exposure to the human body? Three things that a person can do to minimize radiation exposure to themselves is distance, shielding and time. **Distance** – the more distance that is put between the person and the source of radiation the better. **Shielding** – the heavier (more dense) material between the person and the source of radiation the better. **Time** – most of the radioactivity loses its strength fairly quickly.

What are the effects of radiation on materials? The effect of radiation in crystalline solids depends upon the structural type and the nature of the radiation. Light particles such as betas, and gamma ray photons cause very little permanent change in metals. The electrons rapidly lose their excess energy, and it is taken up by the atoms of the metals and manifests itself as vibration energy, such as heat. Heavier particles, such as protons, alpha particles, neutrons, and fission fragments, produce significant changes in the properties of metals. As a result of elastic collisions, appreciable energy can be transferred from the particle of radiation to the atoms of the metal. If the energy is sufficient, these atoms can be displaced from their equilibrium position within the lattice, and permanent physical changes are noted. This effect is known as radiation damage.

What is the difference between design bases and probability risk assessment?

Design Bases vs PRA

Design Bases	PRA
• Single Failure	• **Multiple** failures and **common** cause considered
• Margins used to bound uncertainty	• Best estimate with uncertainty
• Credits only SR equipment	• Considers **both** SR and NSR
• DBEs always shown to reach safe shutdown	• Accidents are evaluated to either safe shutdown **or** **adverse end state**
• Limited set of DBEs are evaluated	• **All accidents** are considered (dependent on scope and quality)

What does bioassay? Bioassay is an assessment of radioactive materials that may be present inside a person's body through analysis of the person's blood, urine, feces, or sweat.

What is the difference in treatments from a part of body exposure as compared to a whole body exposure? In the case of a person who has had only part of their body irradiated then the treatment is easier, as the human body can tolerate very large exposures to the non-vital parts such as hands and feet, without having a global effect on the entire body. For instance, if the hands get a 100Sv dose which results in the body receiving a dose (averaged over your entire body of 5Sv) then the hands may be lost but radiation poisoning would not occur. The resulting injury would be described as localized radiation burn.

What is biological half-life? The time required for one half of the amount of a substance, such as a radionuclide, to be expelled from the body by natural metabolic processes, not counting radioactive decay, once it has been taken in through inhalation, ingestion, or absorption.

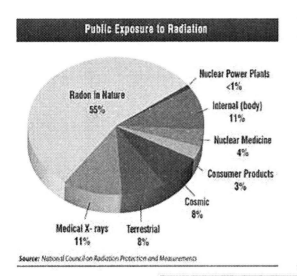

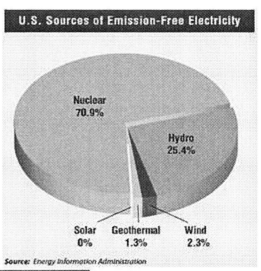

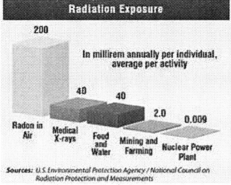

What are some potential sources of radiation in a nuclear plant?

- Reactor Coolant
- Fission and activated corrosion products
- Reactor fuel
- Reactor by-products (neutrons, N-16)
- Plant components (filters, resin, piping, valves,etc.)

160

How do nuclear accidents rate with other fatalities?

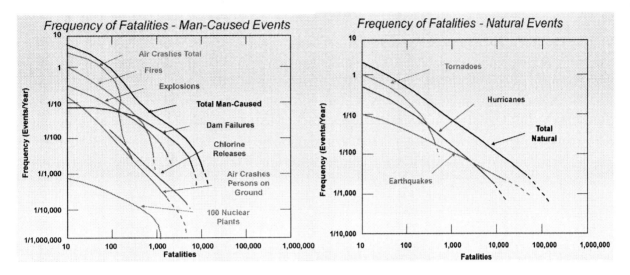

What is meant by dose rate? Dose rate is the amount of radiation dose in a specified period of time. Usually expressed in mr/hr.

What unit of measure is used for contamination? Since contamination is radioactive, it is measured by the level of activity. The more contamination there is, the higher the activity level will be. The units for contamination are counts per minute (cpm) or disintegrations per minute (dpm) per unit of area.

What is meant by total effective dose equivalent (TEDE)? TEDE is an individual's total dose and is determined by adding the external dose to the internal dose. The TEDE is expressed in terms of rem or mr. The rem is a unit used to relate radiation dose to biological effects. A millirem is one thousandth of a rem.

How can radioactive material enter the human body?

- Inhalation – breathing
- Ingestion – eating, drinking, or chewing
- Absorption – absorbing it through the skin
- Open wounds – through an open wound or sore

What are four things that can happen to a cell when exposed to ionizing radiation?

- Nothing
- Cell damage
- Cell death
- Cell Mutation

Who is more sensitive to radiation an adult or a child? Rapidly reproducing cells and organs are more likely to be affected by exposure to radiation. For this reason, a child is much more sensitive to radiation than an adult, and blood forming cells are more sensitive than bone cells.

What is the difference between somatic effects and genetic effects? Somatic effects occur in the individual that received the radiation exposure. Genetic effects appear in the future children of the individual exposed to radiation.

Who sets the limits of how much radiation a worker can receive? The NRC sets these limits low enough to prevent prompt effects, to minimize delayed effects, and to ensure that risks from radiation exposure are comparable with the risks in other industries.

Those federal limits are:

- 5 rem TEDE per year
- 50 rem per year to the skin of the whole body
- 50 rem per year to the extremites (elbows, lower arms, wrists,hands, knees, lower legs, ankles, and feet)
- 15 rem per year to the lens of the eye
- 500 mr for an entire term of pregnancy – this limit only applies to a pregnant worker who declares her pregnancy.
- 50 rem to any internal organ (thyroid)

What happens if these limits are exceeded? Exceeding any of these federal limits is a serious violation of plant and federal policy. Potential consequences of exceeding any federal limit include:

- Increased health risk
- Possible disciplinary action for willful violations
- NRC fines
- Other NRC action against the plant and/or the individual

How do the federal rules and guidelines apply to pregnant women?

- Radiation dose will be limited to 500mr during the term of pregnancy. This is voluntary; and if chosen, requires a pregnant worker to write a statement to plant management stating her desire to limit radiation exposure. Dose should be minimized and spread as evenly as possible throughout the pregnancy.
- It is also the right of the pregnant worker not to limit exposure during her pregnancy. It is the individual worker's choice.
- A pregnant worker may also change her status at any time during the pregnancy.

What are three basic types of contamination?

- **Fixed contamination** – is surface contamination that has become embedded in an object and cannot be removed using normal cleaning methods. Fixed contamination can become airborne or spread through activities such as grinding or welding. Fixed contamination can become loose contamination by leaching.
- **Loose contamination** – is contamination that is loosely adhered to an object or surface. Two problems associated with loose contamination are that it may be transferred to another area or personnel and it can become airborne if disturbed.
- **Discrete or hot particles** – are very small radioactive particles that may be too small to be seen. These particles can be highly radioactive and can cause a dose of several rem to very small localized area on the skin. If ingested, it can give a large dose to an internal organ.

What can be done to prevent the spread of contamination?

- Planning a job and conducting a pre-job brief
- Using protective clothing when working in a contamination area
- Avoiding water that is around or under a contaminated system
- Avoiding skin contact with a contaminated area
- Using step-off pads and warning signs
- Restricting entry to areas that are not routinely monitored for contamination
- Restricting entry into contamination areas
- Using engineering controls such as temporary ventilation with filters and special enclosures
- Wearing protective clothing (PC) to prevent personal contamination

Why is it important to keep liquid and dry radioactive waste separated? If liquid and dry radwaste are mixed, they will require further processing. Mixing of chemicals with either dry or liquid radwaste creates mixed waste. This is even more difficult to dispose of because additional restrictions apply.

Theories, Laws & Measurements

What is the atomic theory? The atomic theory states that all matter is made up of atoms and that each atom consists of a nucleus that is surrounded by electrons.

What is quantum theory? Quantum theory states when radiant energy is released from a body, this radiant energy is transmitted in the form of "bundles" and not as a steady stream.

How big is a gram? A gram is equal to one cubic centimeter of water at 40^0F.

What is a mole? A mole is a unit of measure for the amount of a substance that is equal to the amount of a given substance that has the same number of atoms, molecules, etc., as there are 12 gram atoms of carbon in carbon 12.

What is density? Density is the mass of a substance per unit volume. Density is not weight but how much a substance is concentrated. E.g. Mercury is a very dense material because a great deal of mass is concentrated within a small volume yet when compared to air, air has a very low density even within a large volume.

What is gram molecular weight? It is the mass of a substance that equals in grams its molecular weight.

What is molecular weight? Molecular weight is the mass of a molecule of a compound as compared to $1/12^{th}$ the mass of carbon 12 and is calculated as the sum of the atomic weights of the constituent atoms.

What is the atomic number? The number of protons in the atom's nucleus determines the atomic number.

What is atomic weight? Is the mean of the masses of all the naturally occurring isotopes of a chemical element.

What is pH? The term pH is a unit of measure used to determine the acidity or alkalinity of a solution in terms of the relative concentration of hydrogen ions in a solution. The scale is used to measure a solution ranges from 0 to 14, with 0 being the most acidic and 14 being the most alkalinic and 7 being neutral. With every unit of change in pH represents a tenfold change in the acidity of alkalinity and is the negative logarithm of the effective hydrogen ion concentration in gram equivalents per liter of solution.

The pH Scale

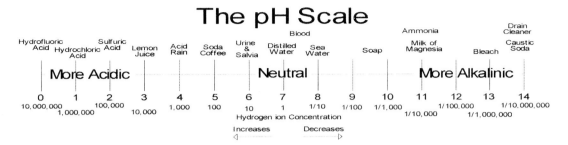

A neutral solution has an equal number of H+ and OH- ions.

If there are more H+ ions than OH- ions, then the solution is an acid.

If there are more OH- ions than H+ ions, then the solution is a base (alkaline).

What is pH$_t$? The symbol pH$_t$ basically means the pH at system temperature. Temperature does effect pH, for example: at 77^0F with a pH of 7.0 will change to a pH of 6.0 when the temperature is increased to 240^0F.

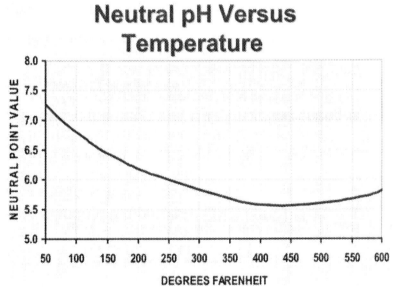

What is equilibrium? Equilibrium is a state of balance. This is a condition in which two opposing forces are equal to each other or exactly balanced.

What is a specific gravity? Specific gravity is a ratio of the weight or mass of a given volume of any substance that is equal to the volume of a substance that is used as a standard, such as water.

What is specific volume? Specific volume is a unit of measure used to measure the amount of space that is occupied by one pound of a substance.

What is viscosity? Viscosity is the property of a fluid, which causes them to resist flowing due to internal friction from the molecules moving against each other in a fluid. It is basically the thickness of a fluid.

What is volume? Volume is the amount of space that is occupied by a body and is measured in three dimensions. Volume is usually expressed in cubic units. Volume can be figured depending on the shape of the object.

Cone	$1/3 (3.14 \times r^2h)$	Sphere	$4/3 (3.14 \times r3)$
Cylinder	$3.14 \times r^2h$	Pyramid	$1/3 (b \times h)$
Cube	$l \times w \times h$	Prism	$b \times h$

What is mass? Mass is the amount of matter that is contained within an object. Mass is not affected or depended on gravity. The mass of an object remains constant regardless of any forces acting upon the object.

What is weight? Weight can be defined as how heavy an object is. Weight can also be defined as the force of gravity acting on a body.

What is weight density? Weight density is the weight of a substance per unit volume.

What is throughput? "Throughput" is a term that is used to express the amount of raw water flowing through a clarifier.

What is atomic mass unit (AMU)? Atomic mass unit is for expressing the mass of individual atoms, based on $1/12^{th}$ of the mass of the Carbon-12 isotope as standard ($1.6605402 \times 10^{-24}$). Masses expressed in AMU are known as relative atomic masses which were formally known as atomic weight. The alternative name for the unit, the Dalton, commemorates the English chemist and physicist John Dalton (1766-1844) who was responsible also for Dalton's Law and for first describing color blindness. (Macmillan) One AMU equals 931 million electron volts (MeV).

What is atomic number? For a particular chemical element, the number of protons in its atomic nucleus is equal to the number of electrons in the un-ionized atom is responsible for the element's chemical identity. (Macmillan)

What is atomic volume? It is the relative atomic mass divided by its density. This quantity varies periodically with increasing atomic number and was one of the factors that led to the idea of the Periodic Table of Elements. (Macmillan)

What is a curie? Curies are used to measure the quantity of radioactivity in a material. In radioactive elements, the rate at which atoms "disintegrate" or emit radiation is measured in "disintegrations per second". 1 curie is equal to 37 billion disintegrations per second, the rate of decay for 1 gram of radium. The quantity of radioactivity in materials such as spent fuel rods might be defined in curies. The quantity of radioactive material in a household smoke detector is measured in millicuries (one thousandth of a curie), while very small amounts of radioactivity are measured in picocuries (one millionth of a curie). Since curies or picocuries are only measures of radioactive decay – one must also consider the: volume of material, radioactive elements present, types of radiation they emit, and dose rate and length of exposure to determine biological impact.

What is a roentgen? Roentgen is a unit that measures radiation exposure in the air. This measurement tells you how much radiation – strictly "X" or gamma radiation – one might be exposed to in a particular location or environment (not what might be absorbed).

What are rads? Rad stands for "radiation absorbed does" and measures the amount of energy deposited in a material – alpha, beta, gamma, neutron or x-ray. Over time, the particular type of radiation absorbed in a nuclear plant component, such as metal pipe, might be expressed in rads.

What are rems? Rem stands for "roentgen equivalent man" and is the standard unit used to measure dose from radiation energy absorbed in human tissue. In this case, the term "dose" includes an allowance or factor for the biological effect of the particular type of radiation absorbed. For "X" and gamma radiation, the roentgen and rem value are essentially the same and are often used interchangeably. Exposure limits for nuclear power plant workers are typically expressed in rem.

What is API gravity? API gravity is a specific scale used to measure the specific gravity of a substance developed by the American Petroleum Institute.

What is meant by half-life? For a radioactive isotope, the time it takes for half of its nuclei to spontaneously decay, equal to $\log_e 2$ divided by the decay constant. It varies, depending on the isotope, from a few seconds to thousands of years and can be used for geological dating.

What are vectors? Vectors are line segments that represent the direction and magnitude of physical quantities like force, distance, acceleration, or velocity.

What is Avogadro's law? At the same temperature and pressure, equal volumes of all gases contain the same number of molecules (or atoms or ions). The number of molecules in 22.4 litres of any gas is known as Avogadro's number. (Macmillan)

What is Avogadro's number? The number of molecules (or atoms or ions) in a mole (gram molecular weight) of a substance, equal (for all substances) to 6.02253×10^{23}. It is known alternatively as Avogadro's constant and was named after the Italian nobleman and physicist Amedeo Avogadro (1776 – 1856). (Macmillan)

What is a barn? A barn is an extremely small unit of area, used for expressing the effective cross-sectional area of the nucleus of an atom. This measure expresses the probability that a particular subatomic particle can be captured by a nucleus. The unit was devised in 1942 by the American nuclear physicists C.P. Baker and H.G. Holloway, and so named because, compared with a subatomic particle, an atomic nucleus is "as big as a barn door": an un-missable target. (Macmillan)

$$1 \text{ barn} = 10^{-28} \text{ square meter}$$

$$= 1.55 \times 10^{-25} \text{ square inch}$$

$$= 10^{-22} \text{ square millimeter}$$

What is a shed? A shed is an extremely small unit of area (nuclear cross-section) that is equal to 10^{-52} square meter. The unit was given this names as a smaller version of the barn.

$$1 \text{ shed} = 10^{-24} \text{ barn}$$

What is a baryon number? A baryon is a member of a group of subatomic particles that are involved in strong interactions with other particles; they include hyperons (short lived particles heavier than a neutron), neutrons themselves, protons, and their respective antiparticles. The baryon number of all baryons is +1; that of antibaryons is -1. Baryon numbers are also ascribed to other particles that do not interact strongly, such as leptons (e.g. electrons, negative munons, tau-minus particles and their neutrinos), Mesons (pi-, K-, and eta-) and gauge bosons (photons, gluons, gravitons, and intermediate vector bosons) all of which have a baryon number of 0. Quarks have a baryon number of +1/3 and for antiquarks it is -1/3. In all reactions between particles, baryon number is conserved. The term derives from the ancient Greek barys 'heavy'. (Macmillan)

What is a Becquerel? A unit of radioactivity in the SI system, equal to the number of nuclei in a radioactive element that disintegrates each second. It has replaced the former unit, the curie.

$$1 \text{ Becquerel} = 2.7 \times 10^{-11} \text{curies}$$

What is Boyle's Law? The volume of a gas, at a given temperature, varies inversely in relation to the pressure it is subjected to.

What is the Bragg rule? The mass stopping power for alpha particles of an element is inversely proportional to the square root of the element's relative atomic mass.

What does a cent measure? A unit of reactivity that is equal to $1/100^{th}$ of a dollar.

What does a dollar measure? A unit of reactivity equal to that contributed by delayed neutrons.

What is a mil? A mil is equal to $1/1000^{th}$ of an inch. (e.g. plastic is measured in mils thick as well as equipment vibration).

What is a chad? A chad is a proposed but seldom used to measure neutron flux, equal to 1 neutron per square centimeter per second. It was named after British physicist James Chadwick (1891 – 1974), who discovered the neutron.

What is a charge? A charge is a property of an object because it has an excess or a deficiency of electrons. It may be termed merely as positive or negative, or it may be quantified in Coulombs.

What is a charge – mass ratio? For an ion or subatomic particle, its charge is divided by its mass.

What is a coulomb? A coulomb is a unit of electric charge, equal to the quantity of electricity carried by 1 ampere of current in 1 second. It was named after the French physicist Charles Coulomb (1736 – 1806).

What is Fajans-Soddy law? For an element undergoing radioactive decay, it's atomic number increases by 1 if it emits BETA PARTICLES, or decreases by 2 if it emits ALPHA PARTICLES. The law is named after Kasimir Fajans and Frederick Soddy.

What is a Fermi? A very small unit of length to express the size of subatomic particles.

What formula is used to find quality (x) of steam? A measure of the amount of vapor in a steam-water mixture.

$$X = lb_m \text{ steam } / lb_m \text{ mixture} = lb_m \text{ steam } / lb_m \text{ steam} + lb_m \text{ water}$$

What is dynamic viscosity? The resistance to shear primarily attributed to the attractive forces between the atoms or molecules of a substance.

What is kinematic viscosity? The absolute viscosity divided by the fluid density.

What are the modes of the reactor? The modes of the reactor describe the activity of the reactor.

MODE	UNIT POSITION	% RATED POWER (THERMAL)
1	POWER OPS	> 5% power
2	STARTUP	$\leq$ 5% power
3	HOT STANDBY	>300^{0}F
4	HOT SHUTDOWN	<300^{0}F
5	COLD SHUTDOWN	<200^{0}F
6	REFUELING	Reactor head taken off
7	DEFUELED	All Fuel Out of Reactor

What is the difference between parts per million (ppm) and parts per billion (ppb)? Parts per million (ppm) is one part for every million parts. E.g. one pound of sugar for every million pounds of water. It is similar with Parts per billion (ppb) in that it is one part for every billion parts. E.g. one pound of sugar for every billion pounds of water.

DEFINITION	AMOUNT OF SOLUTE	AMOUNT OF SOLVENT
Weight Percent	1 part	100 parts
Parts per Thousand	1 part	1,000 parts
Parts per Million (PPM)	1 part	1,000,000 parts
Parts per Billion (PPB)	1 part	1,000,000,000 parts

What does Archimede's principle state? A body wholly or partially immersed in a fluid is buoyed up by a force equal to the weight of the displaced fluid. If a volume's average density is greater than that of the fluid it displaces, it will sink; if its density is less, it will float.

What is the triple point? The triple point of a substance is the temperature and pressure at which three phases of that substance coexist in thermodynamic equilibrium.

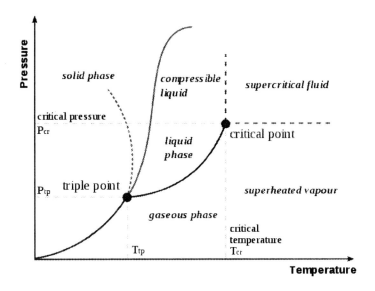

What formula is used to find moisture (y) of water mixture? A measure of the amount of liquid in a steam-water mixture?

$$Y = lb_m \text{ liquid}/lb_m \text{ mixture} = lb_m \text{ liquid} / lb_m \text{ steam} + lb_m \text{ liquid}$$

What is mass flow rate? The mass of fluid passing a reference point per unit time.

What is volumetric flow rate? The total volume of fluid passing a reference point per unit time.

What is the Reynolds Number (N_{re})? A unit-less parameter used to determine the type of flow regime in a system.

What is the Darcy-Weisbach equation? An experimentally derived formula for determining system head loss.

What is quantum mechanics? Quantum mechanics is the study of mechanical systems whose dimensions are close to or below the atomic scale, such as molecules, atoms, electrons, protons and other subatomic particles. The word "quantum" comes from the Latin language meaning "unit of quantity".

Who established quantum mechanics? The foundations of quantum mechanics were established during the first half of the twentieth century by Walter Heisenberg, Max Planck, Louis de Broglie, Albert Einstein, Niels Bohr, Erwin Schrodinger, Max Born, John von Neumann, Paul Dirac, Wolfgang Pauli and numerous others.

What are the differences in mass within an atom?

	(+) PROTON	NEUTRON	(-) ELECTRON
MASS	1.6726×10^{-24} gm	1.6749×10^{-24} gm	0.1085×10^{-28} gm
CHARGE	POSITIVE	NEUTRAL	NEGATIVE
RADIUS	10^{-15}	10^{-15}	UNDERFINED

What is the Pauli Exclusion Principle? This principle states that no two electrons in the same atom can have the same four quantum numbers (identical energies and orbits). The energy level for each electron is unique and can be characterized by a set of four quantum numbers.

- Principal quantum number – main energy level

- Orbital (Azimuth) quantum number – angular orbit around nucleus

- Magnetic quantum number – speed of rotation around nucleus

- Spin quantum number – spin of electron

What is Hooke's Law? A determined relationship which states that strain is proportional to applied stress in the elastic region "only".

What is the Heisenberg uncertainty principle? The exact position and momentum of an electron or other particle cannot be determined at the same moment in time; the product of uncertainties in these two parameters must be greater than the DIRAC CONSTANT.

What is Lenz's law? The electric current induced in a wire moving in a magnetic field generates an additional magnetic field that opposes the movement.

What is rH? A measure of the strength of an oxidizing agent. For a particular oxidation-reduction system, the negative logarithm of the hydrogen gas pressure that would produce the same electrode potential in a solution of the same pH. The larger the rH value, the greater is the oxidizing power of the system.

What does Pascal's law state? In hydrostatics, the pressure in a fluid is the same in all directions, and when pressure is applied to a fluid in a confined space, it is transmitted equally in all directions.

What is the Wigner effect? Hungarian physicist Eugene Wigner discovered that graphite, when bombarded by neutrons, suffers dislocations in its crystalline structure causing a buildup of potential energy. This energy, if allowed to accumulate, could escape spontaneously in a powerful rush of heat.

What are the differences in temperature scales?

Temperature Scale→ Reference Point↓	Kelvin	Celsius	Fahrenheit	Rankine
Absolute Zero	0	-273.15	-459.67	0
Freezing point of water	273.15	0	32	491.67
Triple point of water	273.16	0.01	32.02	491.68
Boiling point of water	373.13	99.98	211.97	671.64

What is the zeroth law? After the first two laws of thermodynamics were created, it was realized that another law should exist if the first two are too exist. The zeroth law states that a body is in thermal equilibrium with its surroundings if the amount of heat flowing into the body is exactly matched by the heat flowing out of the body.

What does the law of the conservation of energy state? The law of the conservation of energy states that energy can neither be created nor destroyed, but only transformed from one form of energy to another.

How can pressure be converted from atmospheres to psi? One atmosphere is equal to 14.7 psi. To convert atmospheres to psi simply multiply the number of atmospheres by 14.7, and there is psi.

What is the Ferranti Rise Effect? The Ferranti Rise Effect is a phenomenon in which a transmission line, with one end closed and one end open, is exposed to its highest voltage magnitude at the open-end of the line. The Ferranti Rise Effect is due to the absorption of a leading charging current when a transmission line is energized but open-ended.

What does Boyle's law state? Boyle's law states that at a constant temperature the volume of a given quantity of gas varies inversely with the pressure to which it is subjected.

How much electricity can the human body withstand? These measurements are with normally dry hands. The more perspiration the hands have the lower the resistance to electricity they become making it easier to get shocked.

1 to 20 ma	little to no sensation
20 to 50 ma	sensation of shock, possibly painful. Reduced muscle control between 10 to 20 ma
50 to 100 ma	same as symptoms above, more severe
100 to 200 ma	ventricular fibrillation may occur resulting in death almost instantly.
200 ma & higher	severe burns and loss of all muscle control. Heart will stop beating for the duration of the shock

This can all occur at less than 1 amp, actually 1/5 of an amp. This is why electricity should be respected.

What formula is used to find density? The formula for density is, P = weight /volume.

How is specific volume figured? Specific volume is the measurement of the amount of space that is occupied by one pound of water or substance. Sv = volume/weight or Sv = 1/density

What is the formula used to find pressure? The height of a fluid or head multiplied by the density of the fluid equals its pressure.

What formula can be used to convert inches of mercury to inches of water? 1" inch of mercury is equal to 13.60" inches of water. Take how many inches of mercury and multiply that number by 13.60 to convert inches of mercury into inches of water.

How much heavier is mercury than water? Mercury is 13.60 times heavier than water of the same volume and gold is 19 times heavier than water.

How is specific gravity figured? For any substance, use the mass of a given volume, divided by the mass of an equal volume of water at it maximum density, 40^0F.

What is the formula used to find work? The formula for work is, Work = distance x force.

What is the formula used to find power? Power is the rate at which work is done. Power = distance x force/time.

What is the difference between kinetic and potential energy? Potential energy is the same as stored energy. The "stored" energy is held within the gravitational field. When you lift a heavy object you exert energy which later will become kinetic energy when the object is dropped. A lift motor from a roller coaster exerts potential energy when lifting the train to the top of the hill. The higher the train is lifted by the motor the more potential energy is produced; thus, forming a greater amount if kinetic energy when the train is dropped. At the top of the hills the train has a huge amount of potential energy, but it has very little kinetic energy.

The word "kinetic" is derived from the Greek word meaning to move, and the word "energy" is the ability to move. Thus, "kinetic energy" is the energy of motion --it's ability to do work. The faster the body moves the more kinetic energy is produced. The greater the mass and speed of an object the more kinetic energy there will be. As the train accelerates down the hill the potential energy is converted into kinetic energy. There is very little potential energy at the bottom of the hill, but there is a great amount of kinetic energy.

What is the formula used to find kinetic energy? The formula for kinetic energy is, KE = velocity^2 x weight / 2g.

What is the formula used to find potential energy? The formula for potential energy is, PE = height x weight.

What are Maxwell's Laws of Electromagnetism? The first law states a changing a magnetic field produces an electric field. The second law is when changing an electrical field will produce a magnetic field. And the third law states, an electric charge is surrounded by an electric field. If the charge should go into motion, it will produce a changing electric field, which in turn generates a magnetic field.

What is a Prandtl number? A Prandtl number is a numerical ratio that is used in calculations of heat transfer in a fluid system, which is defined as the ratio of the viscosity of a substance to its thermal conductivity. This ratio was named after a German physicist Ludwig Prandtl (1875-1953).

What is a curie? A curie is a unit of measure used to measure the intensity of radioactivity. One curie equals precisely 3.7×10^{10} nuclear disintegrations per second. This unit was named after Pierre Curie, a French-Polish physicist.

What is Maxwell's rule? This rule states that every part of an electric current is acted upon by a force tending to move it in such a direction as to enclose the maximum amount of magnetic flux.

What is a calorie? A calorie is the amount of heat required to raise the temperature of one gram of water to 1^0C.

What is the Hall effect? The Hall effect is an electrical effect that is produced when an electrical current flows through a magnetic field. The Hall effect can be used to measure the electrical properties of different materials.

How is Celsius converted to Fahrenheit? The formula for this equation is F = 1.8 x C + 32. First multiply the temperature in Celsius by 1.8 then add 32 to get a temperature in Fahrenheit.

How is Fahrenheit converted to Celsius? The formula for this equation is C = (F- 32) / 1.8. First subtract the temperature in Fahrenheit from 32, and then divide by 1.8 to get a temperature in Celsius.

How is Fahrenheit converted to Rakine? Take the Fahrenheit temperature and add 460^0 to it and the temperature is now in Rakine.

How is Kelvin converted to Celsius? Subtract 273.15 from the temperature to read the temperature in Celsius.

What is absolute zero? Absolute zero is the lowest point on the temperature scale, which is equivalent to 0° K, - 273.16^oC or - 459.69^oF. It is at this temperature is where all molecular motion stops and there is absolutely no heat at all.

What are the Laws of Motion? The first Law of Motion is the tendency of an object to remain at rest or to continue moving in the same direction, unless acted on by some other outside force. The second Law of Motion is a force that moves an object is equal to its mass times its acceleration ($E=mc^2$). The third Law of Motion is for every action there is an equal and opposite reaction.

What is a tesla? A tesla is a unit of magnetic flux or magnetic induction that is equal to one weber per square meter. This unit is named after an American electrical engineer born in Croatia, Nikola Tesla, 1856-1943.

What is a decibel? A decibel (db) is a unit of measure that is used for measuring the intensity or loudness of sounds. At 0 dbs is where human hearing starts, at 40 dbs is a normal conversation, and at 80 dbs is a pneumatic drill. Anything above 70 dbs is harmful to hearing. At 120 dbs or higher, is where the threshold of pain begins.

What are the four states of matter? The four states of matter are liquid, solid, gas, and plasma.

What is a quad? A quad is a unit of energy that is equal to one quadrillion BTUs (10^{15}).

What is the speed of light? The speed of light in a vacuum is approximately 186,282 miles per second.

What is the speed of sound? The speed of sound varies greatly depending on the medium the sound passes through. For instance, the speed of sound through the air at sea level is approximately 768 mph (1 mile in 5 seconds). As the temperature of the air rises, the speed of sound also increases. If sound travels through the air covering one mile in 5 seconds, sound will travel through water covering the same distance in 1 second and travels the same distance through steel at 1/3 of a second.

What is the difference between a short ton, a long ton, and a metric ton? A short ton is equal to 2,000 pounds (commonly used in the US), a long ton is equal to 2,240 pounds (used in Europe and other countries), and a metric ton is equal to 2,204.63 pounds (basically used for comparison).

What units are used to express viscosity? Centistokes (CST) or Saybolt Seconds Universal (SSU) are the two units used to express viscosity.

What is a gram atom? A gram atom is one mole of atoms of an element.

What is a nile? In a nuclear reaction, a unit of reactivity. 1 nile = a reactivity change of 0.01

What is heat rate? The heat rate of a generating unit is a means of measuring the efficiency of the unit by answering a simple question, "How much fuel is required to generate a kWh of electricity?" A smaller heat rate is better, since it means less fuel is being used to create a unit of electricity.

What is the speed of electricity? Electrons in a wire that carry the electrical energy into your home travels at a few millimeter per second, yet the electrical charge that the electrons carry actually travel near the speed of light. If the electricity used in your home is generated from a power plant 50 miles away, the energy will take 50/(186,000 x 95%) or 0.00028 sec (0.28 milliseconds) to reach your home.

What is enthalpy? Enthalpy is the thermal energy or heat content that is in a pound of a substance. Enthalpy is measured in BTUs per pound (BTU/LB).

What is net heat rate? Net heat rate is the heat rate of a generating unit that is calculated by using the electrical power that is fed to the power grid.

What is gross heat rate? Gross heat rate is the heat rate of a generating unit that is calculated by using the total electrical output of the generator.

What is design heat rate? Design heat rate is the theoretical heat rate that is expected based on the construction and design of a unit.

What is specific heat? Specific heat is the amount of heat required to raise the temperature of one pound of a substance one degree Fahrenheit.

What is entropy? Entropy is a calculated property used to evaluate the efficiency of a process.

What is gradient? Gradient is the rate at which a property changes with distance.

What are the different classes of insulation? A classification of insulating materials according to the temperature, which they may be expected to withstand.

A-class insulation- a class of insulating material that is assigned a temperature of 105^0C.

B-class insulation- a class of insulating material that is assigned a temperature of 130^0C.

C-class insulation- a class of insulating material that is assigned a temperature over 180^0C.

E-class insulation- a class of insulating material that is assigned a temperature of 120^0C.

F-class insulation- a class of insulating material that is assigned a temperature of 155^0C.

H-class insulation- a class of insulating material that is assigned a temperature of 180^0C.

Y-class insulation- a class of insulating material that is assigned a temperature of 90^0C.

What is heat flux? Heat flux is the total flow of heat in heat exchange, in appropriate units of area and time.

What is a therm? A therm is a unit of energy used mainly for the sale of natural gas; equals 10^5 Btu or 105.5MJ.

What is R.M.S. power? R.M.S. power is the effective mean power level of an AC electric supply. (Root Mean Square)

What is flux? Flux is the rate of flow of mass, volume, or energy per unit cross-section normal to the direction of flow.

What does the "R" rating mean in insulation? The "R" rating represents the resistances to the flow of heat. The higher the "R" rating the greater the insulating factor. Sheet-rock has a 0.5 insulating value while insulation in the attic can have an "R" value of up to 30 or more for an insulating factor. The better the house is insulated the more energy efficient the house will be.

What is luminous flux? A luminous flux is the total amount of light produced from a light source, measured in lumens.

What units are used to express heat rate? BTU/Kw-hr is the unit used to determine heat rate.

What is heat transfer rate? The heat transfer rate is the amount of heat that is transferred from one body to another in a specific amount of time.

What is the formula used to find heat rate? Heat rate = chemical energy in divided by electrical energy out. Heat rate = amount of fuel (LBS) X heating value (BTU/LB) divided by net generation (Kw-hrs).

What units are used to express specific heat? BTU/LB-^{0}F is the unit used to express specific heat.

What is work? Work is the energy transferred to an object when a force causes the object to move that is equal to the product of the force and the distance. Work is usually measured in joules, 1 joule = 1 watt per second.

What is horsepower? Horsepower is a unit of power used to show the rate at which work is done. One horsepower is a unit of energy needed to lift 550 lbs. the distance of 1 foot in one second. One horsepower is equal to 745.2 watts of power; the heat is equivalent to 2,545 BTUs per hour. One megawatt of power is equal to 1340 horsepower.

What is thermography? Thermography is a method for measuring the slightest variations in temperature of an object by using infrared heat sensors.

What are two units used to express heat? The units used to measure heat energy are in calories or BTU's.

What is torque? Torque is a force that causes rotation or torsion. A measure of the effect that is equal to the product of the force and the perpendicular distance from the axis of rotation or torsion to the line of action of the force.

What is terminal velocity? Terminal velocity is when the maximum velocity of a free falling object has been reached when the resistance of air, water, or some other fluid that has become equal to the force of gravity acting upon the body.

What are eddys? An eddy is a flow of a fluid, gas, etc., that moves against the main current of flow and is usually indicated by a whirling motion like a small whirlpool or a whirlwind depending on the fluid.

What is tensile strength? Tensile strength is the maximum amount of stress that a material can withstand before it breaks. Units used to express tensile strength are pounds per square inch \ psi.

What is an electron volt? An electron volt is a unit of electrical energy that is equal to the energy attained by an electron going through a potential difference of one volt; this is equal to 160.206 x 10^{-21} joules.

What is the BTU rating for different fuels?

FUEL	BTU	UNIT OF MEASURE
FUEL OIL #1	20,000	BTU/GALLON
FUEL OIL #2	19,500	BTU/GALLON
FUEL OIL #4	19,100	BTU/GALLON
FUEL OIL #5	18,950	BTU/GALLON
FUEL OIL #6	18,750	BTU/GALLON
PEAT	7,000	BTU/POUND
SUBITUMINOUS COAL	8300–11,500	BTU/POUND
BITUMINOUS COAL	11,000–14,000	BTU/POUND
ANTHRACITE COAL	15,000	BTU/POUND
NATURAL GAS	950-1150	BTU/CUBIC FOOT
COKE OVEN GAS	500-600	BTU/CUBIC FOOT
BLAST FURNACE GAS	90-100	BTU/CUBIC FOOT
BUTANE	3200-3260	BTU/CUBIC FOOT
PROPANE	2500	BTU/CUBIC FOOT
METHANE GAS	500-750	BTU/CUBIC FOOT
STEAM	1,000	BTU/CUBIC FOOT
ELECTRICITY	3,413	KILOWATT HOUR
GASOLINE	124,000	BTU/GALLON
DIESEL	10,000	BTU/GALLON
LIGNITE	<8300	BTU/POUND
URANIUM	18 MILLION	BTU/ PELLET

What is a magnetic flux? A magnetic flux is the total amount of magnetic lines of force in a magnetic field. Measured in webers (joules per ampere).

What is a magnetic field? A magnetic field is a field that surrounds a magnet or an electric current in which a magnetic force can be felt.

What is centripetal force? Centripetal force is a force that tends to pull an object inward toward the center around which they are turning.

What is a ft-lb? A foot-pound is a unit of work that is equal to the amount of work done by one pound of force that moves an object one foot in the direction that the force is applied.

How many ft-lbs are equal to one BTU? There are approximately 778.26 ft-lbs. per BTU.

What does SCFM stand for? Standard Cubic Foot per Minute.

What does ICFM stand for? International Cubic Foot per Minute.

What does SCFH stand for? Standard Cubic Foot per Hour

What does STP stand for? STP is an abbreviation meaning standard temperature and pressure. This standard is used for comparing gas volumes in which the standard temperature is 0^0 c and a standard pressure of one atmosphere.

How do various energy sources compare to each other?

1 BTU	1 burning match or 250 calories.
1,000 BTU	1 cubic foot of natural gas.
1 million BTU	90 lbs. of coal or 8 gallons of gasoline.
1 quadrillion	45million short tons of coal, 170million barrels of crude oil.
1 barrel of crude oil	5.6 thousand feet of natural gas, 1,700 kilowatt-hours of electricity.
1 short ton of coal	3.8 barrels of crude oil, 6,500 kilowatt-hours of electricity.
1,000 cubic feet of natural gas	300 kilowatt-hours of electricity, 93 lbs of coal.
1,000 kilowatt-hours	310 lbs. of coal, 3,300 cubic feet of natural gas.

*If you leave some of these devices "on" and they are not being used, please shut them "off" and save energy as well as your money.

Physics

What is physics? Physics is a science that deals with the properties and interrelationships of energy and matter. Branches of physics are mechanics, thermodynamics, acoustics, optics, and electronics. Also included in physics is the study of forces, motion, light, heat, sound, electricity, radiation, atomic energy, and magnetism.

What is nuclear physics? Nuclear physics is a branch of physics dealing with the physical properties, structure, and behavior of atomic nucleus and subatomic particles.

What is an atom? An atom is the smallest particle of an element and can take part in a chemical reaction without being permanently changed. An atom is made up of neutrons, protons, and electrons whirling in a variety of orbits. The word atom originated from the Greek word atomos, meaning "uncuttable". The atom was thought to be the smallest thing to exist, until modern science split the atom. The size of an atom is less than one ten-millionth of an inch across. Although atoms are way too small to be seen with the naked eye, scientists are using powerful instruments to study their existence.

What does the structure of an atom tell us?

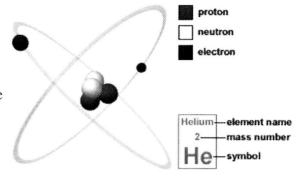

All matter (solid, liquid or gaseous) consists of elements, of which there are more than 100. If, in theory, we cut a block of iron into smaller and smaller pieces, we would finally end up with the smallest piece possible that still has all the characteristics of the iron element. That smallest piece is called an iron atom. An atom is very, very small. In fact, the size of an atom compared to the size of an apple, is like the size of an apple compared to the size of the Earth. Most atoms consist of three basic particles: protons (with a positive electrical charge), electrons (with a negative electrical charge), and neutrons (with no electrical charge). Protons and neutrons are bundled together in the center of the atom, called the nucleus. The electrons move around the nucleus, each in its own orbit like the moon around the earth. Each atom of the same element is characterized by a certain number of protons in the nucleus. That number is called the atomic number. Normally, the atom has the same number of electrons in orbit around the nucleus. This atomic number identifies the elements. The list of elements (ranked according to an increasing number of protons) is called the Periodic Table.

For example, Helium has 2 protons in its nucleus. Its atomic number is therefore 2. Iron has 26 protons in its nucleus. Its atomic number is therefore 26. Uranium has 92 protons. Its atomic number is therefore 92.

Even though the number of protons in the nucleus is the same for all atoms of a particular element, the number of neutrons in the nucleus can differ for different atoms of the same element. Atoms of an element that contain the same number of protons, but different numbers of neutrons, are called isotopes of the element. Isotopes are identified by adding the number of protons and neutrons together -- a number which is referred to as the mass number.

For example, hydrogen: the element hydrogen has 3 isotopes: hydrogen 1 (also called hydrogen), hydrogen 2 (also called deuterium) and hydrogen 3 (also called tritium).

Hydrogen 1 (hydrogen)   Hydrogen 2 (deuterium)   Hydrogen 3 (tritium)

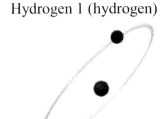
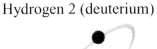

1 proton, 0 neutrons   1 proton, 1 neutron   1 proton, 2 neutrons
Mass number = 1   Mass number = 2   Mass number = 3

Notice that even though the masses of the above atoms are different, each nucleus has only one proton. The one proton identifies all these atoms as hydrogen isotopes. It is the number of neutrons that defines them as different types of hydrogen isotopes. Normally, atoms have the same number of protons and electrons.

The number of positively charged protons is the same as the number of negatively charged electrons so that the atom is electrically neutral. The electrons orbiting at the outside of an atom are the part of the atom that takes part in chemical reactions. They identify the atom chemically. These electrons can be thrown off by the atom, or more can be absorbed. An atom that has lost one or more electrons is positively charged; one that has picked up electrons is negatively charged. These "charged" atoms are called ions.

The nucleus of the atom also contains neutrons. Neutrons are about the same size as protons but have no electric charge. Neutrons are bound very tightly in the atom's nucleus with the protons. When the atom's nucleus contains as many neutrons as protons, the atom is stable. Most atoms are stable. However, when the atom's nucleus contains more neutrons than protons, the nucleus is unstable. The nucleus of such an unstable atom will try to become stable by giving off particles or packets of energy (quanta). These emissions are called radioactivity. The particles and quanta are emitted from the nucleus at high energy. If a particle or quantum hits the electron of another atom, it can knock that electron off of the atom, which makes that atom positively charged and, therefore, an ion. That is why we refer to the particles and quanta emitted by radioactive nuclei as "ionizing radiation."

Very large and heavy atoms that occur in nature are unstable and, therefore, radioactive. These include atoms of the elements uranium (atomic number 92), thorium (atomic number 90), radon (atomic number 86), and radium (atomic number 88), among others. Many smaller atoms are made radioactive artificially for specific uses. Smaller elements like carbon (atomic number 6), often have a stable, non-radioactive form as well as an unstable radioactive form.

Courtesy of the American Nuclear Society

What is a nucleus? A nucleus is the central part of an atom, which consists of protons and neutrons. The size of the nucleus is about one fifty-thousandth of the radius of the whole atom. The nucleus makes up most of the mass or weight of the atom.

What is a molecule? A molecule is the smallest particle of an element of compound that can be divided without affecting its physical and chemical properties. A molecule is made up of two or more atoms.

What is inertia? Inertia is the tendency of all objects and matter in the universe to remain stationary, if stationary of if moving to stay moving in the same direction. Example, riding on a bus, the as the bus begins to slow down you fill yourself leaning forward. This is the inertia in your body, this force can be stopped by an opposing force, and in this case your hand would keep you from moving forward anymore.

What is resonance potential? The potential needed to raise an electron within the atom, from one orbital in one energy level to an orbital at a higher energy level also known as excitation potential

How does atomic mass determine the type of reaction that will be given off? A decrease in mass indicates an exothermic reaction and an increase in mass indicates that the reaction will be endothermic.

What are some different types of decay caused by ejection? If decay occurs by ejection of the same type of particle which initiated the reaction, the process is called scattering. If the emerging particle and recoiling nucleus share all of the available kinetic energy, it is known as elastic scatter. If some of the energy is carried off by gamma radiation, thus reducing the KE to be shared by the emergent particle and the recoiling nucleus, the process is inelastic scatter. If the compound nucleus drops to the ground state solely by gamma emission, the process is called radioactive capture.

What is an omega-minus particle? An elementary particle that carries a negative charge – energy 1,672 million electron volts and is the most massive hyperon.

What is separation energy? The energy needed to completely remove a nucleon from an atom's nucleus, usually measured in electron volts.

What is a sigma particle? A relatively massive, short lived elementary particle, a type of hyperon with a rest mass equivalent to 1,190 mega electron volts.

What is meant by subatomic particle? A particle that is smaller than an atom, often forming parts of the atom, such as an electron, proton or neutron.

What is a tachyon? A hypothetical subatomic particle that can move faster than the speed of light.

What is an electron? An electron is a particle that orbits around the nucleus of an atom at various distances. There is always the same number of electrons orbiting around the nucleus, as there are protons in the nucleus. Electrons carry a negative charge and protons carry a positive charge.

What is a proton? A proton is a subatomic particle that is positively charged. Protons are found in the nucleus of atoms. The number of protons in the nucleus is equal to the same number of electrons that orbit the nucleus. The number of protons in the nucleus of an atom is the atomic number of the element.

What is deuterium? A non-radioactive isotope of the hydrogen atom that contains a neutron in its nucleus in addition to the one proton normally seen in hydrogen. A deuterium atom is twice as heavy as normal hydrogen. Courtesy of the CDC.

What is a photon? A discrete "packet" of pure electromagnetic energy. Photons have no mass and travel at the speed of light. The term "photon" was developed to describe energy when it acts like a particle (causing interactions at the molecular or atomic level), rather than a wave. Gamma rays and x-rays are photons.

Where are the protons, neutrons, and electrons located within the atom? The protons and neutrons are located within the nucleus of the atom while the electrons are located in the outer shells orbiting around the nucleus.

What is an electron volt (eV)? Amount of kinetic energy (eV) gained by an electron when accelerated through an electric potential difference of 1 volt. It is equivalent to 1.603 x 10^{-19} joule. It is a unit of energy or work, not of voltage, and is the common measure of a neutron's energy. Larger multiple units of the electron volt are frequently used: keV for thousand or kilo electron volts, MeV for million electron volts, and BeV for billion electron volts.

What is the distance between the nucleus of an atom and its electrons? The distance between the nucleus of an atom and its electrons is approximately 10^5 times the dimension of the nucleus.

What is a neutron? A neutron is an elementary particle that carries no electrical charge thus remaining neutral. A neutron is located in the nucleus of an atom along with the protons. Neutrons where discovered by James Chadwick in 1932. It can be found in the nucleus of every atom except that of hydrogen which only has a single proton in its nucleus. Its mass is 1.6478 x 10^{-27} kilogram. Free neutrons outside of a nucleus are unstable and have a life cycle that last for about 12 minutes. Outside of the nucleus a neutron cannot exist alone; it is unstable and will decay into a proton and an electron.

What is an isotope? An isotope is two or more forms of an element that has the same chemical properties and the same atomic number but is different in atomic weights. Ex. U-235, U-238, and U-239 are all isotopes of uranium. The term derives from ancient Greek elements meaning of the same place.

What is an ion? An ion is a charged particle (negative or positive) due to the loss or gain of one or more electrons. The term ion originated from the Greek word "ion" meaning "thing going". An ion is an atom or group of atoms that take on an electric charge. This charge can be either negative (-) (anion) or positive (+) (cation).

What is a positron? The positron is a positively charged electron (anti-electron). It is the antimatter (antiparticle) of the electron. The positron has an electric charge of +1, a spin of ½, and the same mass as an electron. In 1930, a graduate student at Caltech, Chung-Yao Chao detected positrons through electron-positron annihilation even though they didn't realize what they were at the time. Paul Dirac in 1928 theorized the positron and Carl Anderson in 1932 actually discovered it.

What is antimatter? Antimatter is the extension of the concept of the antiparticle to matter, where antimatter is composed of antiparticles in the same way that normal matter is composed of particles. For example, an anti-electron (a positron, an electron with a positive charge) and an anti-proton (a proton with a negative charge) could form an anti-hydrogen atom in the same way that an electron and a proton form a *normal matter* hydrogen atom. Furthermore, mixing matter and antimatter would lead to the annihilation of both in the same way that mixing anti-particles and particles does, thus giving rise to high-energy photons (gamma rays) or other particle – anti-particle pairs.

There is considerable speculation both in science and science fiction as to why the observable universe is apparently almost entirely matter, whether other places are almost entirely antimatter instead, and what might be possible if antimatter could be harnessed, but at this time the apparent asymmetry of matter and antimatter in the visible universe is one of the greatest unsolved problems in physics. The process of developing particles and antiparticles is called baryogenesis.

What is a nuclide? A species of atom characterized by the number of protons and number of neutrons contained in the nucleus.

What is an isotope? A nucleus of the same element (same number of protons) with a different number of neutrons.

What is the chart of nuclides? A chart that lists the stable and unstable nuclides in addition to pertinent information about each one. The chart plots a box for each individual nuclide, with the number of protons on the vertical axis and the number of neutrons on the horizontal axis. Each box contains the chemical symbol of the element, the average atomic weight of the naturally occurring substance, and the name of the chemical. The known isotopes of each element are listed to the right. The chart contains much more information for each isotope. By consulting the chart of nuclides, other types of isotopes can be found, such as naturally occurring radioactive types. This chart is color coded to show similar half-lives and neutron cross section for absorption.

What is meant by binding energy? The energy required to remove a particle from an atom or nucleus (e.g. an electron from an atom). Alternatively, it is the difference between the energy of a nucleus and the combined energies of the separate particles (nucleons) of which it is composed (equivalent to the mass decrement or defect). (Macmillan)

What is a boson? A subatomic particle that includes alpha particles, photons, and atomic nuclei with even mass numbers (the total number of protons and neutrons is even).

What is nuclear stability? Radioactive nuclei will undergo radioactive decay in an attempt to reach a ground state. The type of decay is determined by characteristics of the nucleus.

What is a delta particle? A very short lived particle (hyperon) which decays almost instantaneously because of its strong interaction.

What is a gluon? A gluon is a subatomic particle that is the agent for strong interactions between pairs of fundamental particles.

What are gauge boson? Gauge bosons are subatomic particles that are involved in interactions between pairs of fundamental particles. There are four types:

Gluons – strong interactions   **Intermediate vector bosons** – weak interactions

Gravitons – gravitational   **Photons** – electromagnetic

What is G-value? In radiochemistry, a constant that denotes the number of molecules that react after absorbing 100 electron volts of radiation energy.

What are elementary particles? Subatomic particles that are not made up of smaller particles. There are three main kinds:

Leptons (electrons, negative muons, tau-minus particles and their neutrinos)

Quarks (six kinds, and their antiquarks)

Gauge bosons (photons, gluons, gravitons, and intermediate gauge bosons).

What is ground state? The lowest energy level of an atom or molecule that is not excited. Stable condition which exists when an atom is at its lowest energy. Conditions refers to both the nucleus and the orbital electrons in the cloud. All atoms will attempt to reach the ground state.

What is excited state? Unstable condition in which the nucleus, the electron cloud, or both have some finite energy about the ground state. The atom will ultimately release energy in an attempt to reach the ground state.

How are electrons arranged around the nucleus of an atom? Electrons spin very fast, They spin in orbits around the nucleus. The spinning is so fast that they seem to form a solid shell around the nucleus. All atoms except hydrogen and helium atoms have more than one shell. We label the shells of electrons. Each shell is labeled with a capital letter. The first shell is the "K" shell. It is the shell closest to the nucleus. The next shell is the "L" shell. After that comes the "M" shell. And so on. Each shell can hold only a certain number of electrons.

The "K" (1st) shell can hold 2 electrons.

The "L" (2nd) shell can hold 8 electrons.

The "M" (3rd) shell can hold 18 electrons (only up to 8 for the first 18 elements).

The number of shells an atom has depends on the number of electrons for that atom. Each shell must have its full number of electrons before a new shell starts except for the M shell. The M shell is considered full if it has 8 electrons in it for the first 18 elements. For example, Argon [Atomic #18] has 8 valence electrons so it is considered to have a full M shell. If there are more electrons than a shell can hold, a new shell starts. The outer shell of most atoms is not full. Only the atoms in the elements of Group or Family 18 on the Periodic Table have full outer shells.

*Atoms of metals have 3 or less valence electrons.

*Atoms of nonmetals have 5 or more valence electrons.

What is an electron shell? The electron shell is the orbit of electrons rotating around the atom's nucleus.

What is a hyperon? A short life of about 10^{-10} seconds, elementary particle of which the mass is greater than that of a neutron. Hyperons include any Baryon that is not a nucleon.

What is the purpose of an ionization chamber? A closed gas filled tube with a high voltage between a pair of electrodes arranged parallel to each other or coaxially. When ionizing radiation passes through the tube, the gas ionizes and a current flows between the electrodes. The value of the current is a measure of the rate of ion production in the tube.

What are isodiapheres? Two or more atomic nuclei that have the same neutron excess (difference between the numbers of neutrons and protons). The term derives from the ancient Greek elements meaning of equal throughput.

What is kerma? The total kinetic energy transferred to charged particles by ionizing radiation from a given mass of material divided by the mass of the material.

What is a valence? A valence is the combining capacity of an atom or radical, determined by the number of electrons that an atom will add, share, or lose, when it reacts with other atoms. The term valence is derived from the Latin word valent, which means "strength".

Courtesy of Stwertka, Albert

What is a hadron? An elementary particle that undergoes strong interaction with other particles.

What is the Hamiltonian function? A function describing the positions and momentum of a group of particles, in terms of which their motion and energy may be expressed.

What is a harmonic? The frequency of a wave, such as a sound wave, that is an exact multiple of the frequency of the fundamental wave, also known as an overtone.

What are valence electrons? Valence electrons reside in the outermost shell. The number of valence electrons is a significant factor in determining the chemical properties of atoms. Atoms with full valence shells are the most stable while atoms with partially filled shells are more reactive (unstable). The valence electrons determine the chemical "strength" of atoms-their reactivity, or how strongly and in what way they will bind with other atoms.

Courtesy of Stwertka, Albert

What is a valence shell? The valence shell is the outermost shell where electrons rotate around the atom's nucleus.

What determines how many electrons can be in any one shell? The maximum number of electrons per shell can be calculated using the following formula:

$$n = 2N^2$$
where n = number of electrons
and N = shell number (1,2,3, etc)

What are nucleons? A collective name for the particles which represent the main constituents of an atomic nucleus (protons and neutrons).

What is the difference between Isotopes, Isotones, Isobars and Isomers?

Isotopes – one of two or more atoms having the same atomic number (# of protons), but different atomic mass number (# of neutrons).

Isotones – nuclides having the same number of neutrons but different number of protons.

Isobars – nuclides having the same atomic mass number (# of nucleons).

Isomers – a nuclide which can exist for measurable times in different quantum energy states, with different energies and radioactive properties.

What is mass defect? The difference between the mass of an atom and the sum of the masses of its constituent parts. When an atom is assembled from its component parts, the total mass of the nuclide is less than the total mass of the individual particles. The mass difference is called the mass defect. The mass defect is measured in atomic mass units (AMU).

What is coulombic force? Force created as a result of the interaction of charged particles. Charge dependent: like charges repel and unlike charges attract. Has a longer range than nuclear force but weaker than nuclear force.

What is nuclear force? The attractive force between nucleons in an atomic nucleus. It is extremely strong and is charge independent – attract neutrons and protons equally. It has a shorter range and is saturable.

What is electron capture? Occurs in very unstable nuclei with a low neutron/proton ratio. A nucleus captures an electron from the "k" shell and converts a proton into a neutron. The atom neutron/proton ratio increases (found above the line of stability).

What is neutron emission? Neutrons emitted by fissioning nuclei and highly excited fission fragments which have too high a neutron/proton ratio. Neutron emission reduces the neutron/proton ratio (found below the line of stability).

What is rate of decay? Because radioactive decay is a probability function, it is impossible to predict precisely when a single atom will decay. For many radioactive atoms of the same type, a measurable probability can be established.

What is activity? The total number of decays per unit time for a given mass of radioactive material.

What are some types of neutron interactions? Elastic scatter, inelastic scatter, radiative capture and fission.

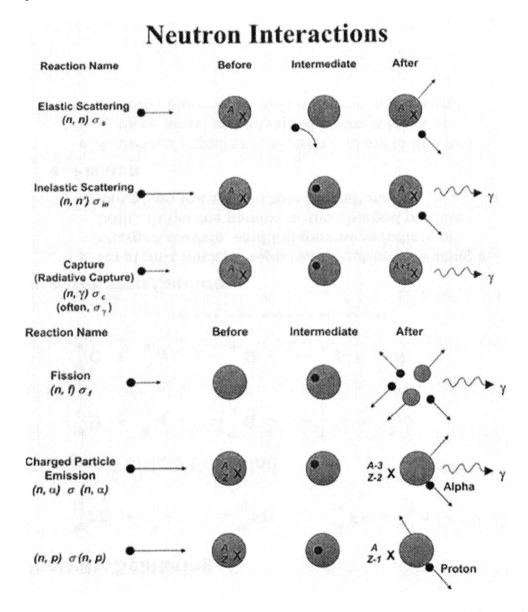

Neutron Interactions

What is Compton scattering? A higher energy photon (.1 – 2 MeV) interacts with an orbital electron, transferring part of its energy. The electron is ejected, and a lower energy photon departs the atom at a scattering angle.

What is elastic scattering? A neutron collides with an atom. The product nucleus is not excited. The neutron transfers kinetic energy and momentum to the nucleus.

What is inelastic scattering? A neutron collides with an atom leaving the nucleus in an excited state. Some kinetic energy and momentum are retained by the nucleus which can later release gamma rays, which can cause further ionization.

What is neutron capture? An atomic nucleus captures a neutron, releasing a charged particle, emitting gamma radiation, or both. Either the gamma or the charged particle can go on to cause further ionizations.

What is fission? Fission occurs when a fissionable nucleus captures a neutron. Capture upsets the internal force between neutrons and protons in the nucleus. The nucleus splits into two lighter nuclei and an average of two or three nucleus is emitted. The resulting mass of products is less than that of the original nucleus plus neutron. The difference in masses appears as energy in an amount determined according to Einstein's formula, $E = mc^2$. If one of the neutrons emitted is captured by another fissionable nucleus, a second fission occurs similar to the first, another neutron may produce a third fission and so on. When the reaction becomes self-sustaining so that one fission triggers at least one more fission, the phenomenon is termed a chain reaction. The device in which this chain reaction is initiated, maintained, and controlled is called a nuclear reactor.

What is spontaneous fission? The splitting of an atom because of its own excess nuclear energy.

What is neutron induced fission? The splitting of an atom due to the addition of energy caused by the absorption of a neutron.

What are prompt fission products?

> **Fission Fragments** – the two or more new nuclides which result from the split of the fissioning nucleus.

> **Prompt Neutrons** – highly energetic neutrons ejected at the time of fission because the fission fragments are too neutron rich.

> **Prompt Gamma Radiation** – high energy photons released at the time of fission due to the high excitation energy of the fission fragments.

What are delayed fission products?

> **Delayed Neutrons** – neutrons emitted sometime after the fission event by a special fission fragment called a delayed neutron precursor.

Decay Betas – emitted by highly excited fission fragments as they undergo radioactive decay toward the line of stability.

Decay Gammas – electromagnetic radiation emitted from the fission fragments as they attempt to reach their ground state.

Neutrinos – charge-less radiation, which accounts for the "rest" of the energy released following a fission event.

What is fissile material? An isotope capable of undergoing a fission reaction upon the absorption of a neutron of any energy level. The binding energy of the neutron is greater than the critical energy.

What is fertile material? Non-fissile material which can be transformed into fissile isotopes by neutron bombardment reactions.

What are fission neutrons? Neutrons liberated during the fission process.

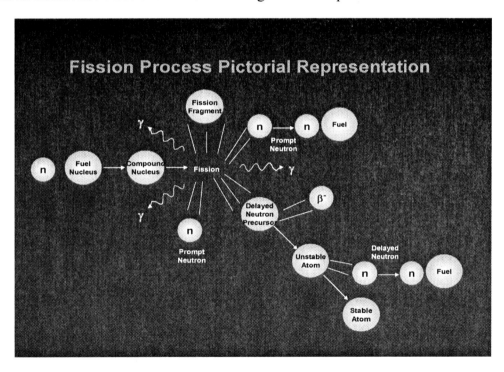

How are fission neutrons classified? By birth time following fission:

Prompt neutrons $< 10^{-14}$ seconds   **Delayed neutrons** $> 10^{-14}$ seconds

What are three types of energy levels for neutrons?

Thermal < 1 Ev   **Epithermal** 1 eV to .1 MeV   **Fast** $> .1$ MeV

What are prompt neutrons? They are neutrons born at the moment of fission. They are highly energetic (approx. 2 MeV). Represent the majority of neutrons from fission of U-235 (99.35%). The number of prompt neutrons produced per fission (nu) depends on the type of nuclide undergoing fission and the energy of the neutron inducing fission.

What are delayed neutrons? Neutrons produced by the B- decay of special fission fragments (delayed neutron precursors) to an excited daughter state, which then emits a delayed neutron. Less energy than prompt neutrons (.5 to .8 MeV). Represent a very small fraction of fission neutrons (~0.65%). Significant in the ability to achieve reactor control since these neutrons do not contribute to the overall neutron population until sometime after the fission event.

What are thermal neutrons? Neutrons that are in thermal equilibrium with their surroundings (kinetic energy < 1 eV). Most probable energy at 68^0F: E_n = .0253 eV and most probable velocity: V_n = 2200 m/s. E_n and V_n increase with increasing reactor temperatures. Thermal fission event induced by the absorption of a neutron in the thermal energy range.

What are epithermal neutrons? Neutrons in the energy range from 1 eV to .1 MeV, also called intermediate neutrons.

What are fast neutrons? Neutrons having energies > .1 MeV. Prompt and delayed fission neutrons are born in the fast energy region. Fast fission induced by the absorption of a neutron in the fast energy range.

What is macroscopic cross section? Macroscopic cross section is a measure of the probability of neutron interaction per centimeter of travel in a given target material.

What is the theory of resonance? Resonances are peaks in the absorption microscopic cross section vs. neutron energy curve, located in the epithermal energy region. These peaks correspond to an abnormally high probability of compound nucleus formation (neutron absorption) when the binding energy of the neutron plus the neutron's kinetic energy equals the excitation energy of a quantum energy level of the compound nucleus.

What is Doppler broadening? The widening and flattening of the resonance absorption peaks of target nuclei due to an increase in temperature (molecular kinetic energy). Major effect is an increase in the probability of neutron absorption. As the temperature of the target material increases, the resonance integral increases. In U-238 atoms, as the reactor fuel is heated, this effect plays a significant role in reactor stability and control.

What is meant by microscopic cross section? The probably of a neutron interacting with a target nucleus. A measure of the effective cross sectional area presented to a neutron by a target nucleus measured in barns (b). $1b = 10^{-24}$ cm^2 The microscopic cross section varies with the type of neutron interaction: Scattering, Capture, Fission   In general, microscopic cross section decreases with increasing neutron energy. Since each type of neutron interaction involves different physical requirements, it follows that the probability of each type of interaction will be different. Therefore, the microscopic cross sections will be different.

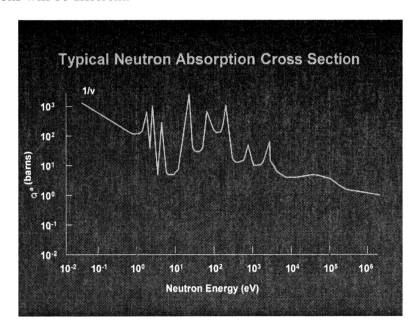

What is the difference between gram atomic weight (GAW) and gram molecular weight(GMW)? Gram atomic weight is the mass, in grams, of a substance that is numerically equivalent to the atomic mass of an atom. Gram molecular weight is the mass, in grams, of a substance that is numerically equivalent to the molecular weight.

What is atom density? The number of atoms per cubic centimeter in a target material.

What two factors help determine neutron interaction with a given target material?

$$\text{Neutron density} \quad n = \text{\# of neutrons} / \text{cm}^3$$
$$\text{Neutron velocity} \quad v = \text{cm} / \text{sec}$$

What is reaction rate? The number of interactions occurring (between an incident neutron flux and a target material) per cubic centimeter per second.

What is power density? The thermal energy output per unit volume of reactor core per second.

Why is it necessary to slow down neutrons? Power reactors rely primarily on thermal neutrons to sustain the fission chain reaction. The probability of a U-235 fission is much greater for neutrons at thermal energies. Nearly all neutrons are born in the fast region. Consequently, the highly energetic fission neutrons must be slowed down to thermal energies to increase the probability of fission. Since neutrons possess no charge, the only means of interaction with matter is by direct collision (scattering). Scattering collisions between neutrons and target atoms results in a transfer of energy from the neutron to the target nuclide.

What takes place in a fission chain reaction? A fission takes place, releasing 2 to 3 neutrons. Released neutrons may go on to induce additional fissions. When each fission produces enough neutrons to cause, on the average, the fission of at least one other fuel atom, a nuclear chain reaction will be sustained. The fission rate depends on the magnitude of the neutron population. The more neutrons present, the more fissions induced in the fuel atoms.

What are some neutron production and removal mechanisms?

Neutron production mechanisms: neutron induced fission, spontaneous fission, cosmic radiation, photo-neutrons and intrinsic sources reactions.

Neutron removal mechanisms: fission capture in fuel, non-fission capture in fuel and parasitic capture in non-fuel materials. Another removal mechanism is actual neutron leakage from the core.

What is meant by k effective?

= # of neutrons produced in present generation / # of neutrons produced in previous generation.

When $K_{eff} = 1$ then Critical
When $K_{eff} < 1$ then Subcritical
When $K_{eff} > 1$ then Supercritical

What factors will affect K_{eff}? Any condition that will affect one or more of the factors in the six factor formula will affect the value of K_{eff}. The four major parameters that affect the six factors are: moderator temperature, fuel temperature, CEA height and soluble Boron concentration.

What is a reproduction factor? The symbol for reproduction factor is η.

η = # of fission neutrons produced by thermal fission / # of thermal fissions absorbed in the specified fuel.

Factors that affect η are: core age (fuel burn-up) and initial fuel composition

What is Fast Fission Factor? The symbol for fast fission factor is ϵ . The formula is:

ϵ = # of fast neutrons produced by all fissions / # of fast neutrons produced by thermal fissions

Factor affecting ϵ is: Core age

As fuel is depleted (U-235), the moderator to fuel ratio increases. This reduces the slowing down time and length, reducing the number of fast fissions. As moderator temperature increases, density decreases N_M / N_F decreases, therefore the moderator is less efficient at thermalizing neutrons, so ϵ increases.

What is the neutron life cycle?

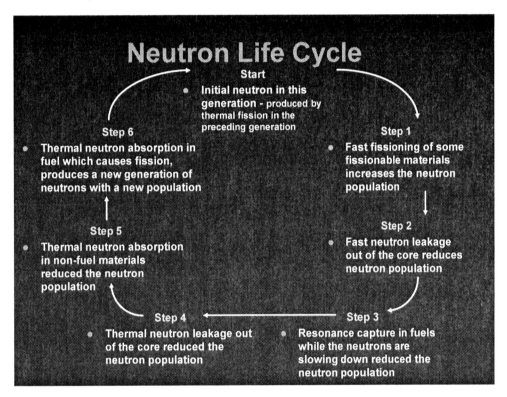

What is the six factor formula?

$$K_{eff} = \epsilon \ L_f \ P \ L_{th} \ f_\eta$$

ϵ = Fast Fission Factor

A measure of the increase in the fast neutron population due to fast fission events

L_f = Fast Non-leakage Factor

The probability that a fast neutron will NOT leak from the core while slowing down to thermal energies.

P = Resonance Escape Probability

The probability that a neutron will NOT be captured in the U-238 resonance region while slowing down to thermal energies.

$$L_{th} = \text{Thermal Non-leakage Factor}$$

The probability that a thermal neutron will NOT leak out of the core prior to absorption in the core.

$$f = \text{Thermal Utilization Factor}$$

The probability that a thermal neutron will be absorbed in fissionable material as opposed to all other core materials.

$$\eta = \text{Reproduction Factor}$$

The average number of fission neutrons produced per thermal neutron ABSORBED in the fuel.

What is Fast Non-leakage Probability (L_f)?

L_f = # of fast neutrons remaining in the core while slowing down / Net # of fast neutrons produced by all fissions

What is Resonance Escape Probability (p)? p = # of fast neutrons which are thermalized / # of fast neutrons remaining in the core while slowing down

Factors affecting p are: as moderator temperature increase – moderator density decreases, neutrons take longer to slow down, spend more time in fast/epithermal ranges, more resonance capture and p decreases. If fuel temperature increases – Doppler broadening occurs, increased resonance capture and p decreases. As core age increases – core fuel composition changes with age and p decreases.

What is Thermal Non-leakage Factor (L_{th})?

L_{th} = # of thermal neutrons remaining in the core / # of fast neutrons which are thermalized

Factors affecting L_{th} are: moderator temperature increase, core age causing L_{th} to decrease.

What is Thermal Utilization Factor (f)?

f = # of thermal neutrons absorbed in the fuel / # of thermal neutrons absorbed in the core

Factors affecting f are: initial enrichment of fuel, core age, moderator density, Boron density, fission product poisons, concentration of burnable absorbers, control rod position, core structural materials.

What is Reactivity (ρ)? A measure of a reactor's departure from criticality. A unitless measure. By performing various operations we can add reactivity to a reactor to change the core conditions. Reactivity can be positive or negative, depending on its effect on the neutron population (fission rate), but is ALWAYS ADDED to the reactor. Numerically equal to the reactivity added by the following six conditions:

1. Reactivity due to fuel (excess reactivity) – large, **positive** reactivity addition which results from loading of nuclear fuel at the beginning of cycle.

2. Reactivity due to control rods – rods insert **negative** reactivity as they are inserted into the core, and insert positive reactivity when withdrawn.

3. Reactivity due to moderator temperature – increasing moderator temperature will add **negative** reactivity except at the beginning of cycle.

4. Reactivity due to fuel temperature – increasing fuel temperature will add **negative** reactivity into the core.

5. Reactivity due to fission product poison buildup – fission product poison buildup will insert **negative** reactivity into the core.

6. Reactivity soluble B^{10} into the core will insert **negative** reactivity

What is reactivity coefficient? The ratio of the change in reactivity divided by the change in a parameter such as: moderator temperature coefficient (MTC), fuel temperature coefficient (FTC), void coefficient (VC) and pressure coefficient (PC).

What is reactivity defect? The total amount of reactivity inserted due to a finite change in a parameter such as: moderator temperature defect, fuel temperature defect and power defect.

What is power defect? The total amount of reactivity added to the reactor when going from one power level to another.

What is ternary fission? Normally, fission is a binary process, in which only two particles (the primary fission fragments) are formed when the fissioning nucleus splits. Much less frequently, more than two particles are formed and if precisely three particles appear, the fission event is classified as a ternary event. A similar effect can be observed when a cylindrical soap-bubble like the one displayed on the main fission-page is chopped up, a small bubble may remain between two big spherical ones.

Ternary fission occurs once every few hundred fission events. Roughly speaking, about 25% more ternary fission is present in spontaneous fission compared to the same fissioning system formed after thermal neutron capture.

What is hydrodynamics? Hydrodynamics is the study of moving liquids and fluids.

What is hydrostatics? Hydrostatics is the study of forces in stationary fluids like water and oil.

What is plasma? Plasma is when a gas is heated so hot that it becomes ionized and can act as a conductor of electricity. There is no greater charge both positive and negative ions are equal in number.

What is a superconductor? A superconductor is anything that can conduct electricity with little or no resistance at temperatures close absolute zero.

What is superconductivity? Superconductivity is the ability of certain materials to conduct an electrical current with little or no resistance at a temperature close to absolute zero. Scientist are working on materials that can be superconductive at room temperature. When this breakthrough occurs technology will take a giant leap forward.

What is sound? Sound is vibrations that travel through the air or some other medium that is caused when a force is applied to a solid, liquid, or gas, thus causing it to move.

What is vibration? Vibration is a reciprocating motion of particles of an elastic body or medium that alternates in opposite directions from the position of equilibrium.

How can vibration be controlled? There are several ways of controlling vibration. Replace worn components, properly balance equipment, check components that may be touching when they shouldn't be, check foundation that the equipment is on, check for errors in manufacturing, check for shaft misalignment and make sure all bolts are tight on component and foundation.

What is vibration displacement? Vibration displacement is the maximum distance between two points that is reached by a vibrating object. This type of displacement is usually measured in mills.

What is noise? Noise is a group of sound waves or vibrations that are not identical to one another and cover a wide range of frequencies. Noise can also be defined as any sound that is not wanted or is unpleasant to the ear.

What is reflected noise? Reflected noise is just that, noise that is reflected off of walls, ceilings, or any other hard surface that does not absorb sound.

How can noise be controlled? There are several ways to control the level of noise: 1) Reduce the noise at the source, like a silencer or muffler at the source, 2) Block the path of the noise, using sound barriers or enclosures, 3) Relocated the equipment causing the noise, and 4) Isolate the noise.

Why are neutrons used to create the chain reaction? Because fissile material is constantly undergoing radioactive decay by spontaneous fission, at least on some small level, there will always be a few neutrons being produced by the material. If enough material is assembled in a "workable" way (speaking to the geometry), critical mass will have been achieved, and the "natural" neutron flux will initiate the nuclear chain reaction and criticality will have been achieved.

In a reactor, which is inside the reactor vessel, we use lots and lots of fissile material. But it is spread out over a large volume and use equipment in there to inhibit or control the nuclear chain reaction. And that means dealing with those neutrons. The chain will begin when the control rods are pulled to a certain height. The rods are neutron absorbers and act to "kill" the chain when they are in the reactor core. By pulling them out, neutrons released by spontaneous natural fission within the fuel are free to cause other fissions (after a bit of slowing down) instead of being absorbed by a control rod. The chain has been initiated and builds according to the physics of the reactor.

Neutrons as have no charge (positive or negative) so they do not discriminate on what target they hit. Protons carrying a positive charge would be attracted to more negative atoms and electrons not having near the mass of a neutron or a proton could not transmit the amount of energy needed to split the atoms as well as having been attracted to other more positive atoms.

What is critical mass? The smallest amount of fissile material that is needed to maintain a nuclear chain reaction.

What is critical potential? The minimum pressure needed to turn an atom into an ion, measured in electron-volts.

What is cross-section? The likelihood that two particles will interact, visualized as an area surrounding the target particle and is measured in barns.

What is fission parameter? For an element that can undergo nuclear fission, the ratio of its atomic number to its relative atomic mass (atomic weight).

What is neutron flux? The total distance traveled (average) by all of the neutrons in one cubic centimeter of a material in one second. Neutron density x neutron velocity.

What is the difference between the thermal, epithermal, and fast region and the effect each has on a neutron? In the "standard" fission reactor, fissile nuclear fuel is "started up" and the neutron chain reaction begins. Neutrons are produced during atomic fission events, and these neutrons are sometimes called "fission energy" or "prompt" or "fast" neutrons. They are the free neutrons that appear as the result of the fission event. And they're moving pretty darn quick when they're "blown out" of the fissioning nucleus. But they're not moving anywhere near the speed of light.

The Boltzman distribution (a fancy way of speaking about the range of energies at which the fast neutrons appear), has a strong peak at close to 2 MeV (20 TJ/kg). That translates into a speed of 28,000 km/s. The speed of light is some 299,792 km/s as we've defined it, and that puts the speed of those fast neutrons at roughly 10% the speed of light. A thermal neutron is a free neutron with a kinetic energy level of less than 0.025 eV (approx. 4.0e-21 J; 2.4 MJ/kg, hence a speed of 2.2 km/s). They are named 'thermal' as this level of kinetic energy, which is similar to the average kinetic energy of a room-temperature gas.

Thermal neutrons have a much larger effective cross-section than fast neutrons, and can therefore be absorbed more easily by any atomic nuclei that they collide with, creating a heavier - and often unstable - isotope of the element as a result.

Most fission reactors use a neutron moderator to slow down, or thermalize the neutrons that are emitted by nuclear fission so that they are more easily captured, causing further fission. Others, called fast breeder reactors, use fast neutrons directly. Fission neutrons are produced at an average energy level of 2 MeV and immediately begin to slow down as the result of numerous scattering reactions with a variety of target nuclei. After a number of collisions with nuclei, the speed of a neutron is reduced to such an extent that it has approximately the same average kinetic energy as the atoms (or molecules) of the medium in which the neutron is undergoing elastic scattering. This energy, which is only a small fraction of an electron volt at ordinary temperatures (0.025 eV at 20 C), is frequently referred to as the thermal energy, since it depends upon the temperature. Neutrons whose energies have been reduced to values in this region (< 1 eV) are designated thermal neutrons. The process of reducing the energy of a neutron to the thermal region by elastic scattering is referred to as thermalization, slowing down, or moderation. The material used for the purpose of thermalizing neutrons is called a moderator. A good moderator reduces the speed of neutrons in a small number of collisions, but does not absorb them to any great extent. Slowing the neutrons in as few collisions as possible is desirable in order to reduce the amount of neutron leakage from the core and also to reduce the number of resonance absorptions in non-fuel materials. Neutron leakage and resonance absorption will be discussed in the next module.

Fast neutrons (>1MeV)
Epithermal neutrons (1MeV – 1 eV)
Thermal neutrons (<1eV)

Control Rods

What is a control rod? A control rod is a cylindrical hollow rod that is loaded with cadmium, boron, or some other absorbing material. The control rods act like sponges that absorb extra neutrons.

How does the control rod system protect the reactor core? The control rod system is designed to absorb neutrons within the core itself. By absorbing neutrons within the core the power level in the core will begin to drop due to the reduction of extra neutrons not being able to cause fission. The control rods are designed to control and/or shutdown the reactor core in the event of an emergency.

What is meant by control rod worth? The reactivity worth of a control rod is a measure of its ability to affect core reactivity. The SL-1 accident created the term control rod worth. Control rod worth is calculated by first calculating the core with the control rod fully inserted to determine its k_{eff}. The core is then similarly calculated with the control rod withdrawn. Control rod worth is defined by the following formula:

$$\frac{k_{eff(i)} - k_{eff(w)}}{k_{eff(w)} \quad k_{eff(i)}}$$

$k_{eff(w)}$ = multiplication factor, control rod withdrawn

$k_{eff(i)}$ = multiplication factor, control rod inserted

How are the control rods arranged in the reactor? The photo below illustrates with circles where the control rods would be located in the core. The difference in color, represents the different groups of control rods. The control rods can be manipulated either, by single control rod, a group of control rods (A,B,C, are shutdown groups while 1,2,3,4,5 are regulating groups), or all control rods (during a reactor trip).

CEA Core Locations

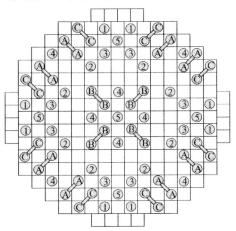

How do the control rods affect the reactor core radial flux profile?

Effect of Control Rods on Radial Flux Profile

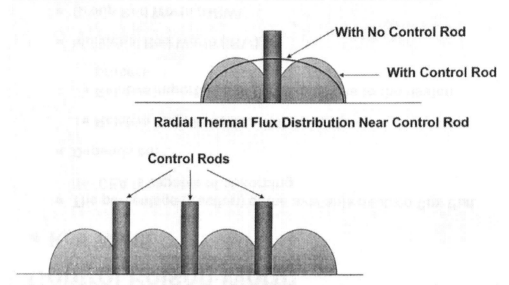

Radial Thermal Flux Distribution Near Control Rod

Radial Thermal Flux Pattern with Several Control Rods

Where are the control rods located? In nuclear reactors it basically depends on the design of the reactor. For example, pressurized water reactors have their control rods located on top of the reactor head. Boiling water reactor has their control rods located on the bottom of the reactor head.

How are the control rods operated? Control rods can be operated either manually or in automatic. They can also be controlled in groups or as a single individual control rod manipulation.

How are the control rods grouped? Control rods are grouped so that when each group is manipulated a balance throughout the reactor will remain. The more balanced the thermal energy is in the reactor core, the more efficient the core can be. If thermal energy within the reactor core becomes unbalanced temperature transients will reduce core efficiency and could possibly damage the core.

What drives the control rods? There are several drive mechanisms that can manipulate the control rods. These drives can be operated by pneumatic, electric, electro-magnetic, hydraulic and/or any combination just mentioned.

What happens to the control rod drive mechanism when they lose power? Depending on the design of the system will determine what might happen to the control rods if they should lose power. In PWRs since the control rods are on top of the reactor vessel, depending on the drive mechanism, if they lost power they may be allowed to just fall into the core by gravity assist. In a BWR since the control rods are located at the bottom of the reactor vessel, gravity assist is not an option so during a loss of power, the control rods will have a backup source of power that continues to allow the control rods to be operated.

How far do the control rods move? Control rods are designed to move in very small increments in order to maintain control of the reactivity within the core.

How fast can the control rods move? Control rods are designed to move quickly in the event of an emergency (inserted into the core), but for startup or increasing unit power control rods are not moved quickly to keep the reactivity in the core from getting out of control. There are numerous control systems that monitor both of these functions to keep the reactor operators aware of what is happening.

What are the control rods made of? Control rods are commonly composed of silver, indium and cadmium. Other elements that can be used include boron, carbon, cobalt, hafnium, gadolinium and europium.

What does the term "SCRAM" mean? Originally the control rods hung above the reactor, suspended by a rope. In an emergency a person assigned to the job would take a fire axe and cut the rope, allowing the rods in fall into the reactor and stop the fission. At some point the title of the person assigned this duty was given as SCRAM, or Safety Control Rod Ax Man. This term continues in use today as the phrase "scram" to describe a shutting down of a reactor by dropping the control rods.

What type of material do control rods use to absorb neutrons? A control rod is made up of chemical elements capable of absorbing many neutrons without fissioning themselves. They are used in nuclear reactors to control the rate of fission of reactor fuel. Because these elements have different capture cross sections for neutrons of varying energies, the compositions of the control rods must be designed for the neutron spectrum of the reactor it is supposed to control. Light water reactors (BWR, PWR) operate with "thermal" neutrons, the breeder reactors operate with "fast" neutrons.

Chemical elements with a sufficiently high capture cross section for neutrons includes silver, indium, cadmium. Other elements that can be used include boron, cobalt, hafnium, dysprosium, gadolinium, samarium, erbium, europium and there are some alloy compounds that can be used like high-boron steel, silver-indium-cadmium alloy, boron carbide, zirconium diboride, titanium diboride, hafnium diboride, gadolinium titanate and dysprosium titanate. The choice of materials is influenced by the energy of neutrons in the reactor, their resistance to neutron-induced swelling, and the required mechanical and lifetime properties. The rods may have the form of stainless steel tubes filled with neutron absorbing pellets or powder. The swelling of the material in the neutron flux can cause rod deformation leading to its premature replacement. The burnup of the absorbing isotopes is another limiting lifetime factor.

What does ARO mean? All Rods Out. This is referring to the control rods being totally out of the reactor core allowing other means to control reactor reactivity.

What do the extension shafts do for the control rods? Connect control element assembly (in core) to its drive mechanism (on the head).

How do control rods affect reactivity? The temperature in the reactor core is monitored carefully. When the core temperature goes down, the control rods are slowly lifted out of the core, and fewer neutrons are absorbed. Therefore, more neutrons are available to cause fission. This releases more energy and heat. When the temperature in the core rises, the rods are slowly lowered into the core and the energy output decreases because fewer neutrons are available for the chain reaction. The control rods absorb neutrons that could otherwise hit reactor fuel atoms and cause them to split. To maintain a controlled nuclear chain reaction, the control rods are manipulated in such a way that each fission will result in just one neutron, since the other neutrons are absorbed by the control rods. The purpose for control rods in controlling reactivity must provide the following: able to shut down reactor at any time in the life of the core under all hypothesized conditions, controls core flux distribution, allows for short term, rapid negative addition and provides for controlled, relatively even reactivity addition during reactor startup.

What is the difference between differential rod worth (DRW) and integral rod worth (IRW)? Differential rod worth is the amount of reactivity inserted per unit length of rod travel. Integral rod worth is the total amount of reactivity inserted upon a rod change between two specified heights.

What is control rod poison worth? The percentage (fraction) of the available neutron flux that the control element assembly is capable of absorbing, depending on relative magnitude flux at the rod tip and relative importance of these neutrons to the fission process. (SL-1)

What are three types of control rods?

1. **Shim rods** – used for making occasional coarse adjustments in neutron density.

2. **Regulating rods** – used for fine adjustments and to maintain the desired power output.

3. **Safety rods** – are capable of shutting down the entire reactor in case of failure of the normal control system. The control rods are normally outside of the reactor core. In the event of a shutdown, they move into the reactor core, absorb neutrons and stop the chain reaction.

Why are control rods positioned on top of a PWR and on bottom of a BWR? In a BWR the control rods are located on the bottom of the reactor vessel. In this type of reactor design the reactor coolant is also converted into steam to power the turbine/generator. With the control rods located at the bottom of the reactor vessel, the control rods remain at the same temperature as the reactor core, this keeps any temperature transients from causing any major stress factors that may cause the control rods to deform thus reducing their ability to be inserted back into the reactor core. For a PWR, since the primary coolant is not converted into steam, the control rods can be located on the top of the reactor vessel.

What about fuel cladding corrosion? Fuel cladding is Zircalloy, a highly corrosion resistant zirconium alloy. It will oxidize to a limited extent from the chemical reaction with water. The oxide layer can cause the cladding to thin and absorb H_2 which can lead to clad embrittlement which could lead to clad failure. Nuclear fuel cores are continuously upgraded to be more efficient which leads to new and different corrosion difficulties.

How long do control rods last? Control Rods have a life expectancy between 10 to 20 years. The basis of their life expectancy is not because the control rods cannot absorb any more neutrons, but because of the cladding that surrounds them…normally breaks down way before the material does.

Chemistry

What is an element? An element is a substance that is made up of atoms that are chemically alike and cannot be broken up into simpler parts by chemical means.

What is a compound? A compound is a substance formed by a chemical combination of two or more elements, which cannot be separated by any physical means.

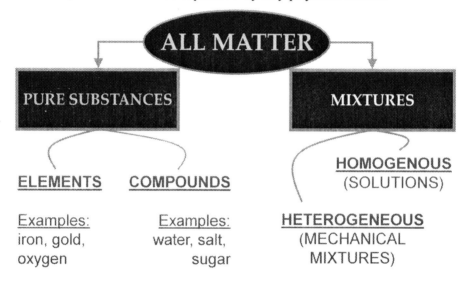

What is a mineral? A mineral is when two or more chemical elements are joined together. It is an inorganic substance found in nature. All minerals are solid, except for two, which are mercury and water.

What are anodes and cathodes? Anodes are negatively charged areas. Cathodes are positively charged areas. A metal surface may have both cathodic and anodic areas.

What is a salt? A salt is created when an acid and a base neutralize each other. Salt is a compound that is ionically bonded and consist of a metal and a non-metal.

What are electrolytes? Electrolytes are a nonmetallic solution that will carry an electrical current by the movement of ions. Electrolytes are liquids that contains ions. Some electrolytes are salts, bases, and acids. E.g. The solution in a car battery is an electrolyte.

How is iron and copper affected by pH? If pH increases, iron corrosion rate decreases. This poses a problem for many power plants that may have both iron and copper alloys in their systems. If pH was optimized for iron corrosion, then copper corrosion was excessive. If pH was optimized for copper then iron corrosion was elevated.

What is the difference between a noble and inert gas? Inert gases are groups of rare gases that include helium, neon, argon, krypton, xenon, and radon. These gases are said to be "inert"; meaning they will not easily react with other substances to make compounds. These rare gases are also called "noble gases".

What are non-metals? A non-metal is an element lacking both the physical and chemical properties of a metal. These non-metals are able to form anions, acidic oxides, acids and stable compounds with hydrogen.

What is an anion? An anion is a negatively charged ion.

What is a cation? A cation is a positively charged ion.

What is an acid? An acid is a chemical compound that is made up of at least two elements, one being hydrogen. Acids are located on the pH scale between 0 and 7. Acids are dangerous; while some will burn skin, others can dissolve metal.

What is a base? A base is a chemical compound that is made up of at least two elements, one being hydroxyl. Bases are located on the pH scale between 7 and 14. Bases are dangerous, while they can burn skin they can also destroy tissue.

What does pH stand for? The term pH is an acronym meaning the **p**otential of **H**ydrogen.

Is pH affected by temperature? YES. At a standard temperature of 77^0F ph 7.0 is neutral. At 240^0F the concentration of H+ ions brings the neutral point of pH from 7.0 to 6.0. In any case where temperature and pH are being measured the symbol would be pH_t.

Why would Oxygen be added to feedwater/condensate? Low concentrations of oxygen means a breakdown in the protective oxide layer. Higher corrosion rates by repeated removal and then replacement of the protective oxide layer. By adding small amounts of oxygen keeps the passive oxide layer intact, reducing the overall corrosion rate. This reduces the amount of corrosion products which migrate into the reactor vessel.

What is hydrazine? Hydrazine is an extremely poisonous chemical used as a fuel for rockets and organic synthesis. This chemical is colorless yet produces potent fumes, which are also corrosive. When used in a water type system, hydrazine acts as an oxygen scavenger, when it combines with oxygen it forms ammonia.

What is a catalyst? A catalyst is a substance which aids in the rate of a chemical reaction without undergoing any permanent changes in itself.

What are polymers? A polymer is a chemical compound formed from the process of polymerization. There are two general types of polymers; there are natural polymers such as starch, cellulose, and proteins. The second type of polymers are synthetic (man-made) such as polyethylene and nylon. A polymer is basically a large molecule that is made up of many smaller molecules being linked together. They bring things together.

What is polymerization? Polymerization is a reaction in which several small molecules combine with the polymer to form a large or more complex molecule with an increased molecular weight and different chemical properties.

What is a solution? It is not just an answer to a problem. A solution is a process in which a solid, liquid or gas is homogeneously mixed together with another solid, liquid or gas.

What is solute? A solute can be a gas, solid, or a liquid that is usually dissolved in a liquid to form a solution.

What is a solvent? A solvent is a substance that is usually a liquid and can dissolve other substances. (Water)

What are coagulants? Coagulants are polymers used to bring together suspended solids < 0.1 milli-microns in size. These products can be positively charged (cationic), or negatively charged (anionic) and carry varying molecular weights. Some examples of coagulants are inorganic salts or polymeric polymers such as polyamines.

What are suspended solids? Suspended solids are solids that will not dissolve in a solution.

What are flocculants? Flocculants are polymers used to bring together suspended solids > 0.1 milli-microns that are comprised of repeating molecular units. These products can be manipulated by varying the molecular weight chains or by changing the charge density on the molecular units.

What is a covalent bond? The term covalent bond is used to describe the bonds in compounds that result from the **<u>sharing</u>** of one or more pairs of electrons. E.g. methane

What is an ionic bond? An ionic bond is when one or more electrons from one atom are removed and attached **<u>(transferred)</u>** to another atom, resulting in positive and negative ions which attract each other. E.g. table salt

What is a diatomic molecule? A molecule that contains two atoms.

What is osmosis? Osmosis is a process in which water on one side of a semi permeable membrane tries to pass through the membrane to dilute a solution on the other side in an attempt to equalize the concentration of salt on both sides of the membrane. The term osmosis derives from the Greek word osmos 'impulse', 'thrust'.

What is osmotic pressure? Osmotic pressure is a pressure that is exerted by a dissolved material in a solution on a semi permeable membrane separating one solution from another solution.

What is an aquifer? An aquifer is a large body of rock that is considered permeable or containing pores and allows water to be stored in it and flows easily through it. The word aquifer is Latin meaning "water to bear". Some examples of permeable rock are sandstone and gravel. The aquifer is replenished from rainfall and filtered as it flows through the rock but can easily be polluted from acid rain, dump sites, spills, etc.

What are two factors that determine the amount of treatment that water receives? The two factors are the source of the water and the purpose for which the water is going to be used for.

What is the periodic table of elements? The periodic table of elements is a table of elements that are arranged in order by their atomic numbers to show specific characteristics of each element.

What is the purpose for using purified water in a power plant? When using water in a power plant for such equipment as the reactor or steam generator, the water must be cleaned or treated to remove impurities in the water that could lead to three major problems: Scale, Corrosion and Carryover.

What are some ways to purify water? Depending on the purpose of the water to be used, will determine how much purifying the water will go through. Some equipment is simple such as trash racks, screens, filters, strainers, clarifiers, etc., other equipment such as polishers, demineralizers and reverse osmosis systems are for water that must be highly purified.

What is the reason to purify water? Remove corrosion causing agents, remove scale forming substances, remove particulate (suspended) matter, remove dissolved oxygen, and remove radioactive contaminants.

What is a demineralizer? A demineralizer is several pieces of equipment working together to purify water for reactor/steam generator use. This process contains several steps but basically there are two types of resin within the demineralizer that create an ion exchange between the hard water coming into the demineralizer and the resin beads themselves. In this ion exchange the hard minerals are removed from the raw water while the resin beads replace the hard minerals with other non-harmful materials that will not hurt the plant equipment when the water is used in the plant cycle.

What systems and/or equipment may use ion exchangers? There are numerous systems that may use ion exchangers like makeup water systems, condensate polisher, steam generator blow-down, stator cooling system, chemical and volume control system, spent fuel pool system, liquid waste system, and sample systems.

What would cause resin bead fouling? Oil and grease can cause major resin bead fouling.

What is acid dosing mean? Acid dosing is a water treatment process that involves the injection of acid into the circulating water system to control the pH and to neutralize any impurities.

What is meant by raw water? Raw water is water that has not been treated for its intended use.

What is the difference between resin beads and powdered resins? Resin beads are small polymeric beads containing active sites and usually used in deep beds. Powdered resin beads are crushed resin beads and usually used in thin layers and are sometimes called a filter demineralizer. Photo below is of resin beads.

At which pH should the reactor and steam generator water pH be maintained? The reactor water for a pressurized water reactor would be around a pH of 5 or 6 while the steam generator feedwater would be closer to a pH of 10. The pH for BWR is around 7.

What two problems can scale cause? In a heat exchanger scale can reduce efficiency of a tube by acting as an insulator on the metal tubing. Second, it can lead to tube failure because the amount of heat transferred from the metal tubing to the water is reduced; the tube will overheat and rupture.

What are hardness salts? Hardness salt is a term that refers to compounds made of calcium or magnesium.

What are impurities? Impurities are foreign substances that hold no positive value or benefit in a substance.

What is activated carbon? Activated carbon is carbon that has been processed to increase its surface area to be used as a filter medium.

What is a filter medium? A filter medium is a material used in filters to trap suspended solids such as gravel, sand or anthracite coal.

What is a carbon filter? A carbon filter is a filter that contains an activated medium to be used to remove any suspended solids, chlorine, and organic material from raw water before it enters the ion exchanger. If a carbon filter was not used before raw water entered the ion exchanger, the ion exchanger would require to be regenerated much sooner than expected.

What are anion resin beads? Anion resin beads are resin beads that attract negatively charged ions in an ion exchanger. Anion resin beads remove anions like chlorides, sulfates and carbonates. These anions in the water are exchanged for hydroxyl ions that are located on the resin beads, thus removing the anions from the water.

What are cation resin beads? Cation resin beads are resin beads that attract positively charged ions in an ion exchanger. Cation resin beads remove cations like calcium magnesium and sodium. These cations in the water are exchanged for hydrogen ions that are located on the resin beads, thus removing the cations from the water.

What are resin sites? Resin sites are charged areas on a resin bead that is occupied by ions. These charged areas could be either positively or negatively charged depending if the resin is for cation or anion.

What is meant by carryover? Carryover is a condition in which suspended and/or dissolved solids that are present in steam generator water that is actually picked up by the steam leaving the steam generator and is moved to other parts of the plant cycle. There are two ways that solids in a steam generator can be picked up by steam. One way is through entrainment in which water containing impurities is physically picked up and carried along with the steam. The second way is through vaporization in which impurities in the water reach their boiling point and change from a solid to a gas and are then carried along with the steam.

How can carryover be prevented? The entrainment of impurities can be prevented through the use of moisture separators and drier panels in the steam generator. However, the vaporization of impurities is harder to control because suspended solids can change from a solid to a vapor quickly under high temperature and pressure conditions and be carried right through the moisture separators and drier panels along with the steam to the turbine. Once these gases reach the turbine they change from a gaseous state into a solid state because of the decrease in temperature and pressure. It is then where the damage starts on the turbine blading.

How are dissolved gases removed from the plant cycle? Gases can be removed by either the condenser vacuum pumps, ion exchange system, waste gas system and/or a deaerator. If gases are not removed from the system, they will buildup and greatly affect heat transfer.

How does heating water affect the amount of gases in a solution? Heating a solution will cause some of the gases within the solution to boil out of the solution. This is why hot water will freeze faster than cold water. The amount of gases removed from a solution is based on several factors; yet the higher the temperature of the solution, the more effective of removing gases from within that solution.

What does volatile mean? Volatile means the ability to evaporate as compared to water. Chemicals that have a high volatility are alcohol and gasoline as compared to water. Chemicals that have a low volatility are motor oil and antifreeze as compared to water.

What are two commonly used chemicals used in an all-volatile treatment system? Two volatile chemicals commonly used are hydrazine and ammonia.

In an all-volatile treatment system, what do these chemicals control when added to the water? These chemicals control the removal of any dissolved oxygen in the water and the pH of the water.

What is the one major disadvantage of using an all-volatile treatment system? Should a major leak develop or impurities that are too excessive to be controlled by either the boiler blow-down or a condensate polisher, the unit must be shut-down for repairs.

What are two methods commonly used to remove excessive amounts of solid impurities in an all-volatile system? One method is the use of the steam generator blow-down to remove solid impurities from the steam generator to the drain. The second method is the use of a condensate polisher, which is a small demineralizer that is tied into the condensate system to remove solid impurities from the water before it returns to the steam generator.

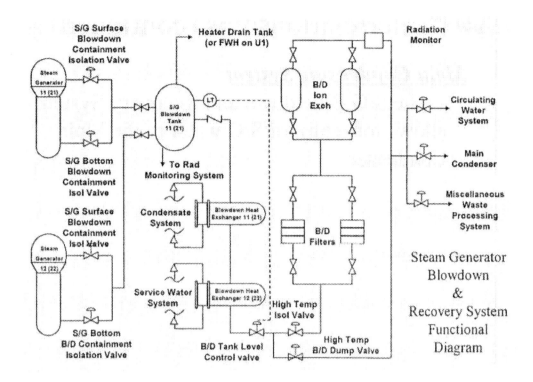

What are sources that could be used to supply coolant to a nuclear plant? Four of these coolants are water based: a river, a lake, an ocean, or a reservoir (cooling tower) could be used to supply coolant. The fifth source is air and is used in air cooled condensers when water is not available.

What is carbohydrazide? Carbohydrazide is similar to hydrazine but more effective as an oxygen scavenger at low temperatures. It is also not a health hazard like hydrazine. It is only used during lay-up conditions and costs 4 to 8 times more than hydrazine.

What are four major problems caused by inadequate treatment to boiler water? Scale, corrosion, erosion, and carryover are caused by poorly treated water. Scale can become attached to the tubes. Corrosion is the chemical reaction within the water and tubes. Erosion is the physical wear and tear on the tubing material. And carryover in which impurities are transferred from one location to another.

What are some procedures for treating raw water that is drawn from a natural source? There are three steps to treat raw water drawn from a natural source. They are: screening to remove large debris such as tree branches, clarifying and filtering to remove any suspended solids with the use of chemicals, and using a demineralizer to purify water for system use through ion exchanger.

What are the two most common methods for removing suspended solids from raw water? Clarification and filtration are two common methods for removing suspended solids from raw water. The clarification process is when chemicals are added to the water to cause the suspended solids to stick or clump together. As the solids clump together, they get heavier causing them to fall out of suspension in the water. The filtration process is when the suspended solids in the water are removed by passing the water through a material that is porous allowing the water to flow through while retaining suspended solids from the water.

What two groups can suspended particles be divided into? The first group is any particles that aid to the turbidity (cloudiness) of the water. The second, is any particles that aid to the color of the water.

What is a clarifier? A clarifier is a device used to remove suspended solids from raw water with the help of chemicals. It consists of several chambers that allow chemicals to mix with the raw water and allows for fallout of these impurities within the water, then blows them out while the clean water flows into a recovery system or is released into the environment.

How does a clarifier work? Raw water enters the clarifier and mixes with chemicals using an agitator which allows the chemicals to react with the suspended solids. Once the chemicals react with the solid separation begins to take place. The now heavier solids fall out of suspension and fall to the bottom of the clarifier. The water leaving the clarifier has had the majority of the suspended solids removed.

What are three stages of clarification? These three stages are coagulation, flocculation and sedimentation. Coagulation is when a chemical coagulant is mixed with the raw water. A coagulant like alum (aluminum sulfate) is mixed into the water and combines with an alkaline material in the water to form aluminum hydroxide precipitate. This precipitate carries a positive charge, which attracts the suspended particles that carry a negative charge in the water. Flocculation is the mixing action in the clarifier, which is slowed down to a gentle agitation. The reason for slowing the agitation process is to allow the negatively charged suspended particles to react with the positively charged precipitate to form heavier particles. These heavier particles are called floc. Sedimentation is the final stage of the clarification process. At this time the water is not agitated at all. As these suspended solids become heavier they settle at the bottom of the clarifier. The buildup of floc particles is called sludge. The sludge is then blown out from the bottom of the clarifier into a holding basin.

What is precipitate? Precipitate is a compound that will not dissolve in the liquid in which it is formed.

What is the purpose of the dilution pit? A dilution pit is a large basin that holds wastewater from the demineralizer until the pH of the water is environmentally safe to pump out.

Why is pH control essential for proper operation of the clarifier? It is essential that pH be controlled to ensure that suspended solids formed by the coagulant in the raw water remains insoluble, the pH must be maintained within a certain pH range. If the pH in the water drops, the suspended solids will dissolve and flocculation will not occur. Also carbonic acid can form when carbon dioxide mixes with the water, which will further lower the pH of the water in the clarifier.

What are some materials that can be used as a filter medium to filter water? Materials like gravel, sand, and anthracite coal can be used as a filter medium to filter raw water.

How does a gravity filter work? A gravity filter is like a large basin or pool that is open to atmosphere and consist of a filter medium, a gravel bed, a water outlet drain pipe, an under drain system, and a raw water inlet pipe. The gravel bed keeps the filter medium from passing through into the water outlet drainpipe. The under drain system is the support system for the gravel bed like a screen or grates, not allowing the gravel to fall through the filter medium where suspended solids will become trapped. The filtered water will then pass through the gravel bed and under drain system to enter the water outlet pipe as filtered water.

What is the purpose of backwashing? As a gravity filter remains in operation the accumulation of suspended solids buildup in the filter medium and restricts flow through the filter. Backwashing is a process of removing suspended solids from the filter medium to restore flow through the filter. This process is accomplished by reversing the flow of water through the filter and washing out the suspended solids from the filter medium. When to backwash a filter can be determined by several methods such as, length of time in operation or by pressure differential depending on plant procedures.

What is the main difference between a pressure filter and a filter demineralizer? The difference between these two filters is the type of medium they use. The pressure filter generally uses a granular filter medium while the filter demineralizer uses a porous metal cylinder or stainless steel mesh for its medium.

How does a reverse osmosis unit work? A reverse osmosis unit uses pressure to force raw water through a porous semi permeable membrane to remove any impurities. The semi permeable membrane is designed to allow water molecules to pass through the membrane, yet traps any dissolved solids in the water. Reverse osmosis depends on the fact that almost all impurities in the water are larger in size than a water molecule.

How do suspended solids affect a reverse osmosis unit? Raw water must be pretreated before it is allowed to enter the reverse osmosis unit. If not, the suspended solids in the water can block the pores of the membrane thus reducing the efficiency of the unit and could possibly damage the membrane.

What is osmotic head? Osmotic head is the liquid level that is higher on the high concentrated solution side.

Why is Carbon 14 used to measure half-life? Radio carbon dating determines the age of ancient objects by means of measuring the amount of carbon-14 there is left in an object. A man called Willard F Libby pioneered it at the University of Chicago in the 50's. In 1960, he won the Nobel Prize for Chemistry. This is now the most widely used method of age estimation in the field of archaeology.

How it works: Certain chemical elements have more than one type of atom. Different atoms of the same element are called isotopes. Carbon has three main isotopes. They are carbon-12, carbon-13 and carbon-14. Carbon-12 makes up 99% of an atom, carbon-13 makes up 1% and carbon-14 - makes up 1 part per million. Carbon-14 is radioactive and it is this radioactivity which is used to measure age.

Radioactive atoms decay into stable atoms by a simple mathematical process. Half of the available atoms will change in a given period of time, known as the half-life. For instance, if 1000 atoms in the year 2000 had a half-life of ten years, then in 2010 there would be 500 left. In 2020, there would be 250 left, and in 2030 there would be 125 left.

By counting how many carbon-14 atoms in any object with carbon in it, we can work out how old the object is - or how long ago it died. So we only have to know two things, the half-life of carbon-14 and how many carbon-14 atoms the object had before it died. The half-life of carbon-14 is 5,730 years. However knowing how many carbon-14 atoms something had before it died can only be guessed at. The assumption is that the proportion of carbon-14 in any living organism is constant. It can be deduced then that today's readings would be the same as those many years ago. When a particular fossil was alive, it had the same amount of carbon-14 as the same living organism today.

The fact that carbon-14 has a half-life of 5,730 years helps archaeologists date artifacts. Dates derived from carbon samples can be carried back to about 50,000 years. Potassium or uranium isotopes which have much longer half-lives are used to date very ancient geological events that have to be measured in millions or billions of years.

What is boiling point? The temperature at which a liquid boils, when its vapor pressure equals the external (atmospheric pressure) and the liquid freely turns into a vapor. It varies with external pressure.

What is homolog? Homolog are elements in the same periodic table group that tend to exhibit similar, but not identical, chemical properties.

What is hormesis? A controversial theory that argues that there is a benefit to health, or decrease in biological damage from radiation as dose is increased (valid only for very small doses).

What impurities in the primary system coolant must be monitored? The primary system coolant is monitored for anions, O2, Zinc, Lithium, pH, Nickel, Boron, and radiation (Gamma, Alpha).

What is the difference between an exothermic and endothermic reaction? An exothermic reaction is a reaction where heat generated from the reaction is given off to the surrounding atmosphere. An endothermic reaction is a reaction where heat generated from the reaction remains within the reaction itself.

What is N-16? Radioisotope N^{16} is the dominant radionuclide in the coolant of pressurized water reactors during normal operation. It is produced from O^{16} (in water) via (neutron/proton) reaction. It has a short half-life of about 7.1 seconds, but during its decay back to O^{16} produces high-energy gamma radiation (5 to 7 MeV). Because of this, the access to the primary coolant piping must be restricted during reactor power operation. N^{16} is one of the main means used to immediately detect even small leaks from the primary coolant to the secondary steam cycle. The secondary system of a PWR is monitored for N^{16} which would give an indication that there is a primary to secondary system leak.

Why is the letdown fluid cooled before entering the ion exchanger? The resin beads in the ion exchanger do not tolerate heat very well so the letdown fluid is cooled to a safe and functional temperature that the ion exchanger can effectively operate. If the temperature gets too high the ion exchanger is bypassed in order to keep the resin beads from damage.

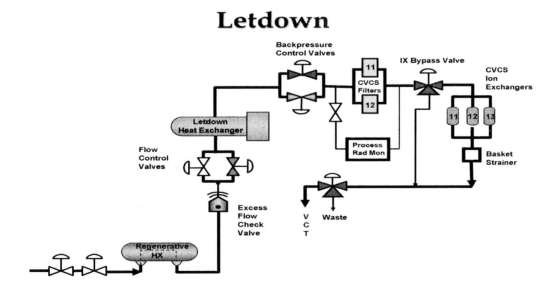

Letdown

How is the reactor coolant purified? The CVCS bleeds the reactor coolant off of the primary system through the letdown. From here it flows through a heat exchanger and then to the ion exchanger to remove unwanted material from the coolant. After leaving the ion exchanger the coolant is routed back to the reactor through a series of charging pumps that pump the coolant back into the primary system.

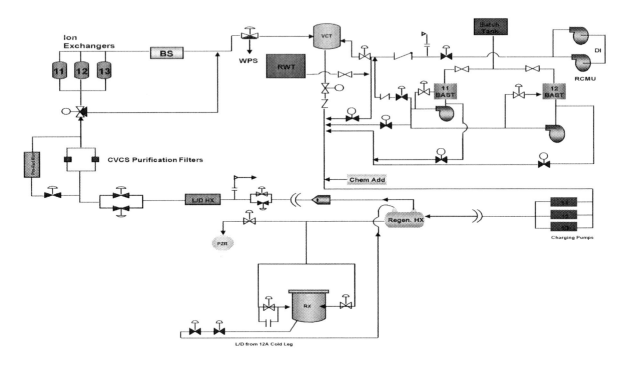

Is conductivity of a fluid affected by temperature? YES. Conductivity increases as temperature increases.

What is the difference between oxidation and reduction? Oxidation is corrosion where metals form metal oxides (corrosion products). Reduction is just the opposite of oxidation.

What is radiolysis? Radiolysis is the radiation induced splitting of water; meaning that water molecules H_2O are broken up by radiation into individual atoms. This separation can lead to a number of problems like corrosion and even worse a possible explosion. By injecting hydrogen gas into the water creates an overabundance of hydrogen in the fluid which greatly reduces the chance of the water molecules braking down into individual atoms.

What is gamma spectroscopy? Gamma spectroscopy involves the spectroscopy of radionuclides. While a Geiger counter determines only the count rate, a gamma spectrometer also determines the energy of gamma rays emitted by radioactive substances.

What is tritium? Tritium is a radioactive isotope of hydrogen, known as Hydrogen-3. The nucleus of tritium contains one proton and two neutrons, whereas the nucleus of protium contains no neutrons and one proton. It has a very low decay energy and is a pure Beta emitter. Its half-life is 12.3 years.

What is the purpose of the sample cooler? The purpose of a sample cooler is to cool the sample flow below its boiling temperature to keep the fluid from boiling when it leaves the sample line.

What is a chemistry hold? A chemistry hold is a procedure that the chemistry department implements when the chemistry in the plant system is abnormal. This procedure is in place to help reduce any further problems in the plant systems. If the chemistry was not controlled, the plant systems could be severely damaged.

Why is condensate polisher resin, different sizes in diameter? The resin inside a condensate polisher is of different sizes so when it is time to regenerate the resin it can be separated through backwashing allowing a more efficient method of regeneration.

In a condensate polisher, what does cross contamination mean? The amount of cation resin trapped in the anion resin or the amount of anion resin trapped in the cation resin.

What is resin throw? Resin throw is when resin is somehow escapes the demineralizer and becomes free floating in the system that it is designed to protect.

What is resin selectivity? A measure of how strongly different ions are held by ion exchange resins. This affects demineralizer performance.

Cations	**Anions**
Very Weak Bonds	
Lithium	Fluoride
H+	OH-
Sodium	Silica
Ammonia	Carbonate
Rubidium	Chloride
Magnesium	Sulfate
Potassium	Bromide
Silver	Nitrate
Calcium	Iodine
Strontium	
Barium	
Very Strong Bonds	

What is meant by hideout? Hideout is a term used to describe the loss of boiler water phosphate residual that accompanies this solid phase formation problem. The phosphate hideout phenomenon can pose a significant chemistry control problem in high pressure boilers that experience load variations. Typically, phosphate tends to "disappear" as load (steaming rate load) is increased towards maximum. The pH will also vary, usually in an upward direction, although one would expect that the change in pH would be consistent with the change in Na:PO4 ratio of the solution. The upward trend in pH may be explained by an interaction between phosphate and magnetite that is also known to exist. Just as phosphate disappears with increasing load, so does it reappear with decreasing load, with concurrent impact on the system pH. Dissolution of the solid phase occurs as load is reduced. Historically, "hide-out" had been regarded as an operating nuisance, which continually needed additions of tri-, di-, and mono-sodium phosphates to keep the boiler water in the CPT control range. However, it has been recognized recently that serious corrosion (termed "acid phosphate corrosion") can result when these chemicals are added to boilers where "hide-out" and deposition are occurring.

Why is hideout bad? Hideout is bad whenever there is a condition change in a system. It is this change that causes chemicals that may have become hidden in the system to come back out into fluid suspension causing fluctuations in pH, suspended solids, etc.

How does this build up affect steam generator performance? Buildup inside the steam generator reduces heat transfer between the primary and secondary system by insulating the tubes. This buildup also creates temperature differential stresses between the clean tubes and the dirty tubes. The more buildup a steam generator has an increase in reduced efficiency will be found.

What is the purpose of using three different ion exchangers in the CVCS? One is used to remove lithium, one is used to remove boron and the third one is used to remove everything else.

What is Crud Induced Power Shift (CIPS)? CIPS is where boron rich deposits in hotter portion of the core shifts flux downward. At end of cycle (EOC), flux may be shifted upwards (deposits dissolve, less burn up in the upper core). Deposits may contain lithium, boron or zeolite elements. The axial offset anomaly crud induced power shift mainly affects cores that have more boiling. This occurrence has significantly restricted power operation at some plants.

What is matter? Anything that has mass and occupies space. Has specific chemical and physical properties.

What is meant by hardness of water? Hardness of water is caused mainly b the presence of calcium and magnesium salts. Temporary hardness is caused by bicarbonates and is removed by boiling the water. But for permanent hardness caused by carbonates, it remains even after boiling.

What are heavy metals? Heavy metals are chemical elements with a specific gravity that is at least 5 times the specific gravity of water. The specific gravity of water is 1 at 39^0F. There are 23 heavy metals: antimony, arsenic, bismuth, cadmium, cerium, chromium, cobalt, copper, gallium, gold, iron, lead, manganese, mercury, nickel, platinum, silver, tellurium, thallium, tin, uranium, vanadium and zinc.

How are heavy metals dangerous? In minute amounts, some of these elements are needed for good health; however in large doses they can quickly become toxic to living organisms. Heavy metal toxicity can result in damaged or reduced mental and central nervous system function, lower energy levels, damage blood composition, lungs, kidneys, liver and many other vital organs. Long term exposure may result in slowly progressing physical, muscular and neurological degenerative processes that mimic Alzheimer's disease, muscular dystrophy and multiple sclerosis. Allergies are not uncommon and repeated long term contact with some metals or their compounds may even cause cancer. When heavy metals enter the body, the body is unable to remove them through the body's waste system. Through continuous consumption of these heavy metals will cause the body to increase levels in heavy metals, eventually becoming more and more toxic in the human body thus leading to medical issues later on. The most commonly encountered toxic heavy metals are: arsenic, lead, mercury, cadmium and iron.

What is radiation chemistry? Radiation chemistry is the study of the chemical effects of radiation on matter. E.g. The conversion of water into hydrogen gas and hydrogen peroxide.

What part do HEPA and charcoal filters play in the nuclear industry? HEPA and charcoal filters are used to catch and trap air borne particles that may be radioactive. These high quality filters ensure that our environment inside does not impact our outside environment.

What is a zirc/water reaction? During normal reactor operation, the surface temperature of the Zircaloy is approximately 650°-700° F. This causes limited surface oxidation and is accounted for in the typical reactor design. As the surface temperature increases, the oxidation rate increases. If the Zircaloy temperatures approach approximately 2000° F., which can occur during an accident, the Zircaloy reacts rapidly with the steam in the reactor in what is generally referred to as a Zirc-water reaction. The reaction results in rapid oxidation of Zircaloy and the rapid formation of hydrogen gas. The oxidized high temperature Zircaloy lacks structural strength and will slump under its own weight and release fission gases contained in the fuel rods.

What is plutonium? A heavy, man-made, radioactive metallic element. The most important isotope is Pu-239, which has a half-life of 24,000 years. Pu-239 can be used in reactor fuel and is the primary isotope in weapons. One kilogram is equivalent to about 22 million kilo-watt hours of heat energy. The complete detonation of a kilogram of plutonium produces an explosion equal to about 20,000 tons of chemical explosive. All isotopes of plutonium are readily absorbed by the bones and can be lethal depending on the dose and exposure time.

What is meant by transuranium elements? Transuranium elements are those chemical elements with atomic numbers greater than 92 (Uranium).

All these elements that have an atomic number greater than 92 have been first discovered artificially and other than plutonium and neptunium, none occur naturally on earth. They are all radioactive with half-lives showing a trend to decrease with atomic number. Trace amounts of neptunium and plutonium form in some uranium-rich rock.

What was the first transuranium element? Neptunium was the first transuranium to be created followed by the second transuranium element…plutonium.

What are super-heavy atoms? Super-heavy atoms are the transactinide elements beginning with rutherfordium. They are artificially created and currently do not pose a useful purpose because of their short half-lives causing them decay from just a few minutes to milli-seconds.

What does off-gas mean? Off-Gas is the un-dissolved ozonated air collected from the reaction tanks or de-ozonizing filters that can cover various manufactured parts, such as some computer parts. It is also a term used to describe the process of removing such ozonated air. It is potentially dangerous and is advised that people air out such contaminated items.

What gases are lighter or heavier than air? The term specific gravity is used to determine the weight of a gas when compared to air. If a gas is heavier than air it will have a value that is greater than 1 and if the gas is lighter than air it will have a value that is less than 1. Here are some gas values:

Acetylene	0.9	Helium	0.14
Air	1	Hydrogen	0.07
Ammonia	0.59	Hydrogen chloride	1.26
Argon	1.38	Hydrogen sulfide	1.18
Benzene	2.7	Methane	0.55
Butane	2	Natural gas	0.65
Carbon dioxide	1.52	Oxygen	1.1
Carbon monoxide	0.96	Ozone	1.66
Chlorine	2.48	Propane	1.52
Deuterium	0.07	Sulfur Dioxide	2.26

What is meant by chemisorptions? Chemisorption is a classification of adsorption characterized by a strong interaction between an adsorbate and a substrate surface, as opposed to physisorption which is characterized by a weak Van der Waals force. A distinction between the two can be difficult and it is conventionally accepted that it is around 0.5 eV of binding energy per atom or molecule. The types of strong interactions include chemical bonds of the ionic or covalent variety, depending on the species involved. It is characterized by:

- High temperatures.
- Type of interaction: strong; covalent bond between adsorbate and surface.
- High enthalpy: -50 kJ/mol $>\Delta H>$ -800 kJ/mol
- Adsorption takes place only in a monolayer.
- High activation energy
- Increase in electron density in the adsorbent-adsorbate interface.
- Reversible only at high temperature.

Due to specificity, the nature of chemisorption can greatly differ from system to system, depending on the chemical identity and the surface structure.

What is a radical? A radical can be an atom or a group of atoms that function together as a unit in a chemical reaction.

What is rust? Rust is a brittle coating that forms on iron or steel due to the oxidation or chemical attack to the surface metal when it is exposed to moisture and oxygen. This form of corrosion is a reddish brown or orange color.

How can rust be controlled? Rust can be controlled through the use of chemicals such as hydrazine. When hydrazine mixes with oxygen it forms ammonia and the ammonia can be removed safely by other means. The problem here is that if ammonia is created, care must be taken that ammonia will not be as harmful as oxygen is to creating rust. Ammonia is extremely corrosive to copper and copper alloys.

What liquids are lighter or heavier than water? The term specific gravity is used to determine the weight of a liquid when compared to water. If a liquid is heavier than water it will have a value that is greater than 1 and if the liquid is lighter than water it will have a value that is less than 1. Here are some liquid values:

Acetone	1.05	Gasoline	0.74
Alcohol, propyl	0.8	Hydrazine	0.8
Antifreeze	1.05	Hydrochloric acid	1.19
Chloride (liquid)	1.56	Mercury	13.63
Creosote	1.07	Milk	1.04
Crude oil	0.918	Sea water	1.03
Formaldehyde	0.82	Saline	0.72

What are electrolytes? Electrolytes are a nonmetallic solution that will carry an electrical current by the movement of ions. Electrolytes are liquids that contain ions. Some electrolytes are salts, bases, and acids. E.g. Electrolytes like Sodium, Potassium, Calcium, Magnesium, Chloride, Hydrogen phosphate and Hydrogen carbonate may be found in everyday drinks like Gatorade.

What would be the purpose for hydrogen peroxide injection? For a PWR, going into an outage hydrogen peroxide is injected into the primary system to create an oxygen rich environment thus allowing CRUD to break loose to be removed within the system. The hydrogen peroxide is introduced into the chemical volume control system (CVCS). CAUTION must be used because this process is very likely to increase dose due to more radioactive material now present in the primary coolant system.

What is Cobalt 58 & Cobalt 60? Cobalt can also exist in radioactive forms. A radioactive isotope of an element constantly gives off radiation, which can change it into an isotope of a different element or a different isotope of the same element. This newly formed nuclide may be stable or radioactive. This process is called radioactive decay. Co-60 is the most important radioisotope of cobalt. It is produced by bombarding natural cobalt, Co-59, with neutrons in a nuclear reactor. Co-60 decays by giving off a beta ray (or electron), and is changed into a stable nuclide of nickel (atomic number 28). The half-life of Co-60 is 5.27 years. The decay is accompanied by the emission of high energy radiation called gamma rays. Co-60 is used as a source of gamma rays for sterilizing medical equipment and consumer products, radiation therapy for treating cancer patients, and for manufacturing plastics. Co-60 has also been used for food irradiation; depending on the radiation dose, this process may be used to sterilize food, destroy pathogens, extend the shelf-life of food, disinfest fruits and grain, delay ripening, and retard sprouting (e.g., potatoes and onions). Co-57 is used in medical and scientific research and has a half-life of 272 days. Co-57 undergoes a decay process called electron capture to form a stable isotope of iron (Fe-57). Another important cobalt isotope, Co-58, is produced when nickel is exposed to a source of neutrons. Since nickel is used in nuclear reactors, Co-58 may be unintentionally produced and appear as a contaminant in cooling water released by nuclear reactors. Co-58 also decays by electron capture, forming another stable isotope of iron (Fe-58). Co-60 may be similarly produced from cobalt alloys in nuclear reactors and released as a contaminant in cooling water. Co-58 has a half-life of 71 days and gives off beta and gamma radiation in the decay process.

What is the purpose of zinc injection? Zinc will mitigate primary water stress corrosion cracking of Alloy 600. Forms a thin, tough, protective oxide film on reactor coolant system piping and steam generator tubes. Zinc oxide layer will absorb less radioactivity and increases corrosion protection. Reduces the amount of nickel activation which decays into cobalt. Cobalt is the biggest contributor of radiation to personnel during an outage. Zinc addition is the addition of Zinc Oxide into the Feedwater (BWR) to reduce primary system piping and component dose rates by reducing the amount of Co-60. Zinc acts to lower Co-60 by suppressing the corrosion release rate for in-core alloys and incorporating into the iron-based fuel deposits resulting in an oxide which releases Co-60. Zinc is also used in many PWRs (as zinc acetate) to reduce dose.

What is the purpose of using three different ion exchangers in the CVCS? One is used to remove lithium, one is used to remove boron and the third one is used to remove everything else.

What is lithium and Boron used for? Lithium is used to control ph and Boron is used to control reactivity.

Why is boron monitored in a BWR? BWRs measure for boron to detect problems inside the reactor…coming from control rod leaking or possibly from the liquid poison system.

What happens if the lithium is too high or too low in the reactor vessel? High lithium levels will raise pH. It can also increase stress corrosion cracking or clad oxidation. If lithium levels are too low, it can increase corrosion rates of pipes and tubes, increase deposits on fuel surface, increase out-of-core radiation buildup and potential fuel cladding damage.

What is biocide? Biocide is a substance that is poisonous and destructive to living organisms. E.g. pesticides, herbicides, chlorine, etc.

What is radiation chemistry? Radiation chemistry is the study of the chemical effects of radiation on matter. E.g. The conversion of water into hydrogen gas and hydrogen peroxide.

Materials

What properties should be accounted for when determining material usage?

Material Properties

Mechanical	Thermal	Electrical
Toughness	Conductivity	Conductivity
Ductility	Heat Capacity	Resistivity
Elastic Modulus	Expansion	Polarization
Strength		Dielectric Constant

What is intergranular attack? Intergranular attack (IGA) is a form of corrosion where attack occurs along microscopic grain boundaries. Metal grain boundaries are found in many metals and can be seen under a microscope. These grains have a different chemistry than the bulk metal. Many metals are composed of fine grains that can be seen under a microscope. The grains are a consequence of manufacturing. When a molten alloy is cooled, some species may precipitate out of the melt. The chemistry at the grain boundaries is slightly different than that of the gains themselves. This leads to differences in corrosion behavior.

What is corrosion current? When two dissimilar metals are coupled in a corrosion cell, a current will be produced. The magnitude of the current depends on the metals being used. The corrosion current is sometimes called the galvanic potential. There are related measures of corrosion called oxidation-reduction potential and electrochemical potential. These are very similar to galvanic potential.

ANODIC END	Magnesium	**MOST CORRODED**
	Magnesium alloys	
	Zinc	
	Aluminum	
	Cast Iron	
	Copper	
	400 Series Stainless Steels	
	Lead	
	Alloy 600 (Inconel)	
	300 Series Stainless Steels	
	Titanium	
	Platinum	
CATHOIC END	Graphite	**LEAST CORRODED**

When two dissimilar metals are coupled in a corrosion cell a current will be produced. The magnitude of the current depends on the metals being used. The corrosion current is sometimes called the galvanic potential. There are related measures of corrosion called oxidation-reduction potential and electrochemical potential.

The galvanic series list various metals and alloys according to the currents they will produce. If a cathodic metal is coupled with an anodic metal in a corrosion cell, then the anodic metal will preferentially corrode. All that it may take for this corrosion cell is contact (or near contact) and an electrolyte.

How can galvanic corrosion be controlled? By selecting proper materials, insulation, chemistry control (keep ionic strength low), the use of sacrificial anodes and/or cathodic protection.

What is the difference between an axial crack and a circumferential crack? An axial crack is a crack that runs along (parallel) with the pipe. Circumferential crack is a crack that runs across the pipe (perpendicular). An axial crack is easy to detect, even if the crack is large strength in the pipe is not affected, this type of crack will leak before it breaks and is considered less serious than a circumferential crack. A circumferential crack is harder to detect, this type of crack is more of a concern since it is likely to rupture than leak and is generally taken more serious.

What is Microbiological Influenced Corrosion (MIC)? Microbiological influenced corrosion is the attack of metal that is directly associated with microbiological organisms. Attack occurs at the metal-organism contact zone. MIC is mostly prevalent under stagnant or low flow conditions in waters with high nutrient loading. MIC bacteria can be anaerobic or aerobic. MIC bacteria can be sulfate-reducing or iron-metabolizing. MIC often occurs at weld seams, as the heat during welding can make the metal structure more sensitive to attack. MIC remedies include biocides and dispersants.

What is the problem with tube denting? Tube deformation leads to stress in the tube. Denting is associated with high levels of contaminants.

What is a metal? A metal is any class of elements that give up electrons when it creates compounds. Metals usually have a shiny surface and are great conductors of electricity and heat. Metals can be melted or fused and hammered into thin metal sheets or made into wire.

What are alloys? An alloy is a metal that is made by mixing two or more elements, one of them being a metal melted together. Alloys are used to create a desirable metal such as hardness, toughness, reduced weight, thermal conductivity and strength. Ex. Copper and tin make bronze or copper and zinc make brass.

What is the difference between magnetite and hematite? Magnetite is a black iron oxide and is a protective coating for the inside of pipes and hematite is a red iron oxide which indicates the corrosive effects of oxidation of the inside of pipe. Hematite forms at lower temperatures in the presence of oxygen and flow. Hematite film if tightly adherent, it can protect piping surface from further corrosion. At very low oxygen levels hematite is not stable and therefore the corrosion rate increases. Magnetite forms under high pH and reducing (low oxygen) conditions.

What about fuel cladding corrosion? Fuel cladding is Zircalloy, a highly corrosion resistant zirconium alloy. It will oxidize to a limited extent from the chemical reaction with water. The oxide layer can cause the cladding to thin and absorb H_2 which can lead to clad embrittlement which could lead to clad failure. Nuclear fuel cores are continuously upgraded to be more efficient which leads to new and different corrosion difficulties.

How is iron and copper affected by pH? If pH increases, iron corrosion rate decreases. This poses a problem for many power plants that may have both iron and copper alloys in their systems. If pH was optimized for iron corrosion, then copper corrosion was excessive. If pH was optimized for copper then iron corrosion was elevated.

What are non-metals? A non-metal is an element lacking both the physical and chemical properties of a metal. These non-metals are able to form anions, acidic oxides, acids and stable compounds with hydrogen.

What is corrosion? Corrosion is the electrochemical reaction that causes destructive alteration or loss of metal.

What are some types of corrosion? Galvanic corrosion, stress corrosion cracking, crevice corrosion, pitting corrosion, hydrogen embrittlement, boric acid corrosion and flow assisted corrosion.

What is crevice corrosion? Crevice corrosion is a corrosion occurring in spaces to which the access of the working fluid from the environment is limited. These spaces are generally called crevices. Examples of crevices are gaps and contact areas between parts, under gaskets or seals, inside cracks and seams, spaces filled with deposits and under sludge piles.

The susceptibility to crevice corrosion varies widely from one material-environment system to another. In general, crevice corrosion is of greatest concern for materials which are normally passive metals, like stainless steel or aluminum. Crevice corrosion tends to be of greatest significance to components built of highly corrosion-resistant superalloys and operating with the purest-available water chemistry. For example, steam generators in nuclear power plants degrade largely by crevice corrosion.

Crevice corrosion is extremely dangerous because it is localized and can lead to component failure while the overall material loss is minimal. The initiation and progress of crevice corrosion can be difficult to detect.

What is flow accelerated corrosion (FAC)? Flow accelerated corrosion occurs where abrasion increases the chemical corrosion. This occurs readily where there is high velocity and two phase flow. The way to correct flow assisted corrosion is basically to replace the piping with a more resistant material. Increasing pH in the liquid phase will decrease the corrosion rate.

What are passive films? A reactive metal may form a tightly adherent protective oxide layer that will protect the underlying metal. Example: aluminum is reactive but under the right conditions a protective oxide film will form to protect the underlying or base metal. On cast iron or mild steels, magnetite or hematite can form. If these are tightly adherent, they can protect base metal. Ferric hydroxide is not a protective film.

What are some facts about corrosion? All metals and alloys are susceptible to corrosion. Corrosion is a natural process and is thermodynamically driven.

What are four requirements for corrosion to occur? Anode, Cathode, Electroyte, Electronic conduction path.

What factors affect the corrosion of steel? The type of material can determine how fast corrosion will affect steel. E.g. Chromium or Nickel can reduce corrosion rate. The pH strength will determine how fast corrosion will occur. The lower the pH, the faster the corrosion rate. The amount of oxygen controls the cathodic and overall reaction rate, but can also facilitate formation of protective films. If the temperature increases, the corrosion rate will increase. And fluid velocity, especially high flow rates can strip protective film layers, exposing bare metal.

What is the purpose of nitrogen blanketing? Nitrogen blanketing prevents air contact with layup solutions or metal surfaces. This blanketing is effective but can pose other safety hazards, like confined space entries.

How does oxygen affect corrosion? Oxygen can have either a beneficial or a detrimental effect on corrosion rates. If the presence of oxygen stabilizes oxide films, then the effect can be positive. Oxygen or oxidants can raise the electrochemical potential (corrosion potential) and greatly increase stress corrosion cracking.

What is a covalent bond? The term covalent bond is used to describe the bonds in compounds that result from the **sharing** of one or more pairs of electrons. E.g. methane

What is an ionic bond? An ionic bond is when one or more electrons from one atom are removed and attached **(transferred)** to another atom, resulting in positive and negative ions which attract each other. E.g. table salt

What is a diatomic molecule? A molecule that contains two atoms.

What is localized attack? Localized attack is a form of corrosion that takes place only at specific areas on a material.

What is pitting? Pitting is a form of localized attack where pits are formed on a material's surface.

What is selective corrosion? Selective corrosion is a type of corrosion that affects a specific type of material and where all other materials are not affected by corrosion.

What is scale? Scale is a buildup of impurities on a material's surface. Scale can cause a reduction in heat transfer, effecting heat transfer thus effecting heat transfer. Secondly, it increases the risk of material failure since there is not enough heat transfer from the surface causing the material to overheat increasing the risk of failure.

What is the Rockwell number? A scale created to represent the hardness of metal. Under standard loading conditions a diamond cone is pressed into the metal. The Rockwell number is related to the depth of penetration into the material. Invented by Hugh and Stanley Rockwell in 1908.

What is elasticity? A property of a material being deformed that causes it to take on its orginial shape again after the stress is removed.

What is fatigue strength? For a metal, the maximum stress that it can tolerate without weakening or snapping.

What is a metalloid? A metalloid is an element which possesses certain properties characteristics of metals but are generally classified as nonmetals.

What is the difference between a metal and a non-metal? A metal is a element which exhibits metallic properties such as high thermal and electrical conductivity, luster, malleability, opaqueness, high strength, hardness and ductility. A non-metal is an element which exhibits the properties of low thermal, low electrical conductivity, low strength, brittleness and dullness.

What are metallic bonds? Metallic bonds occur when a group of atoms achieve a more stable electron configuration by sharing the electrons in their valence shells with many atoms. Metallic bonds exhibit the following characteristics: easily deform, high thermal conductivity and high electrical conductivity. E.g. Sodium

What is stress? The measure of the force per cross section area exerted on a material.
Stress = Force / Area   measured in psi, ksi

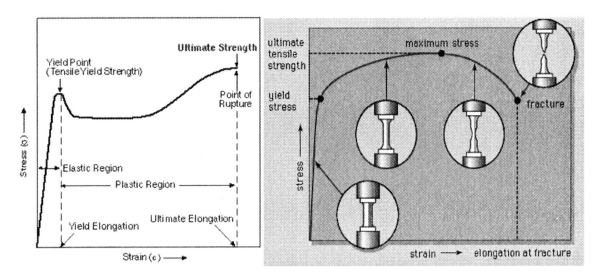

Courtesy of Britannica Encyclopedia

What is a Van der Waals bond? A Van der Waals bond develops between atoms with filled electron shells. Atoms attain symmetry in electron charge (dipole) distribution. An attractive force is established between the dipole in adjacent atoms. It is the weakest of all atomic bonds. Characteristics of this type of bond are low melting point and it is easily compressed.

What is strain? The measure of the fractional deformation of a material due to an applied stress.
Strain = Change in length / Original length

What is tensile stress? Exerted when a force acts perpendicular to a cross-sectional area, tending to stretch or elongate the material. Sign convention: Positive

What is compressive stress? Exerted when a force acts perpendicular to a cross-sectional area, tending to squeeze the material. Sign convention: Negative

What is shear stress? Exerted when a force acts parallel to a cross-sectional area, tending to slide one section of the material past another.

What is residual stress? Caused by manufacturing process which leaves stresses in the material. (rolling, machining, welding, etc)

What is reaction stress? Also called structure stress. Caused by the weight of the material which is supported.

What is pressure stress? Stress induced in vessels or piping due to contents under pressure.

What is flow stress? Stress which occurs when a mass of flowing fluid induces a dynamic pressure on a conduit wall.

What is thermal stress? Stress induced by temperature gradients in a material or when a constrained material is heated or cooled.

What is fatigue stress? Stress induced due to cyclic loading of a material.

What is the difference between elastic strain and plastic strain? Elastic strain is a transitory dimensional change which exist only when an initiating stress is present. The change disappears immediately upon the removal of the stress also called elastic deformation. Plastic strain is a permanent dimensional change induced in a material due to the application of a stress, also called permanent strain or plastic deformation.

What is Modulus of Elasticity? The numerical relationship between stress and strain at any point within the elastic range.

What is Modulus of Rigidity? Is defined as the relationship between shear stress and shear strain in the linear elastic region of the stress-strain curve.

What is yield point? It is the point at which there is an increase in strain with no increase in stress.

What is yield stress? Is the amount of stress applied that, when removed, leaves a permanent deformation.

What is the purpose of tempering metal? Tempering is a process of heating metal to a specific temperature for a given period of time. This improves the metal's toughness with minimal loss of hardness and strength.

What is the purpose of quenching? Quenching is a process of rapidly cooling hot metal to control grain size and pattern. Grain boundaries arrest dislocation movement. The faster the material is cooled, the smaller the grain size. Phase composition and distribution is also affected. Hardness and tensile strength increase while toughness and ductility decrease.

What is the difference between austenitic, martensitic and ferritic steels?

Austenitic steel – good for high temperature applications, cannot be hardened by heat treatment, better weldability.

Martensitic steel – magnetic, very hard and strong and reduced corrosion resistance.

Ferritic steel – relatively soft (very ductile), easily machined, used for common industrial uses, good corrosion resistance.

What is cement? Cement is a material that has adhesive and cohesive properties enabling it to bond mineral fragments into a solid mass. Cement is made up of lime, iron, silica and alumina (LISA).

What is compressive strength? The strength of an object subjected to a compressive load. E.g. concrete.

What is tensile strength? The strength of an object subjected to a tensile load. E.g. rebar

What is heat of hydration? The heat generated when cement and water react.

How many types of cement are there? There are numerous types of cement, the ones listed are the more popular ones.

> Type l – normal
> Type ll – moderate sulfate resistance
> Type lll – high early strength
> Type lV – low heat of hydration
> Type V – high sulfate resistance

How does concrete cure? Concrete cures by formation of a crystalline structure which binds aggregate and sand together – typically 28 days for full cure. Concrete does not harden or cure by drying. Concrete needs moisture to hydrate and harden. When concrete dries it ceases to gain strength. Concrete must be kept moist until it has undergone sufficient hydration.

What is the main factor in curing concrete? The water – cement ratio is the main factor for determining concrete strength. The lower the water – cement ratio the higher the strength.

Why is it not good to paint concrete? Painting concrete with an oil-based alkyd will cause the oil in the paint to saponify (convert to soap) and the paint will detach and peel off.

What is the pH of concrete? Concrete is very alkaline in nature with a pH between 12 and 13 due to its lime content.

What determines the porosity of concrete? The porosity of concrete is dependent on a number of factors, including cure temperatures, mineral composition, chemicals added, water/cement ratio and curing time. Adding fly ash, fumed silica or slag helps to decrease the porosity of concrete.

Why is rebar added to a concrete structure? Concrete is a strong substance when under compressive loads but becomes weak when under tensile loads. Rebar is a strong material when under tensile loads and under compressive loads. By combining rebar to concrete a strong, inexpensive structure can be built that will support compressive and tensile loads.

If rebar can handle compressive and tension loads why isn't used by itself instead of mixing it with concrete? Using all steel construction greatly increases the cost. The shape considerations is another factor since concrete is easier to form than all steel. Concrete requires lower maintenance than steel and has a better fire rating.

What is fracture mechanics? Area of mechanics used to predict interaction between applied loads and material response in the presence of flaws.

How steel cracks up? The structural steels used in reactor pressure vessels exhibit a characteristic known as ductile-brittle transition. At relatively high temperatures the material is ductile: it retains its strength when stresses alter its shape, and a large amount of energy is needed to break it. At low temperatures the material becomes brittle, and relatively little energy is required to break it.

A common way of measuring this property is the Charpy impact test. A weighted pendulum breaks a standard notched specimen, fixed at both ends. The follow-through of the pendulum allows the energy absorbed in breaking the specimen to be measured. Specimens are tested over a range of temperatures, and the temperature at which the energy absorbed is 40 joules is taken as a reference point.

The temperature range over which the energy needed to break the specimen increases is known as the transition region. In this region the material can show brittle and ductile behavior. The region at higher temperature where the energy is high and relatively constant is known as the upper shelf. Normally the 40-joule temperature is below the freezing point of water and is only a concern in arctic conditions in which vehicle parts can snap.

The effect of radiation is to shift the transition from ductile to brittle behavior to a higher temperature. The degree of shift depends on the accumulated displacement dose, the material composition (particularly the levels of copper and nickel) and to a lesser extent the rate at which the damage accumulates. In addition, radiation tends to lengthen the transition region and reduce the upper shelf energy.

A radiation-hardened material no longer deforms at high temperatures in response to loads applied to it, which would give some warning of incipient failure. Instead, the loads cause undetected cracks to grow in the brittle material, which means failure is unexpected.

Hardening of the material arises from two sources. The first comes from the debris left behind from the displacement damage process - the formation of clusters of defects, both vacancies and interstitials. The second comes from tiny precipitates of copper in the steel. Radiation damage hastens the precipitation and then, through interactions with the cascades of defects, keeps the precipitate small, preserving them as effective obstacles. The clusters of point defects continue to accumulate as the displacement dose increases, but the rate of accumulation decreases with time.

Reactors

What is a thermal neutron reactor?

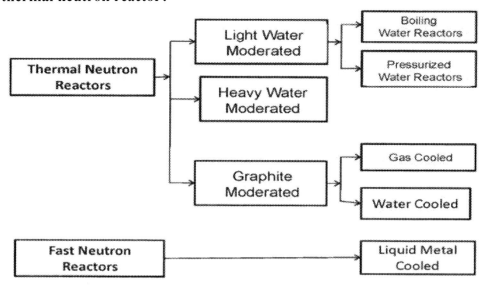

When and where was the first man-made nuclear reactor created? On 2 December 1942, another team led by Enrico Fermi was able to initiate the first artificial nuclear chain reaction. Working in a lab below the stands of Stagg Field at the University of Chicago, the team created the conditions needed for such a reaction by piling together 400 tons of graphite, 58 tons of uranium oxide, and six tons of uranium metal. Later researchers found that such a chain reaction could either be controlled to produce usable energy or could be allowed to go out of control to produce an explosion more violent than anything possible using chemical explosives.

Why was CP-1, 2 & 3 known as the Chicago Pile? 'CP' denotes 'Chicago Pile'. It was called a "pile" because it was, literally, a pile of bricks of uranium, uranium oxide, and graphite. The term "pile" was a neutral term that would not betray the actual nature of the structure if outsiders heard of it. After the war, "atomic pile" continued to be used for a short while and then gave way to much more appropriate term, nuclear reactor. Asimov, I.

What did the CP-1 look like? When CP-1 was created, it was 30 feet wide, 32 feet long and 21 ½ feet high. It weighed 1400 tons, of which 52 tons of it was uranium. The uranium, uranium oxide, and graphite were arranged in alternate layers with, here and there, holes into which long rods of cadmium could be fitted. Asimov, I.

What did the telegraph say to Washington after CP-1 became self-sustaining? News of the success was announced to Washington by a cautious telegram reading: "The Italian navigator (Enrico Fermi) has entered the new world". There came a questioning wire in return: "How were the natives?" The answer was sent off at once: "Very friendly". Asimov, I.

When did the first reactors take place?

Date	Reactor	Place	Power	First
1942	CP-1 Chicago Pile 1	Chicago, Ill	low	first reactor
1943	Oak Ridge X-10	Oak Ridge, Tn	3.8 MW	first megawatt range reactor
1944	Y-Boiler	Los Alamos, NM	low	first enriched fuel reactor
1944	CP-3 Chicago Pile 3	Chicago, Ill	300 KW	first heavy water reactor
1945	ZEEP	Chalk River, Ont	low	first Canadian reactor
1945	Hanford	Richland, Wash	>100 MW	first high power reactor
1946	Clementine	Los Alamos, NM	25 KW	first fast neutron reactor
1947	NRX	Chalk River, Ont	42 MW	first high flux reactor
1947	GLEEP	Harwell, England	low	first British reactor
1948	ZOE	Chatillon, France	150 KW	first French reactor
1950	LITR	Oak Ridge, Tn	3 MW	first plate fuel reactor
1951	EBR-1	Idaho Falls, Idaho	1.4 MW	first breeder reactor
1951	JEEP 1	Kjeller, Norway	350 KW	first international reactor
1954	APS-1	Obinsk, Russia	5 MW	first Russian reactor
1954	PWR	USS Nautilus	1 MW	First propulsion reactor
1955	BORAX-III	Idaho Falls, Idaho	3.5 MW	first U.S. reactor capable of significant electric power
1956	Calder Hall A	Calder Hall, England	20 MW	world's first large scale commercial reactor for power production
1956	EBWR	Argonne, Ill	5 MW	First boiling water reactor
1959	Dresden 1	Morris, Ill		First commercial boiling water reactor in U.S.
1974	Phenix	France	250 MW	First commercial fast breeder reactor

What are the generations for nuclear reactors?

Generation 1 – one of a kind like Fermi 1

Generation 2 – include PWR, CANDU, BWR and AGR

Generation 3 – incorporated evolutionary improvements in design which have been developed during the lifetime of the generation 2 reactor design, such as improved fuel technology, passive safety systems and standardized design.

Generation 4 – set of theoretical designs currently being researched and are not expected to be available for commercial construction before 2030. The primary goals being to improve nuclear safety, improve proliferation resistance, minimize waste and natural resource utilization and to decrease the cost to build and operate such plants.

Generation 5 – designs which are theoretically possible, but which are not being actively considered or researched at present. Such reactors could be built with current or near term technology, they trigger little interest for reasons of economics, practicality, or safety.

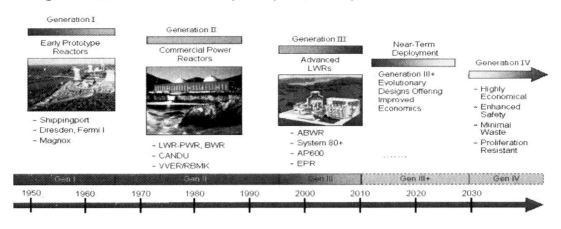

Courtesy of NRC

What is the oldest boiling water reactor still in service? Oyster Creek Nuclear Power Station is the oldest operating nuclear power plant in the U.S. producing 610 MWe. Nine Mile Point nuclear station unit 1 is the second oldest BWR in the U.S. producing 621 MWe.

What are the advantages of a pressurized water reactor?

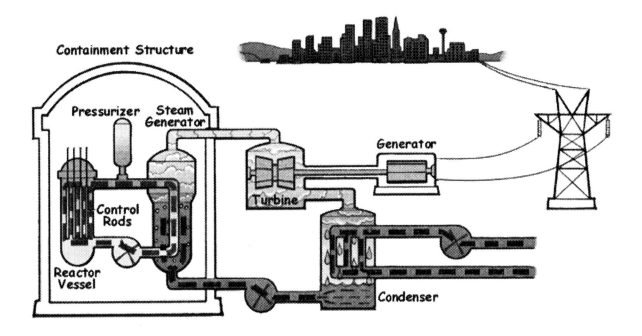

Advantages

- PWR reactors are very stable due to their tendency to produce less power as temperatures increase; this makes the reactor easier to operate from a stability standpoint.

- PWR reactors can be operated with a core containing less fissile material than is required for them to go prompt critical. This significantly reduces the chance that the reactor will run out of control and makes PWR designs relatively safe from criticality accidents.

- Because PWR reactors use enriched uranium as fuel, they can use ordinary water as a moderator rather than the much more expensive heavy water as used in a pressurized heavy water reactor.

- PWR turbine cycle loop is separate from the primary loop, so the water in the secondary loop is not contaminated by radioactive materials.

What are the disadvantages of a pressurized water reactor?

Disadvantages

- The coolant water must be highly pressurized to remain liquid at high temperatures. This requires high strength piping and a heavy pressure vessel and hence increases construction costs. The higher pressure can increase the consequences of a loss of coolant accident.

- Most pressurized water reactors cannot be refueled while operating. This decreases the availability of the reactor- it has to go offline for comparably long periods of time (some weeks).

- The high temperature water coolant with boric acid dissolved in it is corrosive to carbon steel (but not stainless steel), this can cause radioactive corrosion products to circulate in the primary coolant loop. This not only limits the lifetime of the reactor, but the systems that filter out the corrosion products and adjust the boric acid concentration add significantly to the overall cost of the reactor and radiation exposure. Occasionally, this has resulted in severe corrosion to control rod drive mechanisms when the boric acid solution leaked through the seal between the mechanism itself and the primary system.[4][5]

- Natural uranium is only 0.7% Uranium-235, the isotope necessary for thermal reactors. This makes it necessary to enrich the uranium fuel, which increases the costs of fuel production. If heavy water is used it is possible to operate the reactor with natural uranium, but the production of heavy water requires large amounts of energy and is hence expensive.

- Because water acts as a neutron moderator it is not possible to build a fast neutron reactor with a PWR design. A reduced moderation water reactor may however achieve breeding ratio greater than unity, though these have disadvantages of their own.

What are the advantages of a boiling water reactor?

Advantages

- The reactor vessel and associated components operate at a substantially lower pressure (about 75 times atmospheric pressure) compared to a PWR (about 158 times atmospheric pressure).
- Pressure vessel is subject to significantly less irradiation compared to a PWR, and so does not become as brittle with age.
- Operates at a lower nuclear fuel temperature.
- Fewer components due to no steam generators and no pressurizer vessel. (Older BWRs have external recirculation loops, but even this piping is eliminated in modern BWRs, such as the ABWR.)
- Lower risk (probability) of a rupture causing loss of coolant compared to a PWR, and lower risk of a severe accident should such a rupture occur. This is due to fewer pipes, fewer large diameter pipes, fewer welds and no steam generator tubes.
- Measuring the water level in the pressure vessel is the same for both normal and emergency operations, which results in easy and intuitive assessment of emergency conditions.
- Can operate at lower core power density levels using natural circulation without forced flow.
- A BWR may be designed to operate using only natural circulation so that recirculation pumps are eliminated entirely. (The new ESBWR design uses natural circulation.)

- BWRs do not use boric acid to control fission burn-up, leading to less possibility of corrosion within the reactor vessel and piping. (Corrosion from boric acid must be carefully monitored in PWRs; it has been demonstrated that dangerous reactor vessel head corrosion can occur if the reactor vessel head is not properly maintained. See Davis-Besse. Since BWRs do not utilize boric acid, these contingencies are eliminated.)

What are the disadvantages of a boiling water reactor?

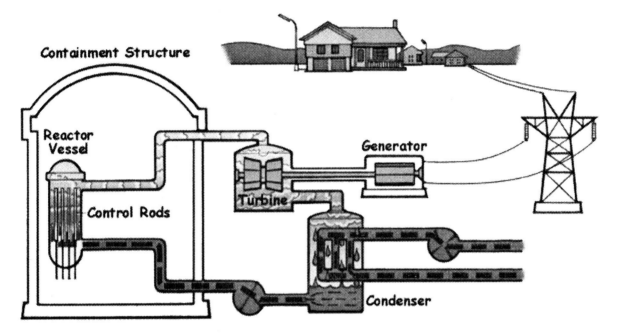

Disadvantages

- Complex calculations for managing consumption of nuclear fuel during operation due to "two phase (water and steam) fluid flow" in the upper part of the core. This requires more instrumentation in the reactor core. The innovation of computers, however, makes this less of an issue.
- Much larger pressure vessel than for a PWR of similar power, with correspondingly higher cost. (However, the overall cost is reduced because a modern BWR has no main steam generators and associated piping.)
- Contamination of the turbine by short-lived activation products. This means that shielding and access control around the steam turbine are required during normal operations due to the radiation levels arising from the steam entering directly from the reactor core.
- Control rods are inserted from below for current BWR designs. There are two available hydraulic power sources that can drive the control rods into the core for a BWR under emergency conditions. There is a dedicated high pressure hydraulic accumulator and also the pressure inside of the reactor pressure vessel available to each control rod. Either the dedicated accumulator (one per rod) or reactor pressure is capable of fully inserting each rod. Most other reactor types use top entry control rods that are held up in the withdrawn

position by electromagnets, causing them to fall into the reactor by gravity if power is lost.

What is meant by adding negative reactivity? Negative reactivity is a term referring to a reduction in reactor reactivity by whatever means.

What is meant by adding positive reactivity? Positive reactivity is a term referring to an increase in reactor reactivity by whatever means.

What is the purpose of the reactor vessel? Supports fuel element assemblies (the core), absorbs dynamic loads, contains reactor coolant, supports control elements, acts as a pressure boundary, directs coolant through the core and contacts reactor coolant pumps.

What is the purpose of the reactor head? The reactor head is removal when the unit needs to be refueled. Depending on the design of the reactor vessel the reactor head may also hold the control rod assemblies for example in a pressurizer water reactor.

What is the purpose of the reactor head vent? The reactor head vent allows (on certain reactor designs) to remove any air trapped in the reactor vessel after the reactor head has been re-installed on the reactor vessel. By removing the air, allows the primary reactor coolant to completely cover the reactor core and upper internal equipment to ensure equal temperature distribution throughout the reactor vessel.

What is the purpose of the in-core instrumentation? The in-core instrumentation gives continuous indication of the reactor core to the control room operators. These indications help the reactor operators to make adjustments that help keep the core reactivity maintained.

Why are nuclear power plants base loaded? Nuclear power plants are essentially base-load generators, running continuously. This is because their power output cannot readily be ramped up and down on a daily and weekly basis, and in this respect they are similar to most coal-fired plants. (It is also uneconomic to run them at less than full capacity, since they are expensive to build but cheap to run.) However, in some situations it is necessary to vary the output according to daily and weekly load cycles on a regular basis, for instance in France, where there is a very high reliance on nuclear power.

The ability of a power reactor to run at less than full power for much of the time depends on whether it is in the early part of its 18 to 24-month refueling cycle or late in it, and whether it is designed with special control rods which diminish power levels throughout the core without shutting it down. Thus, though the ability on any individual reactor to run on a sustained basis at low power decreases markedly as it progresses through the refueling cycle, there is considerable scope for running a fleet of reactors in load-following mode.   www.world-nuclear.org

How is reactivity affected? Through reactivity management plant personnel must know which variables can affect reactivity. Some variables have a direct impact on core reactivity while others may have an indirect impact on core reactivity. Regardless, all of these variables must be accounted for in order to mitigate the likelihood of an event or even an accident.

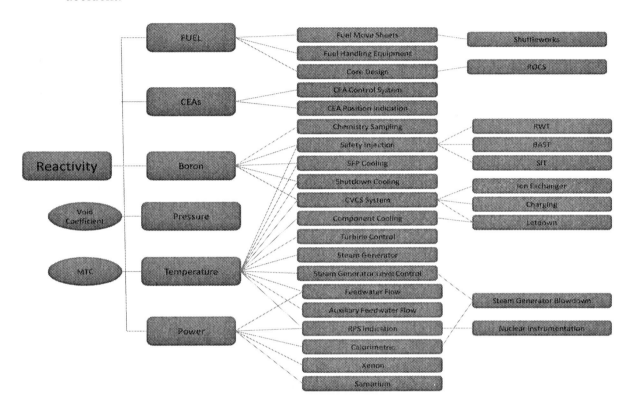

When the core is up graded with different, new type material what are some other considerations need to be taken into account? If a new material is introduced into the core, it is important to determine that the material can handle the environment of a nuclear reactor. Primarily, the material must be able to withstand whatever temperature and pressure it will experience in its location in the core. Also, the material's properties after being irradiated are important. Most materials become brittle after becoming irradiated (i.e. bombarded by neutrons), so it's important to know that the material will be strong enough to support its designed function.

What is the purpose of the reactor vessel flange seals? The reactor vessel flange seals are the seals between the reactor vessel and the reactor vessel head. There are two seals a primary seal and a secondary seal. There is instrumentation that monitors the pressure in-between both of these seals. These seals are continuously monitored at all times, except during a unit outage. During the outage the reactor vessel head is removed to perform refueling and component inspections. If the primary seal leaks, the temperature and pressure indicator will notify the control room. This informs the control room that the primary seal has failed and the unit must be shut down.

At what elevation does the reactor vessel sit? The reactor vessel sits at the lowest level in the primary system. This setup is designed in such a way that in the event of a LOCA, all the primary fluid would head in the direction of the lowest point…being the reactor vessel. This should help maintain fluid in the reactor vessel in order to keep the core covered, contained and cooled.

What is the thimble support plate? Guides instruments from nozzles (on the head) to the upper guide structure or into the fuel.

What do the extension shafts do for the control rods? Connect control element assembly (in core) to its drive mechanism (on the head).

What does the fuel alignment plate do? Aligns the top portion of the fuel assembly inside the reactor vessel.

The core support assembly does what? Supports weight of core and transmits load to bottom flange of core support barrel.

What is the purpose of the flow skirt? Reduce inequalities of core inlet flow distributions and prevent formation of vortices in the lower plenum.

What does the core stops do? The reactor core stops prevent vertical motion of the core. They are located at the bottom of the reactor vessel and is normally not in contact with the core support barrel. They are designed to support the entire weight of the reactor core.

What does the reactor coolant system do? Removes heat generated in the core, controls reactivity, controls pressure using the pressurizer. Removes heat from the nuclear reactor core and transfer it to the steam generators. A barrier is to prevent release of fission products into the environment.

What is the purpose of the chemical volume and control system (CVCS)? The CVCS conditions, purifies, and supplements the volume of the reactor coolant system by receiving a portion of the RCS flow, passing it through filters and ion exchangers, adding chemicals as required, and receiving or discharging water to the RCS from the volume control tank depending on pressurizer level. The CVCS also furnishes water for the auxiliary pressurizer spray.

What is the purpose of the safety injection tanks? The safety injection tanks are located at the discharge of each reactor coolant pump. They are designed to hold a limited amount of borated water at primary system pressure and is released into the reactor vessel in the event of a LOCA. A safety measure designed to reduce the chance that the reactor core would become exposed, causing major reactivity issues and temperature problems leading to worse conditions. By covering the reactor core at all times, the safety of the plant and public is protected.

What fills the safety injection tanks? The high pressure safety injection pumps supply borated water to the safety injection tanks in order to maintain their level.

Where do the safety injection tanks discharge to? The safety injection tanks are attached to the primary system at the reactor coolant pump discharge piping leading into the reactor vessel. This setup will allow the safety injection tanks to discharge directly into the reactor core providing immediate coverage to the reactor core.

Are there limits to the safety injection tanks in the amount of borated water should be in the fluid? The borated water within the safety injection tanks should be maintained between 2300 ppm and 2700 ppm in order to ensure that the concentration would be affective in the event of a LOCA. *CAUTION:* Boric acid, in concentrations could precipitate out of solution when the solution temperature is too low. If the boron does precipitate it will form rock like crystal formations.

How is the recirculation ratio affect by reactor power output? For pressurized water reactor there is a recirculation ratio in the steam generator. This ratio is dependent on the steam generator load to the turbine. At very low loads the recirculation ratio is extremely high in the steam generator since the flow of steam to the turbine is very low. As steam load to the turbine increases, the amount of steam flow from the steam generator increases, therefore, the amount of recirculation ratio continues to decrease until the unit (steam turbine) reaches 100%. (e.g. 40% ratio at 5% load, 4% ratio at 100% load).

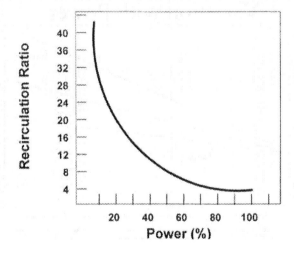

Variation of Recirculation Ratio with Power

How much heat can the reactor coolant pumps provide to the reactor coolant? A reactor coolant pump can provide up to 13 MWt (thermal) to heat the reactor coolant system from a cold shutdown up to over 5000F prior to reactor startup.

What is the purpose of the reactor coolant pumps? Provide forced circulation of the reactor coolant water, transferring heat from the reactor core to the steam generator and back again into the reactor core to be reheated.

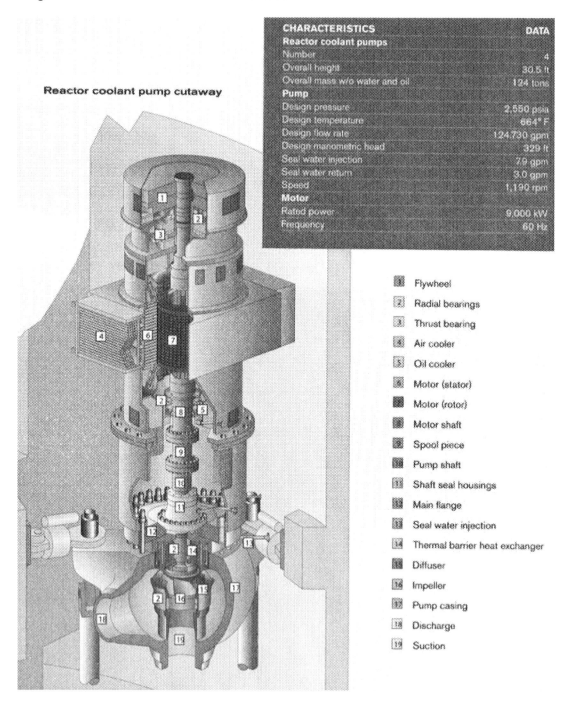

CHARACTERISTICS	DATA
Reactor coolant pumps	
Number	4
Overall height	30.5 ft
Overall mass w/o water and oil	124 tons
Pump	
Design pressure	2,550 psia
Design temperature	664° F
Design flow rate	124,730 gpm
Design manometric head	329 ft
Seal water injection	7.9 gpm
Seal water return	3.0 gpm
Speed	1,190 rpm
Motor	
Rated power	9,000 kW
Frequency	60 Hz

Reactor coolant pump cutaway

1 Flywheel
2 Radial bearings
3 Thrust bearing
4 Air cooler
5 Oil cooler
6 Motor (stator)
7 Motor (rotor)
8 Motor shaft
9 Spool piece
10 Pump shaft
11 Shaft seal housings
12 Main flange
13 Seal water injection
14 Thermal barrier heat exchanger
15 Diffuser
16 Impeller
17 Pump casing
18 Discharge
19 Suction

Courtesy of AREVA

How much pressure does the reactor coolant pumps add to the fluid flow in the primary system? The reactor coolant pumps add approximately 35 psi to the reactor coolant through the reactor coolant pump discharge piping.

How can you verify that there is natural circulation in the primary system? The driving force for natural circulation is a differential temperature between the primary system and the secondary system.

- The steaming rate affects RCS temperatures
- T_{HOT} is constant or decreasing
- T_{COLD} is constant or decreasing
- Loop Delta T between 10^0F and 50^0F
- Core exit thermocouple temperature consistent with T_{HOT}

What is the function of the reactor coolant system leakage detection? To determine if there might be a pressure boundary leak. There are limits to how much leakage a unit can have and remain in-service. E.g.

- 1 GPM unidentified leakage (through-wall leak)
- 10 GPM identified leakage
- 100 GPM gallons per day primary to secondary side leakage through any one steam generator.

What is the purpose of the shutdown heat exchangers (SDC HX)? Designed to cool reactor coolant when shutting down the reactor. Even when the reactor is shutdown, there is always decay heat being generated by the reactor fuel. If this heat is not removed, it would continue to build heat and overtime could eventually lead to a meltdown.

What is the difference between a homogeneous and heterogeneous reactor? A homogeneous reactor the core materials are distributed in such a manner that the neutron characteristics can be accurately described by the assumption of homogeneous distribution of materials throughout the core. A heterogeneous reactor the core materials are segregated to such an extent that the neutron characteristics cannot be accurately described by the assumption of homogeneous distributions of materials throughout the core.

What does a BWR reactor look like?

GEHitachi
NuclearEnergy

Advanced Boiling Water Reactor Assembly

1. Vessel Flange and Closure Head
2. Vent and Head Spray
3. Steam Outlet Flow Restrictor
4. RPV Stabilizer
5. Feedwater Nozzle
6. Forged Shell Rings
7. Vessel Support Skirt
8. Vessel Bottom Head
9. RIP Penetrations
10. Thermal Insulation
11. Core Shroud
12. Core Plate
13. Top Guide
14. Fuel Supports
15. Control Rod Drive Housings
16. Control Rod Guide Tubes
17. In Core Housing
18. In-Core Instrument Guide Tubes and Stabilizers

19. Feedwater Sparger
20. High Pressure Core Flooder (HPCF) Sparger
21. HPCF Coupling
22. Low Pressure Flooder (LPFL)
23. Shutdown Cooling Outlet
24. Steam Separators
25. Steam Dryer
26. Reactor Internal Pumps (RIP)
27. RIP Motor Casing
28. Core and RIP Differential Pressure Line
29. Fine Motion Control Rod Drives
30. Fuel Assemblies
31. Control Rods
32. Local Power Range Monitor

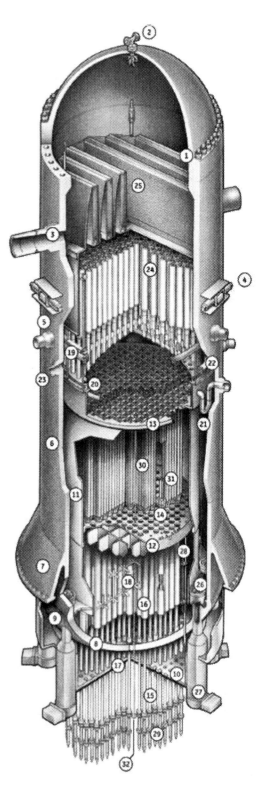

What are the internal structures of a pressurized water reactor vessel?

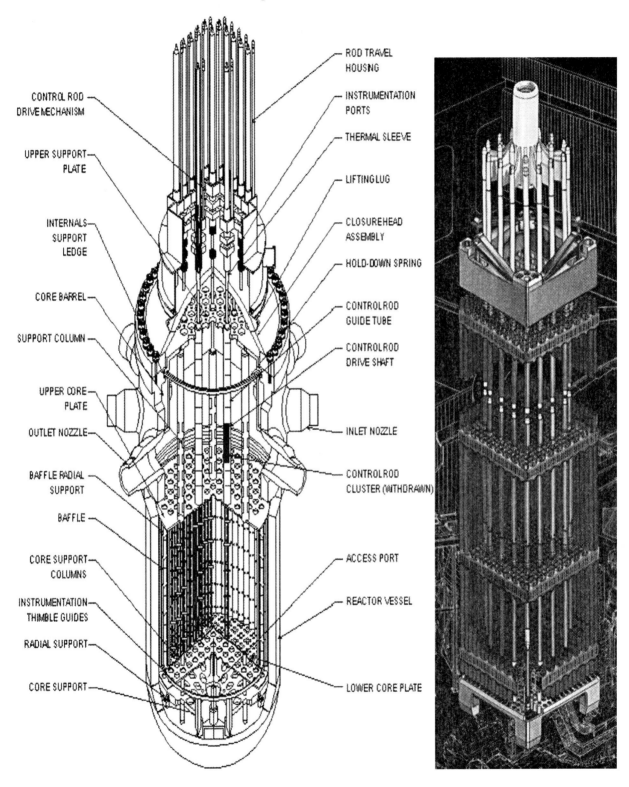

CONTROL ROD DRIVE MECHANISM

UPPER SUPPORT PLATE

INTERNALS SUPPORT LEDGE

CORE BARREL

SUPPORT COLUMN

UPPER CORE PLATE

OUTLET NOZZLE

BAFFLE RADIAL SUPPORT

BAFFLE

CORE SUPPORT COLUMNS

INSTRUMENTATION THIMBLE GUIDES

RADIAL SUPPORT

CORE SUPPORT

ROD TRAVEL HOUSING

INSTRUMENTATION PORTS

THERMAL SLEEVE

LIFTING LUG

CLOSURE HEAD ASSEMBLY

HOLD-DOWN SPRING

CONTROL ROD GUIDE TUBE

CONTROL ROD DRIVE SHAFT

INLET NOZZLE

CONTROL ROD CLUSTER (WITHDRAWN)

ACCESS PORT

REACTOR VESSEL

LOWER CORE PLATE

What does a PWR and a BWR look like? The pictures below give you an indication of how the systems are tied together.

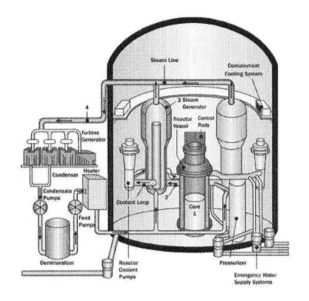 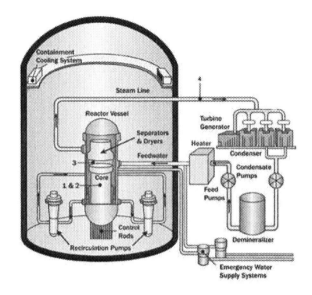

Pressurized Water Reactor (PWR)   Boiling Water Reactor (BWR)

How are reactors classified? Reactors may be classified in a number of ways: the use to which the neutrons produced by fission are put, the energy spectrum of the neutron population, the degree of conversion of fertile material, the dispersion of the core materials, and by the types of materials selected for fuel, moderator, cladding, and control. Here is a list of rector components and their functions:

Component	Material	Function
Fuel	U^{233}, U^{235}, Pu^{239}, Pu^{241}	Fission reaction
Moderator	Light water, heavy water, carbon, beryllium	To reduce energy of fast neutrons to thermal neutrons
Coolant	Light water, heavy water, air CO_2, He, sodium, bismuth, sodium, potassium, organic	To remove heat
Reflector	Same as moderator	To minimize neutron leakage
Shielding	Concrete, water, steel, lead, Polyethylene	To provide protection from radiation
Control Rods	Cadmium, boron, hafnium	To control neutron production rate

Structure	aluminum, steel, zirconium,	To provide physical support of reactor structure and components, containment of fuel elements.

What is a fast breeder reactor? A fast breeder reactor (FBR) is a fast neutron reactor designed to breed fuel by producing more fissile material than it consumes.

Liquid Metal cooled Fast Breeder Reactors (LMFBR)

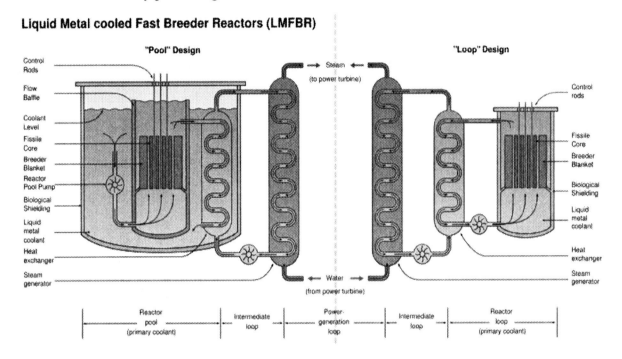

What is a fast reactor? A category of nuclear reactors in which the fission chain reaction is sustained by fast neutrons. Such a reactor needs no neutron moderator, but must use fuel that is relatively rich in fissile material when compared to that required for a thermal reactor.

On average, more neutrons per fission are produced from fissions caused by fast neutrons than from those caused by thermal neutrons. Therefore, there is a much larger excess of neutrons not required to sustain the chain reaction. These neutrons can be used to produce extra fuel, or to transmute long half-life waste to less troublesome isotopes, such as the Phenix reactor near Cadarache in France, or some can be used for each purpose. Through conventional thermal reactors also produce excess neutrons, fast reactors can produce enough of them to breed more fuel than they consume. Such designs are known as fast breeder reactors. Fast neutrons also have an advantage in the transmutation of nuclear waste. The reason for this is that the ratio between the fission cross-section and the absorption cross-section in plutonium and minor actinides are higher in a fast spectrum.

What is a Magnox reactor? A now obsolete type of nuclear reactor which was designed as a producer of plutonium for nuclear weapons. The name magnox comes from the alloy used to clad the fuel rods inside the reactor. Magnox is also the name of an alloy mainly composed of magnesium with small amounts of aluminum and other metals which is used in cladding unenriched uranium metal fuel with a non-oxidizing covering to contain fission products. Magnox is short for Magnesium non-oxidizing. This material has the advantage of a low neutron capture cross-section but is limited in temperature and that it reacts with water. The first Magnox reactor was in Calder Hall, England which operated for 47 years.

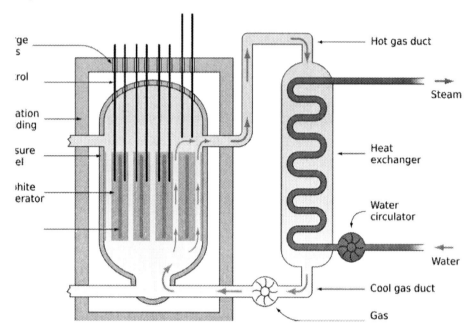

What is an advanced gas-cooled reactor? These are the second Generation of British gas-cooled reactors, using graphite as the neutron moderator and carbon dioxide as coolant. The AGR was developed from the Magnox reactor, operating at a higher gas temperature for improved efficiency, and using enriched uranium fuel so requiring less frequent refueling. The first proto-type AGR became operational in 1962 but the first commercial AGR did not come on-line until 1976.

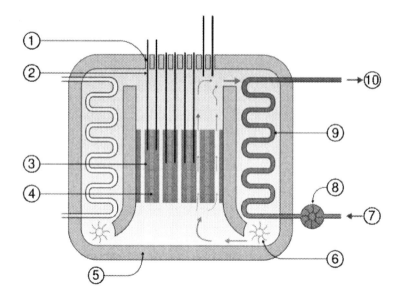

Schematic diagram of an Advanced Gas-cooled Reactor type nuclear reactor:

1. Charge tubes
2. Control rods
3. Graphite moderator
4. Fuel assemblies
5. Concrete pressure vessel and radiation shielding
6. Gas circulator
7. Water
8. Water circulator
9. Heat exchanger
10. Steam

What is a gas-cooled fast reactor? A reactor design which is currently in development. Classified as a Generation IV reactor, it features a fast neutron spectrum and closed fuel cycle for efficient conversion of fertile uranium and management of actinides. The reference reactor design is a helium cooled system operating with an outlet temperature of 8500C using a direct Brayton cycle gas turbine for high thermal efficiency. Several fuel forms are being considered for their potential to operate at very high temperatures and to ensure an excellent retention of fission products: composite ceramic fuel, advanced fuel particles, or ceramic clad elements of actinide compounds. Core configurations are being considered based on pin or plate-based fuel assemblies or prismatic blocks.

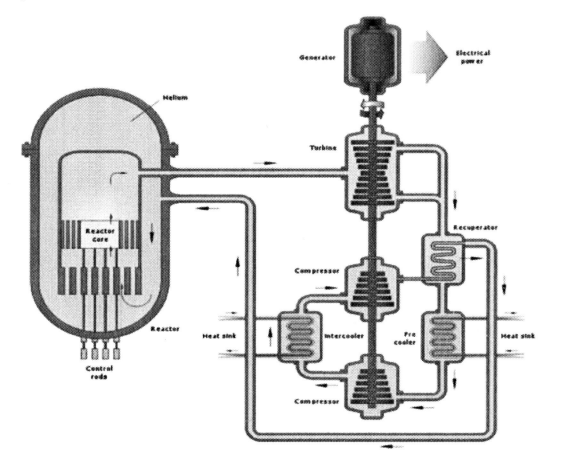

What is a molten salt reactor? A molten salt reactor (MSR) is a type of nuclear reactor where the primary coolant is a molten salt. There have been many designs put forward for use of this type of reactor as a nuclear power plant and a few prototypes built. The concept is one of those proposed for development as a generation IV reactor.

In many designs the nuclear fuel is dissolved in the molten fluoride salt coolant as uranium tetrafluoride (UF_4). The fluid becomes critical in a graphite core which serves as the moderator. Many modern designs rely on ceramic fuel dispersed in a graphite matrix, with the molten salt providing low pressure, high temperature cooling.

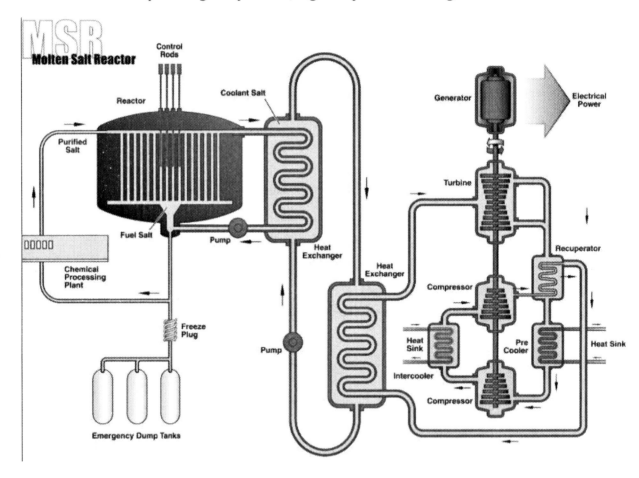

What is a liquid metal cooled reactor? All liquid metal cooled reactors are fast neutron reactors, and to date most fast neutron reactors have been liquid metal cooled fast breeder reactors (LMFBRs), or naval propulsion units. The liquid metals used typically need good heat transfer characteristics. Fast neutron reactor cores tend to generate a lot of heat in a small space when compared to reactors of other classes. A low neutron absorption is desirable in any reactor coolant, but especially important for a fast reactor, as the good neutron economy of a fast reactor is one of its main advantages. Since slower neutrons are more easily absorbed, the coolant should ideally have a low moderation of neutrons. It is also important that the coolant does not cause excessive corrosion of the structural materials, and that its melting and boiling points are suitable for the reactor's operating temperature.

Ideally the coolant should never boil as that would make it more likely to leak out of the system, resulting in a loss of coolant accident. Conversely, if the coolant can be prevented from boiling this allows the pressure in the cooling system to remain at neutral levels, and this dramatically reduces the probability of an accident. Some designs immerse the entire reactor and heat exchangers into a pool of coolant, virtually eliminating the risk that cooling will be lost.

Liquid metal cooled reactors were first adapted for nuclear submarine use but have also been extensively studied for power generation applications. They have safety advantages because the reactor doesn't need to be kept under pressure, and they allow a much higher power density than traditional coolants. Disadvantages include difficulties associated with inspection and repair of a reactor immersed in opaque molten metal, and depending on the choice of metal, corrosion and/or production of radioactive activation products may be an issue.

What is a lead cooled fast reactor? The lead-cooled fast reactor is a nuclear power Generation IV reactor that features a fast neutron spectrum, molten lead or lead-bismuth eutectic coolant, and a closed fuel cycle. Options include a range of plant ratings, including a number of 50 to 150 MWe (megawatts electric) units featuring long-life, pre-manufactured cores. Plans include modular arrangements rated at 300 to 400 MW, and a large monolithic plant rated at 1,200 MW. The fuel is metal or nitride-based containing fertile uranium and transuranics. The LFR is cooled by natural convection with a reactor outlet coolant temperature of 550 °C, possibly ranging over 800 °C with advanced materials. Temperatures higher than 830 °C are high enough to support thermo-chemical production of hydrogen.

The LFR battery is a small factory-built turnkey plant operating on a closed fuel cycle with very long refueling interval (15 to 20 years) cassette cores or replaceable reactor modules. Its features are designed to meet market opportunities for electricity production on small grids, and for developing countries that may not wish to deploy an indigenous fuel cycle infrastructure to support their nuclear energy systems. The modular "battery" system (ie consisting of a number of identical elements, not "battery" in the sense of an electro-chemical energy storage system), is designed for distributed generation of electricity and other energy products, including hydrogen and potable water.

What is a sodium-cooled fast reactor? The Sodium-cooled fast reactor or SFR is a Generation IV reactor project to design an advanced fast neutron reactor.

It builds on two closely related existing projects, the LMFBR and the Integral Fast Reactor, with the objective of producing a fast-spectrum, sodium-cooled reactor and a closed fuel cycle for efficient management of actinides and conversion of fertile uranium.

The operating temperature should not exceed the melting temperature of the fuel. It has been found that the melting point of a fuel called SFR-MOX (20% tranuranic oxides and 80% uranium oxide). Fuel-to-cladding chemical interaction (FCCI) has to be designed against. FCCI is eutectic melting between the fuel and the cladding; uranium, plutonium, and lanthanum (a fission product) inter-diffuse with the iron of the cladding. The alloy that forms has a low eutectic melting temperature. FCCI causes the cladding to reduce in strength and could eventually rupture. The amount of tranuranic transmutation is limited by the production of plutonium from uranium. A design work-around has been proposed to have an inert matrix. Magnesium oxide has been proposed as the inert matrix. Magnesium oxide has an entire order of magnitude smaller probability or interacting with neutrons (thermal and fast) than the elements like iron.

The SFR is designed for management of high-level wastes and, in particular, management of plutonium and other actinides. Important safety features of the system include a long thermal response time, a large margin to coolant boiling, a primary system that operates near atmospheric pressure, and intermediate sodium system between the radioactive sodium in the primary system and the water and steam in the power plant. With innovations to reduce capital cost, such as making a modular design, removing a primary loop, integrating the pump and intermediate heat exchanger, or simply find better materials for construction, the SFR can be a viable technology for electricity generation.

The SFR's fast spectrum also makes it possible to use available fissile and fertile materials (including depleted uranium) considerably more efficiently than thermal spectrum reactors with once-through fuel cycles.

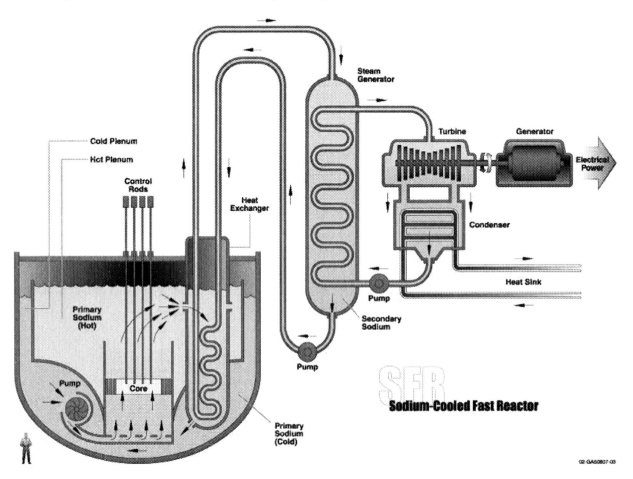

What is a supercritical water reactor? The Supercritical water reactor (SCWR) is a Generation IV reactor concept that uses supercritical water as the working fluid. SCWRs are basically LWRs operating at higher pressure and temperatures with a direct, once-through cycle. As most commonly envisioned, it would operate on a direct cycle, much like a BWR, but since it uses supercritical water (not to be confused with critical mass) as the working fluid, would have only one phase present, like the PWR. It could operate at much higher temperatures and pressure than both current PWRs and BWRs.

Supercritical water-cooled reactors (SCWRs) are promising advanced nuclear systems because of their high thermal efficiency (i.e., about 45% vs. about 33% efficiency for current light water reactors (LWR) and considerable plant simplification.

A key issue in natural circulation is constituted by the stability of the flow mainly when two phase conditions are concerned and when the feedback with neutron kinetics is possible.

The main mission of the SCWR is generation of low-cost electricity. It is built upon two proven technologies, LWRs, which are the most commonly deployed power generating reactors in the world, and supercritical fossil fuel fired boilers, a large number of which are also in use around the world. The SCWR concept is being investigated by 32 organizations in 13 countries.

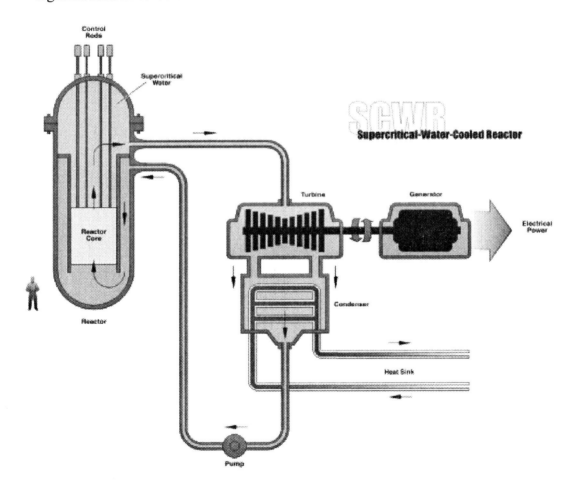

What is a very high temperature reactor? The Very High Temperature Reactor is a
   Generation IV reactor concept that uses a graphite-moderated nuclear reactor with a
   once-through uranium fuel cycle. This reactor design envisions an outlet temperature of
   1000°C. The reactor core can be either a "prismatic block" or a "pebble-bed" core. The
   high temperatures enable applications such as process heat or hydrogen production via
   the thermo-chemical sulfur-iodine cycle.

This design was formerly known as "High Temperature Gas-cooled Reactor" or HTGR,
originally developed in the 1950's. The Fort St. Vrain Generating Station was one
example of this design that operated as an HTGR from 1979 to 1989; though the reactor
was beset by some problems which led to its decommissioning due to economic factors,
it served as proof of the HTGR concept in the United States (though no new commercial
HTGRs have been developed there since).[1] HTGRs have also existed in Germany, Japan
and China, and are promoted in several countries by reactor designers.[2] More recently,
this reactor design type has been updated and is now more commonly known as the Very
High Temperature Reactor.

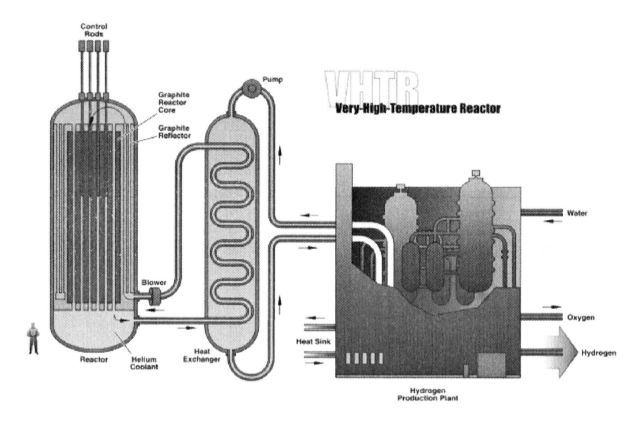

How are test reactors regulated? Research and Test Reactors – also called RTRs or "non-power" reactors – are low-power nuclear reactors that are primarily used for research, training and development. There are 32 operating NRC-licensed RTRs around the country that are used to study almost every field of science. Regulating the safety and security of RTRs is one of NRC's jobs.

RTRs are designed and operated so that material is not easily handled or dispersed. This protects the public and environment against potential radiological exposure or theft of the material. RTRs are licensed to have only small amounts of radioactive material on site. The NRC evaluates and inspects each RTR's security plans, procedures and systems to verify that effective security measures are in place to protect the reactors.

- Size Matters

NRC-licensed RTRs range in size from 20 Megawatts (MW) to 5 Watts (about the size of a child's nightlight). In comparison, the typical operating nuclear power plant is 3,000MW and can power over 1 million homes.

Rules of Regulation

Because NRC-licensed RTRs operate at significantly lower power levels than their power plant cousins and have a limited amount of radioactive material on site, the standard for regulating these reactors is different. In fact, the NRC is federally mandated to apply the minimum regulation needed to protect the public health and safety at RTRs so they can effectively conduct education and research.

After 9/11, the NRC established additional security measures and inspected RTRs to ensure the measures were followed. The NRC identified several potential enhancements and RTRs around the country voluntarily implemented many of the improvements. With these security measures in place, the NRC has determined that these reactors pose minimal risk to public health and safety.

Today, the NRC continues to monitor RTR security through our regulatory processes. If threat conditions change, such that they could potentially affect public health and safety, the NRC will act promptly to further enhance security at RTRs.   Courtesy of the NRC.

Does the reactor have to be shutdown in order to reload the reactor core? It depends on the design of the reactor core. In a CANDU type reactor, fuel assemblies can be changed out while the unit remains online. In most reactors, the reactor has to be shut-down and secured in order to refuel the reactor core.

How is the reactor core configured? The picture on the left gives you an indication of the fuel layout inside the reactor core and the picture on the right sides gives you an indication where the control rods might be located within the reactor core assemblies. The different colors in the right picture is to illustrate the groups of control rods and how they can be moved in groups, all at once and/or one at a time…depending on what is needed. The

CEAs are labeled as such for example: A,B,C are shutdown groups while 1,2,3,4,5 are regulating groups.

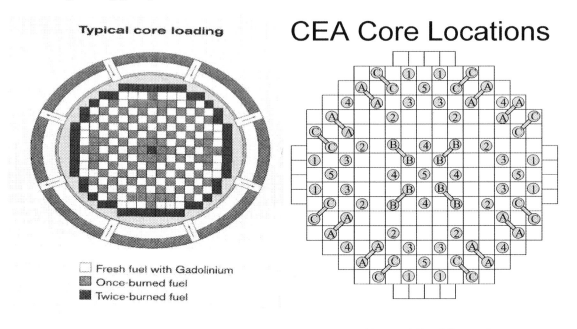

Typical core loading

☐ Fresh fuel with Gadolinium
▦ Once-burned fuel
■ Twice-burned fuel

Fuel Locations

CEA Core Locations

Control Rod Locations

How long does it normally take for a reactor to come up to 100% power? During a startup, there are multiple steps in being the reactor core up to power. There are numerous variables that must be continuously monitored in order to ensure that the reactor is operating within its design parameters. There is never a rush to force a nuclear reactor online…safety is the number 1 priority.

How long does it normally take for a nuclear to get to 0% power? Reactors are designed to trip should the unit trip and would be shut-down within seconds by admitting all control rods into the core at once. Normally, groups of control rods are slowly inserted into the core while reducing steam flow to the turbine until the reactor is no longer able to generate enough heat to generate steam for the turbine.

How does coolant flow through the PWR reactor core? Approximately 90% of the coolant travels the following path through the reactor vessel – into the vessel through the four inlet nozzles, one per reactor coolant pump; downward in the outer annulus, between the outside surfaces of the core support shield and thermal shield and the inside surface of the reactor vessel, to the bottom of the reactor vessel; upward through the flow holes in the plenum cylinder to the two outlet nozzles, one per steam generator. Some of the flow traveling the core continues upward within the control rod guide tubes to the upper plenum, above the plenum assembly cover. This fluid rejoins the main flow through holes in the periphery of the plenum cover. Whereas 90% of the coolant follows the paths just described, the remainder bypasses the primary heat transfer surfaces of the core. There are three major components of this bypass flow. First, a portion of the flow at the inlet nozzles proceeds directly to the outlet nozzles through residual clearances between the outlet nozzles of the reactor vessel and the nozzle openings of the core support shield.

Second, a portion of the flow in the lower plenum is directed into the gap between the core barrel and the thermal shield, primarily to cool the thermal shield. Finally, bypass flow also occurs in the control rod and instrument guide tubes. The amount of bypass flow through these tubes varies with the configuration of the control components within the core.   Courtesy of Babcock & Wilcox

What is meant by positive void coefficient and negative void coefficient? The use of water as a moderator is an important safety feature of PWRs, as any increase in temperature causes the water to expand and become less dense; thereby reducing the extent to which neutrons are slowed down and hence reducing the reactivity in the reactor. Therefore, if reactivity increases beyond normal, the reduced moderation of neutrons will cause the chain reaction to slow down, producing less heat. This property, known as the negative temperature coefficient of reactivity, makes PWR reactors very stable.

In contrast, the RBMK reactor design used at Chernobyl, which uses graphite instead of water as the moderator and uses boiling water as the coolant, has a high positive coefficient of reactivity that increases heat generation when coolant water temperatures increase. This makes the RBMK design less stable than pressurized water reactors. In addition to its property of slowing down neutrons when serving as a moderator, water also has a property of absorbing neutrons, albeit to a lesser degree. When the coolant water temperature increases, the boiling increases, which creates voids. Thus there is less water to absorb thermal neutrons that have already been slowed down by the graphite moderator, causing an increase in reactivity. This property is called the void coefficient of reactivity, and in an RBMK reactor like Chernobyl, the void coefficient is positive, and fairly large, causing rapid transients. This design characteristic of the RBMK reactor is generally seen as one of several causes of the Chernobyl accident.[11]

Heavy water has very low neutron absorption, so heavy water reactors such as CANDU reactors also have a positive void coefficient, though it is not as large as that of an RBMK like Chernobyl; these reactors are designed with a number of safety systems not found in the original RBMK design, which are designed to handle or react to this as needed.

PWRs are designed to be maintained in an under-moderated state, meaning that there is room for increased water volume or density to further increase moderation, because if moderation were near saturation, then a reduction in density of the moderator/coolant could reduce neutron absorption significantly while reducing moderation only slightly, making the void coefficient positive. Also, light water is actually a somewhat stronger moderator of neutrons than heavy water, though heavy water's neutron absorption is much lower. Because of these two facts, light water reactors have a relatively small moderator volume and therefore have compact cores. One next generation design, the supercritical water reactor, is even less moderated. A less moderated neutron energy spectrum does worsen the capture/fission ratio for ^{235}U and especially ^{239}Pu, meaning that more fissile nuclei fail to fission on neutron absorption and instead capture the neutron to become a heavier non-fissile isotope, wasting one or more neutrons and increasing accumulation of heavy transuranic actinides, some of which have long half-lives.

What is a dry well? In a BWR, the containment strategy is a bit different. A BWR's containment consists of a drywell where the reactor and associated cooling equipment is located and a wetwell. The drywell is much smaller than a PWR containment and plays a larger role. During the theoretical leakage design basis accident the reactor coolant flashes to steam in the drywell, pressurizing it rapidly. Vent pipes or tubes from the drywell direct the steam below the water level maintained in the wetwell (also known as a torus or suppression pool), condensing the steam, limiting the pressure ultimately reached. Both the drywell and the wetwell are enclosed by a secondary containment building, maintained at a slight sub-atmospheric or negative pressure during normal operation and refueling operations. The containment designs are referred to by the names Mark I (oldest; drywell/torus), Mark II, and Mark III (newest). All three types house also use the large body of water in the suppression pools to quench steam released from the reactor system during transients.   Courtesy of Wikipedia

What is a torus or suppression pool? The torus or suppression pool is used to remove heat released if an event occurs in which large quantities of steam are released from the reactor or the reactor recirculation system, used to circulate water through the reactor.

Which units have been decommissioned?

Decommissioning Status for Shut Down Power Reactors (As of Jan. 2008)

Reactor	Type	Thermal Power	Location	Shutdown	Status	Fuel Onsite
Big Rock Point	BWR	67 MW	Charlevoix, MI	8/97	ISFSI Only	Yes
CVTR	Pressure Tube, Heavy Water	65 MW	Parr, SC	1/67	License Terminated	No
Dresden I	BWR	700 MW	Morris, IL	10/31/78	SAFSTOR	Yes
Fermi I	Fast Breeder	200 MW	Monroe Co., MI	9/22/72	SAFSTOR/DECON	No
Fort St. Vrain	HTGR	842 MW	Platteville, CO	8/18/89	License Terminated	Yes
GE VBWR	BWR	50 MW	Alameda Co., CA	12/9/63	SAFSTOR	No
Haddam Neck	PWR	1825 MW	Haddam Neck, CT	7/22/96	ISFSI Only	Yes
Humboldt Bay 3	BWR	200 MW	Eureka, CA	7/02/76	DECON	Yes
Indian Point I	PWR	615 MW	Buchanan, NY	10/31/74	SAFSTOR	Yes
LaCrosse	BWR	165 MW	LaCrosse, WI	4/30/87	SAFSTOR	Yes
Main	PWR	2772 MW	Bath, ME	12/96	ISFSI Only	Yes

Reactor	Type	Thermal Power	Location	Shutdown	Status	Fuel Onsite
Yankee						
Millstone I	BWR	2011 MW	Waterford, CT	11/04/95	SAFSTOR	Yes
N.S. Savannah	PWR	80 MW	Norfolk, VA	1970	SAFSTOR	No
Pathfinder	Superheat BWR	190 MW	Sioux Falls, SD	9/16/67	DECON NRC Part 30	No
Peach Bottom I	HTGR	115 MW	York Co., PA	10/31/74	SAFSTOR	No
Rancho Seco	PWR	2772 MW	Sacramento, CA	6/7/89	DECON	Yes
San Onofre I	PWR	1347 MW	San Clemente, CA	11/30/92	DECON	Yes
Saxton	PWR	28 MW	Saxton, PA	5/72	License Terminated	No
Shoreham	BWR	2436 MW	Suffolk Co., NY	6/28/89	License Terminated	No
Three Mile Island 2	PWR	2772 MW	Middletown, PA	3/28/79	SAFSTOR*	No
Trojan	PWR	3411 MW	Portland, OR	11/9/92	ISFSI Only	Yes
Yankee Rowe	PWR	600 MW	Franklin Co., MA	10/1/91	ISFSI Only	Yes
Zion 2	PWR	3250 MW	Zion, IL	2/98	SAFSTOR	Yes
Zion I	PWR	3250 MW	Zion, IL	2/98	SAFSTOR	Yes

Courtesy of the NRC

* Post-defueling monitored storage (PDMS).

Note: An independent spent fuel storage installation (ISFSI) is a stand-alone facility within the plant boundary constructed for the interim storage of spent nuclear fuel.

What is the purpose of the reactor coolant pump? The purpose of the reactor coolant pump is to circulate primary coolant around and through the reactor core to transfer heat from the core to the secondary system. These pumps are the prime movers of the fluid for the reactor coolant system. They must provide the motive force necessary to move a large volume of coolant through the reactor, the coolant piping and the steam generators.

What is the flow rate of a reactor coolant pump? The flow rate for the reactor coolant pumps range from tens to hundreds of thousands of gallons per minute through the reactor coolant system.

What is the purpose of the letdown line? The letdown line is a part of a big system called the chemical volume and control system (CVCS). The letdown line allows fluid from the reactor primary coolant to be removed from the primary system so that it can be purified and conditioned before having it pumped back into the primary system.

How long does a reactor core (fuel) last? Depending on the type of reactor and its usage will determine how long a reactors core will last. Commercial reactors normally last between 18 to 24 months. This is also dependent upon the percentage of enrichment the fuel is at.

What is a moderator? In nuclear fission reactors, the moderator is a substance that is used to reduce the speed of neutrons, making them more likely to split atomic nuclei (fission).

What is heavy water? Heavy water is water which contains a higher proportion than normal of the isotope deuterium, as deuterium oxide, D_2O or ^2H_2O, or as deuterium protium oxide, HDO or $^1H^2HO$. Its physical and chemical properties are somewhat similar to those of water, H_2O. Heavy water may contain as much as 100% D_2O, and usually the term refers to water which is highly enriched in deuterium. The isotopic substitution with deuterium alters the bond energy of the hydrogen-oxygen bond in water, altering the physical, chemical, and especially biological properties of the pure or highly-enriched substance to a larger degree than is found in most isotope-substituted chemical compounds.

Heavy water should not be confused with hard water or with tritiated water.

What is reactor coolant? Reactor coolant is a fluid (gas or liquid) that removes heat from the reactor core.

What is the purpose of the containment air coolers? Containment air coolers are used to basically keep the environment inside the containment building at a given temperature. The air coolers help to keep the equipment inside working properly keeping them from overheating.

What is the purpose of the penetration coolers? There are numerous penetrations leading into and out of the containment building. These penetrations are tightly sealed to prevent the inside containment atmosphere to leak out, protecting the environment. These penetrations need to be cooled in order to maintain the strength of the concrete structure.

What is the purpose of the containment purge air system? The containment purge air system is used to purge the air within the containment building when work is to be performed such as an outage. There is a system to bring fresh filtered air into the containment building another system to remove air from the containment building, filter the air before it is released into the outside environment. This system is monitored when in service.

What is the purpose of the iodine removal unit? The iodine removal unit is a giant filter that uses charcoal to capture iodine particles that were created during the fission process. The unit is placed in service when the unit is being shut-down and the reactor head is being removed or in the event of a LOCA so that personnel working in the containment building will not be exposed to these particles and gases.

What is the purpose of maintaining H₂ pressure in the VCT? The purpose for maintaining hydrogen pressure on the volume control tank is to reduce and/or mitigate the separator of hydrogen and oxygen within the water.

What is the purpose of the RVLMS? Reactor Vessel Level Monitoring System, measures the water level in the head and upper guide structure. Indication range from fuel alignment plate to near the top of the vessel head. Designed primarily to aid in detecting the approach to inadequate core cooling.

How does the reactor vessel level monitoring system work? If a sensor is surrounded by water, the temperature difference between the heated and the unheated thermocouple is minimal because water absorbs the heat. If a sensor is in steam or non-condensable gas, the temperature difference between the heated and unheated thermocouple increases due to the poor heat conduction of the vapor (steam).

Where does the reactor head vent to? The reactor head vent, vents to the pressurizer quench tank (post-accident) and to refueling pool for normal startups.

When is the reactor head vented? The reactor head is vented whenever the reactor head has been removed (outage) and re-installed to ensure that the reactor vessel is completely full of coolant.

What is meant by sub-cooled margin? By sub-cooled, means to be below a certain temperature to ensure that a liquid/gas/solid does not phase change when progressing through different stages of a process. E.g. Feedwater heater drains are sub-cooled to ensure that when the drains cascade into a lower pressure heater they do not flash back into steam causing severe pressure and thermal shock. By sub-cooling, the water should remain mostly in one phase (liquid).

What is the purpose of the quench tank? A reservoir where numerous system can be discharge safely without releasing hazardous materials into the environment. Systems like relief and safety valves, safety injection system recirculation relief valve, reactor coolant drain tank relief valves and reactor coolant head vent. They discharge into the quench tank below the water level to maximize steam condensing thus reducing pressure buildup inside the tank.

What kind of protective devices does the quench tank have? The tank is supplied with a rupture disc and a vent relief valve.

What is the purpose of the loose parts monitoring system? A loose material within the reactor vessel can become high velocity projectiles within this system due to the mass flow rate going through the reactor vessel. The flow rates are high enough to not only pick up loose pieces of material but can give them such energy that they become high impact objects. A single of wire or tie wrap could puncher a single fuel rod causing the fuel inside the fuel rod to leak outward thus creating a more hazardous work environment. When the sound waves generated by a loose part strike the transducer, an electrical output is produced which is proportional to the instantaneous acceleration of the sensor.

What is the function of the acoustic monitors? The acoustic monitors provide a position indication and determine if liquid and/or steam is flowing through the relief valve discharge pipe.

What is meant by hot leg and cold leg? The hot leg is the piping that is transporting the hot water leaving the reactor core, heading to the steam generators. The cold leg is the piping that is transporting the cold water leaving the steam generators, heading to the reactor core.

Where does the letdown system remove coolant from? The letdown fluid is removed from the primary system through the suction piping to the reactor coolant pump.

Where does the charging pumps supply coolant back into the reactor? The charging pumps discharge the letdown fluid back into the primary system at the discharge of the reactor coolant pumps.

How is the reactor kept full of coolant? In a pressurized water reactor (PWR) there is a vent system on the very top of the reactor vessel head. After the reactor vessel head is put back on and bolted down, the vessel is vented to remove a vast majority of the air they may have become trapped. A makeup system adds coolant to the reactor to bring the coolant level to the very top of the reactor vessel head. This also ensures that the maximum amount of coolant is available in the vessel at all times.

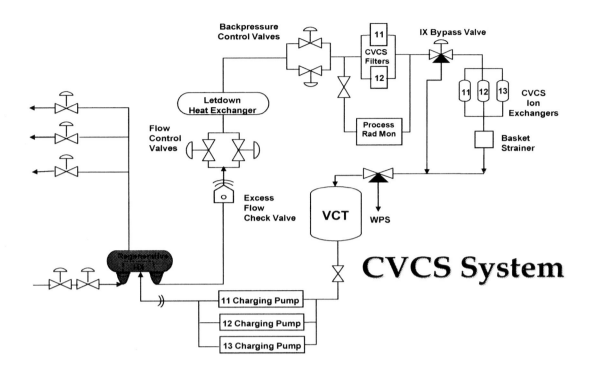

CVCS System

What is the purpose of the radiation monitoring system on the CVCS? This system is one of the first indications that there is a fuel pin rupture and where fuel material is leaking into direct contact with the primary coolant. It measures the letdown fluid before it flows into the ion exchanger.

What is the function of the volume control tank (VCT)? It is a surge tank for the reactor coolant system volume changes due to variations in temperature. Collects reactor coolant pump seal control bleed-off and provides net positive suction pressure for the charging pumps.

What is the makeup system for the CVCS? The makeup system allows for plant personnel to run in four different modes depending on the circumstances. (Adding deminerialized water & chemicals, borate – if raising the boron concentration, diluting – adding deminerialized water by itself and in auto to work with the control room for whatever they need for the system.

Makeup System Overview

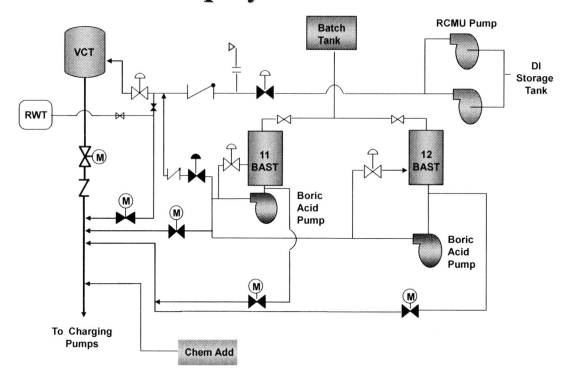

What could cause a loss of pressure control? Many failures in a reactor plant or its supporting auxiliaries could cause a loss of pressure control, including:

- Inadvertent isolation of the pressurizing vessel from the reactor plant, via the closing of an isolation valve or mechanically clogged piping. Because of this possibility, no commercial nuclear power plant has any kind of valve in the connection between the pressurizer and the reactor coolant circuit. To avoid clogging anywhere in the primary circuit, the coolant is kept very clean, and the connecting pipe between the pressurizer and the reactor coolant circuit is short and large diameter.

- A rupture in the pressurizer vessel, which would also be a loss of coolant accident. In most reactor plant designs, however, this would not limit flow rate through the core and therefore would behave like a loss of pressure control accident rather than a loss of coolant accident.

- Failure of either the spray nozzles (failing open would inhibit raising pressure as the relatively cool spray collapses the pressurizer vessel bubble) or the heaters of the pressurizing system.

- Thermal Stratification of the liquid portion of the pressurizer. When the liquid portion of the pressurizer becomes stratified, the lower layers of water (furthest from the steam bubble) are sub-cooled and as the steam bubble slowly condenses, pressurizer pressure will appear relatively constant but actually will be slowly lowering. When the operator energizes pressurizer heaters to maintain or raise pressure, pressure will continue to drop until the sub-cooled water is heated up by the pressurizer heaters to the saturation temperature corresponding to the pressure of the steam (bubble) portion of the pressurizer. During this reheating period, pressure control will be lost, since pressure will still be dropping when it is desired to raise pressure.

What could result in a loss of pressure control in a PWR? When pressure control is lost in a reactor, depending on the level of heat being generated by the reactor plant, the heat being removed by the steam or other auxiliary systems, the initial pressure, and the normal operating temperature of the plant, it could take minutes or even hours for operators to see significant trends in core behavior.

For whatever power level the reactor is currently operating at, a certain amount of enthalpy is present in the coolant. This enthalpy is proportional to temperature, therefore, the hotter the plant, the higher the pressure must be maintained to prevent boiling. When pressure drops to the saturation point, dry-out in the coolant channels will occur.

As the reactor heats the water flowing through coolant channels, sub-cooled nucleate boiling takes place, in which some of the water becomes small bubbles of steam on the cladding of the fuel rods. These are then stripped from the fuel cladding and into the coolant channel by the flow of water. Normally, these bubbles collapse in the channel, transferring enthalpy to the surrounding coolant. When the pressure is below the saturation pressure for the given temperature, the bubbles will not collapse. As more bubbles accumulate in the channel and combine, the steam space within the channel becomes larger and larger until steam blankets the fuel cell walls. Once the fuel cell walls are blanketed with steam, the rate of heat transfer lowers significantly. Heat is not transferred out of the fuel rods as fast as it is being generated, potentially causing a nuclear meltdown. Because of this potential, all nuclear power plants have reactor protection systems that automatically shut down the reactor if the pressure in the primary circuit falls below a safe level, or if the sub-cooling margin falls below a safe level. Once the reactor is shut down, the rate at which residual heat is generated in the fuel rods is similar to that of an electric kettle, and the fuel rods can be safely cooled just by being submerged in water at normal atmospheric pressure.

What would cause a reactor to trip? There are numerous events that could cause the reactor to trip, here are 11 such events: variable overpower trip, high rate of change of power, low reactor coolant flow, low steam generator water level, low steam generator pressure, reactor coolant system high pressure, thermal margin low pressure, loss of load, high containment pressure, Asymmetric S/G tilt and axial power distribution.

What is the reactor made of and why? Solid carbon steel and clad with stainless steel inside. The carbon steel provides the strength while the stainless steel cladding protects the carbon steel from corrosion. (E.g. Davis Besse)

What is Axial Shape Index? The axial shape index is the ratio of the lower half of the reactor core power minus the upper half of the reactor core power to the total power.

Why would low reactor coolant flow trip the reactor? Low reactor coolant flow trip protects the core against departure from nucleate boiling if reactor coolant system flow decreases.

What is meant of high rate of change of power (SUR)? Hi startup rate (SUR) trip is designed to prevent core damage due to an uncontrolled withdrawal of control rod event and/or boron dilution event when the reactor is critical at low power.

What is meant by thermal margin low pressure? The thermal margin low pressure trip is designed to prevent overheating of the fuel due to voiding caused by low reactor coolant system pressure. This is normally what would trip the reactor in the case of a small to medium size line break releasing reactor coolant (LOCA).

Why must the reactor trip due to loss of load? The loss of load reactor trip is designed to trip the reactor when the turbine/generator is tripped so that it would limit the increase in reactor coolant system stored energy and pressure caused by the suspension of normal steam demand from the steam generators.

Why would high containment pressure trip the reactor? The reactor vessel is a massive amount of space. The only event that could cause the containment pressure to increase is a break in the primary system and/or the reactor vessel…releasing coolant into the containment building and the fluid instantly flashing into steam (raising pressure).

Why would axial power distribution trip the reactor? The axial power distribution trip is designed to ensure excessive axial peaking caused by xenon oscillations or control rod movement will not cause localized overheating of the fuel and thus cause fuel damage.

Why would asymmetric steam/generator tilt trip the reactor? The asymmetric steam generator tilt trip is designed to protect against events that affect only one of the steam generators and would therefore cause a non-uniform core inlet temperature distribution.

What is the purpose of the containment spray system? The containment spray system sprays borated water into the containment building through a spray header located at the top of the reactor vessel to reduce containment internal pressure and temperature by condensing the steam being released into the containment building during a LOCA. Aids in the removal of radioactive material from the containment building atmosphere.

What is the purpose of the high pressure and low pressure safety injection systems? To inject borated water into the core for reactivity control, to flood and cool the core. Provides long term core cooling following a loss of coolant accident (LOCA). The safety injection system also provides heat removal from the reactor coolant system during normal cool down and refueling, transfer water between the refueling water tank and the refueling pool.

Where do the high and low pressure safety injection pumps discharge into? Both high and low pressure safety injection pumps discharge into the reactor coolant pumps discharge piping leading into the reactor vessel.

What is the purpose of the reactor protective system? The reactor protective system is designed to remove the reactor from service in order to maximize safe shutdown efficiency to protect the public and environment.

What is a variable overpower trip (VOPT)? The variable overpower trip is designed to prevent the reactor core from exceeding rated core power and limit the margin reduction to safety limits. The trip is effective in overpower transients such as excess load or CEA withdrawal accidents.

What is CSAS? The Containment Spray Actuation Signal (CSAS) provides a reliable means of automatic initiation of equipment to maintain containment pressure within design structural limits by providing atmospheric cooling Protects from over pressure and over temperature and reduces leakage of radioactivity from containment.

What is RAS? The Recirculation Actuation Signal (RAS) occurs when water sources have become depleted it provides a signal allowing the emergency pumps to pull suction from the containment sump causing the pumps to continuously pump the same fluid into the containment building and reactor core, maintaining core safety and containing radioactive contaminants. Provides a reliable means of automatic initiation of actions to provide continuous source of emergency core and containment cooling water during long-term recovery.

What is SIAS? The Safety Injection Actuation Signal (SIAS) used to ensure that the core is covered, the core is kept cool and the core is kept NOT critical. Provides a reliable means of automatic initiation of actions to remove core decay heat and ensure adequate shutdown in the event of a LOCA.

What is CIS? The Containment Isolation Signal (CIS) provide a reliable means of automatically isolating the containment, and operating ventilation systems in a manner that reduces the potential for radioactive release to the environment during a LOCA.

What is CRS? The Containment Radiation Signal (CRS) provides a reliable means of automatic initiation of actions to limit the release of radioactive fission products during refueling & maintenance periods.

What is SGIS? The Steam Generator Isolation Signal (SGIS) provides a reliable means of automatic initiation of actions to minimize the effect of a main steam line break upon personnel and plant equipment (including containment overpressure).

What is UV? Under-voltage signal provides automatic actions to provide safe, reliable power from emergency diesel generators to their respective bus.

What is SASB? Sequential actuation system blocking (SASB) restores certain safeguards equipment in a controlled manner to prevent overloading the emergency diesels.

What is CVCIS? The Chemical Volume Control Isolation Signal (CVCIS) provides a reliable means of automatically initiation of actions to minimize the effect of a letdown line break, outside of containment and limit release of radioactive fission products to the public.

What is RTT? The Reactor Trip-Turbine Trip signal is designed to ensure that if the turbine or reactor should trip that the turbine or reactor still in service will be tripped as well. This trip signal will prevent excessive cool down of the primary system.

What is DSS? The Diverse Scram System (DSS) provides a means of automatically tripping the reactor due to high pressurizer pressure conditions in the event of an Anticipated Transient without Scram (ATWS).

What is SGHLTT? The Steam Generator High Level Turbine Trip Signal (SGHLTT) provides a reliable means of automatically initiation of actions to prevent steam generator water carryover from reaching the main turbine and damaging blading.

What is AFAS? The Auxiliary Feedwater Actuation System (AFAS) provides a means of electronically detecting the need for auxiliary feedwater and other protection functions. Example: may start the auxiliary feedwater system on a low level in the steam generator or alarms in the event of a high level in the steam generator.

What is the purpose of the reactor regulating system? The purpose of the reactor regulating system is to sense the operating conditions of the reactor and provide numerous control signals in the event of a problem.

What is the purpose of the atmospheric dump valves? Used for rapid removal of NSSS stored energy and sensible heat following a turbine/reactor trip by dumping steam. This allows the operator the ability to maintain positive control steam generator pressure and reactor coolant system temperature during normal heat-up / cool-down. The use of these valves help to reduce the use of safety valves. Since safety valves do not have any way of being isolated, if they fail to close off and continue to leak through, the unit may have to be shut down.

How is reactor reactivity controlled in PWRs and BWRs? For BWRs, reactivity control for routine operation is implemented through a combination of control rods and coolant flow adjustments. Bottom mounted control rods are made of long boron carbide filled pins in a cruciform shape that fits between four fuel assemblies.

Flow adjustment can provide another effective control method, since the water changes density with temperature. At low temperature, the dense water is very effective at moderating neutrons and thereby encourages fission. With increased temperature, the density decreases; or, equivalently, the void content increases as steam is being produced, causing a reduction in moderation and fission rate. If flow is increased, energy removal can be increased without a net change in coolant temperature with a resulting increase in power generation.

As for PWRs, reactivity control is accomplished mainly with soluble poison in the form of boric acid assisted by control rods. The boron concentration is adjusted to match general changes from fuel burn-up, conversion of fertile material, and depletion of burnable poisons.

What is the purpose of the containment liner? The containment liner serves as a containment leakage boundary.

What is the function of the containment personnel airlock? The containment airlock is design to allow plant personnel entrance into the reactor area while the unit remains online. This design helps to keep the plant safe as well as the personnel working around the area. This airlock is another protective device designed to keep the containment environment from escaping outside to the atmosphere around the plant.

What are charging pumps? The charging pumps are pumps designed to replace fluids in the primary system of a pressurized water reactor. As reactor coolant is letdown from the primary system the charging pumps allow chemicals and other fluids to be pumped right back into the primary system thus maintaining a proper amount of fluid to remain in the reactor vessel and primary system.

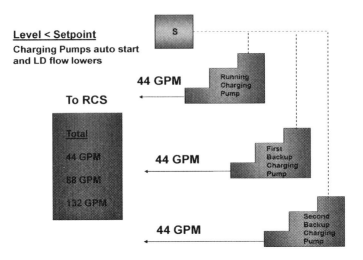

What is the operating pressure and temperature of a pressurized water reactor? The operating pressure and temperature for a PWR is between 2200 and 2300 psi and approximately 600^0F.

What is the operating pressure and temperature of a boiling water reactor? The operating pressure and temperature of a BWR is between 1000 and 1100 psi and approximately 550^0F.

What is the primary system? All nuclear reactors have some type of primary system used to help cool the reactor and to generate steam in one form or another. A primary system for a PWR will consist of several reactor coolant pumps, a letdown system, a charging system along with piping, valves and instrumentation.

What are good characteristics of a moderator? Good characteristics of a moderator are inexpensive, plentiful, and compatible with plant material, low absorption cross section and has a high scattering ability.

What is moderator temperature defect (MTD)? The amount of reactivity added to the reactor (due to moderator temperature changes) when going from one power level to another.

What is fuel temperature defect (FTD)? The amount of reactivity added to the reactor (due to a fuel temperature changes) when going from one power level to another.

What is startup rate (SUR)? A measure of the rate of change in reactor power, in decades per minute. Since neutron population increases geometrically, it is important to be able to determine the rate of change, particularly at low power levels. Startup rate is read directly from the reactivity control panel during startup and can be calculated using this equation:
$$P = P_0 10^{(SUR)(t)}$$

What is meant by point of adding heat (POAH)? The power level at which temperature of the core will begin to increase due to addition of nuclear heat.

What is shutdown margin (SDM)? The instantaneous amount of reactivity by which a reactor is subcritical, or would be made subcritical from its present condition.

What is heat capacity? The amount of heat needed to raise the temperature of an object by 1 degree, expressed in joules per Kelvin. The heat capacity is known also as thermal capacity.

What is reaction rate? In a nuclear reactor, it is the rate at which fission takes place.

What is the purpose of the recirculation pumps for a BWR? Basically, the recirc pumps help natural circulation within the reactor vessel by forcing water downward toward the bottom of the core, allowing it to flow up through the reactor core to become steam. The forced recirculation flow is very useful in controlling power. Thermal power levels are easily varied by simply increasing or decreasing the speed of the recirculation pumps.

What two ways are there to controlling power levels in a BWR? Positioning (withdrawing or inserting) control rods is the normal method for controlling power when starting up a BWR. As control rods are withdrawn, neutron absorption decreases in the control material and increases in the fuel, so reactor power increases. As control rods are inserted, neutron absorption increases in the control material and decreases in the fuel, so reactor power decreases. Some early BWRs and the proposed ESBWR designs use only natural circulation with control rod positioning to control power from zero to 100% because they do not have reactor recirculation systems.

Is all the reactor fuel changed out during an outage? No, for a PWR approximately 1/3 of the fuel is replaced and for a BWR approximately ¼ of the fuel is replaced.

Why not replace the entire core during every outage? The fuel in the core does not burn up at the same rate. There are different power levels within the core. The areas with a higher power level burn up more fuel that those with lower power levels. In order to maintain the highest efficiency of the nuclear fuel…there are calculations that determine what fuel assemblies will be replaced and how the reactor core will be re-organized to maintain its high efficiency usage.

What makes the best moderator? Materials consisting of low-mass-number atoms usually make the best moderators. In this respect hydrogen makes an ideal moderator (except that hydrogen absorbs some neutrons). Light water is desirable because it is cheap and plentiful. Light water can be used as both a moderator and a coolant, however, the water must be completely free of impurities in order to avoid neutron absorption and possible radioactivity. Water also has a relatively low boiling point, so pressures must be high if high temperatures are desired. Heavy water, D_2O is also an excellent moderator and coolant. Heavy water has a smaller probability for neutron absorption than light water, but is not quite as effective as light water in slowing down neutrons. Its major disadvantage is the high cost. Carbon is another good moderator because it does not absorb many neutrons and does scatter neutrons well. Carbon is readily available in the form of graphite. One disadvantage is that graphite may oxidize at high temperatures. Beryllium is one of the best solid moderators and is used either as metallic beryllium or as beryllium oxide. Beryllium has a low absorption cross section, a high scattering cross section, and a high melting point, 1158^0K.

What is ESFAS? The ESFAS system monitors various parameters throughout the plant and provides automatic signals to control the engineered safety features.

What is ECCS? Emergency Core Cooling System (ECCS) is a component in nuclear power plants designed to deal with a loss of coolant accident (LOCA) by providing massive backup sources of coolant. An ECCS also may be used after a "partial" (incomplete) SCRAM to help bring a runaway reaction under control. Each power plant has multiple independent ECCS systems, any one of which should be adequate to cool the core. ECCS systems are nuclear safety-grade components. ECCS systems can be powered by plant power (as long as the generator is on-line), offsite power, or the plant's Emergency Diesel Generators (another nuclear safety-grade system). ECCS systems typically activate automatically upon occurrence of a LOCA (along with other automatic plant actions), to restore cooling as fast as possible so as to prevent a nuclear meltdown.

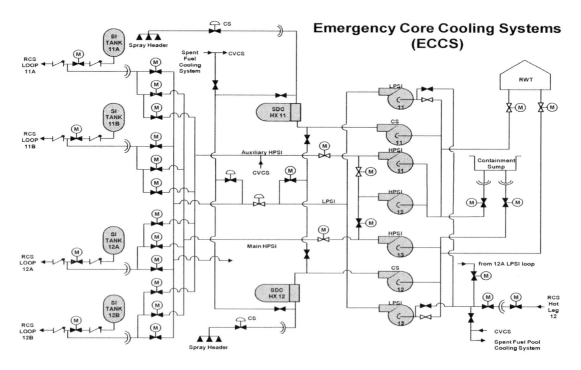

What is the purpose of the reactor core? Provides a means of heat generation to produce steam to drive the turbine/generator to produce electricity.

What is the purpose of the control rod guide tubes? The purpose of the control rod guide tubes is to keep the control rods in line as they are being moved into and out of the reactor core.

What is meant by reduced inventory? Reduced inventory is a term used to describe a condition that happens during a unit refueling. Reduced inventory happens when the amount of coolant available to cool the core during the refueling process is reduced in order to complete other overhaul maintenance.

What is meant by time to boil? As with reduced inventory, time to boil is a term used by plant personnel to determine how long it will take the reactor core coolant to boil without any other heat removal process. Since the core is continuously giving off heat regardless if the unit is online or offline, the core must be continuously cooled and monitored. Once the coolant begins to boil, the cooling ability of the coolant is no longer able to remove the heat away from the core causing the core to increase in temperature. If the core temperature is not controlled, the core could reach a condition where it would begin to meltdown.

Poisons

Why use boron? The isotopes that might be used for neutron capture therapy are those that have a very good ability to absorb neutrons but normally not radioactive. Boron is suitable because the B_{10} nucleus has a very large neutron absorption cross section for slow neutrons Boron's ability to absorb these neutrons is several thousand times better than that of the elements constituting living tissues (such as hydrogen, oxygen, carbon etc.). Therefore, if the concentration in cancerous tissues is sufficiently high, boron will absorb the vast majority of neutrons. Boron is never found as a free element in nature. The B_{10} isotope is good at capturing thermal neutrons. Natural boron is about 20% B_{10} and 80% B_{11}. The nuclear industry enriches natural boron to nearly a pure state. The waste product or depleted boron is nearly pure B_{11}. B_{11} is a candidate as a fuel for aneutronic fusion and is used in the semiconductor industry. Enriched boron or B_{10} is used in both radiation shielding and in boron neutron capture therapy.

In PWR nuclear reactors, B_{10} is used for reactivity control and in emergency shutdown systems. It can serve either function in the form of borosilicate control rods or as boric acid. In pressurized water reactors, boric acid is added to the reactor coolant when the plant is shut down for refueling. It is then slowly filtered out and diluted with water over many months as fissile material is used up and the fuel becomes less reactive.

What is a neutron poison? A neutron poison is a substance with a large neutron absorption cross section in applications are inserted into some types of reactors in order to lower the high reactivity of their initial fresh fuel load in the reactor core. Some of these poisons deplete as they absorb neutrons during reactor operation, while others remain relatively constant.

What is Xenon-135? ^{135}Xe is an unstable isotope of xenon with a half-life about 9.2 hours. ^{135}Xe is a fission product of uranium (yield 6.3%) and is the most powerful known neutron-absorbing nuclear poison (2 million barns), with a significant effect on nuclear reactor operation.

The inability of a reactor to be started due to the effects of Xe-135 is sometimes referred to as xenon precluded start-up. During periods of steady state operation at a constant neutron flux level, the Xe-135 concentration builds up to its equilibrium value for that reactor power in about 40 to 50 hours. When the reactor power is increased, Xe-135 concentration initially decreases because the burn up is increased at the new higher power level. Because 95% of the Xe-135 production is from decay of iodine-135, which has a 6 to 7 hour half-life, the production of Xe-135 remains constant; at this point, the Xe-135 concentration reaches a minimum. The concentration then increases to the new equilibrium level for the new power level in roughly 40 to 50 hours. During the initial 4 to 6 hours following the power change, the magnitude and the rate of change of concentration is dependent upon the initial power level and on the amount of change in power level; the Xe-135 concentration change is greater for a larger change in power level. When reactor power is decreased, the process is reversed.

Iodine-135 is a fission product of uranium with a yield of about 6%. This ^{135}I decays with a 6.7 hour half-life to ^{135}Xe. Thus, in an operating nuclear reactor, ^{135}Xe is being continuously produced. ^{135}Xe has a very large neutron absorption cross-section, so in the high neutron flux environment of a nuclear reactor core, the ^{135}Xe soon absorbs a neutron and becomes stable ^{136}Xe. Thus, in about 50 hours, the ^{135}Xe concentration reaches equilibrium where its creation by ^{135}I decay is balanced with its destruction by neutron absorption.

When reactor power is decreased or shut down by inserting neutron absorbing control rods, the reactor neutron flux is reduced and the equilibrium shifts initially towards higher ^{135}Xe concentration. The ^{135}Xe concentration peaks about 11.1 hours after reactor power is decreased. Since ^{135}Xe has a 9.2 hour half-life, the ^{135}Xe concentration gradually decays back to low levels over 72 hours.

The temporarily high level of ^{135}Xe with its high neutron absorption cross-section makes it difficult to restart the reactor for several hours. The neutron absorbing ^{135}Xe acts like a control rod reducing reactivity. The inability of a reactor to be started due to the effects of Xe-135 is sometimes referred to as xenon precluded start-up. The period of time where the reactor is unable to override the effects of Xe-135 is called the xenon dead time.

If sufficient reactivity control authority is available, the reactor *can* be restarted, but a xenon burn-out transient must be carefully managed. As the control rods are extracted and criticality is reached, neutron flux increases many orders of magnitude and the ^{135}Xe begins to absorb neutrons and be transmuted to ^{136}Xe. The reactor burns off the nuclear poison. As this happens, the reactivity increases and the control rods must be gradually re-inserted or reactor power will increase. The time constant for this burn-off transient depends on the reactor design, power level history of the reactor for the past several days, and the new power setting. For a typical step up from 50% power to 100% power, ^{135}Xe concentration falls for about 3 hours. Failing to manage this xenon transient properly caused the Chernobyl reactor power to overshoot ~100x normal causing a steam explosion. The xenon burn-out rate is proportional to neutron flux and thus reactor power. If reactor power doubles, the xenon burns out twice as quickly. The larger the rate of increase in reactor power, the faster the xenon burns out and the more quickly reactor power increases.

Reactors using continuous reprocessing like many molten salt reactor designs might be able to extract Xe-135 from the fuel and avoid these effects.

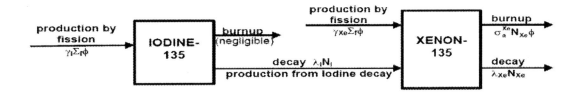

How was the effect of Xenon-135 discovered during reactor operation? A Tuesday in late September 1944, the scene is the "B" pile (graphite-moderated nuclear reactor) at the Hanford Reservation in Washington State. The dramatic discovery of the effect of Xe-135 is described this way (Rhodes 1986):

The pile went critical at a few minutes past midnight; by 2am it was operating at a higher power level than any previous chain reaction. For the space of an hour all was well. The operating engineers were discussing with each other, adjusting control rods, knowing something was wrong. The pile reactivity was steadily decreasing with time; the control rods had to be withdrawn continuously from the pile to hold it at 100 MWs. The time came when the rods were completely withdrawn. The reactor power began to drop down and continued to decrease. By early Wednesday evening the "B" pile died... Early Thursday morning the pile came back to life. It was running again. But twelve hours later it began another decline. The culprit was Xe-135.   Courtesy of Paul L. Roggenkamp

What are transient fission product poisons? Some of the fission products generated during a nuclear reaction have a high neutron absorption capacity, such as xenon-135 (Xe-135) and samarium-149 (Sm-149). Because these two fission product poisons remove neutrons from the reactor, they will have an impact on the thermal utilization factor and thus the reactivity. The poisoning of a reactor core by these fission products may become so serious that the chain reaction comes to a standstill.

Xe-135 in particular has a tremendous impact on the operation of a nuclear reactor. The inability of a reactor to be started due to the effects of Xe-135 is sometimes referred to as xenon precluded start-up. The period of time where the reactor is unable to override the effects of Xe-135 is called the xenon dead time. During periods of steady state operation, at a constant neutron flux level, the Xe-135 concentration builds up to its equilibrium value for that reactor power in about 40 to 50 hours. When the reactor power is increased, Xe-135 concentration initially decreases because the burn up is increased at the new higher power level. Because 95% of the Xe-135 production is from iodine-135 decay, which has a 6 to 7 hour half-life, the production of Xe-135 remains constant, at this point, the Xe-135 concentration reaches a minimum. The concentration then increases to the new equilibrium level for the new power level in again roughly 40 to 50 hours. The magnitude and the rate of change of concentration during the initial 4 to 6 hours following the power change is dependent upon the initial power level and on the amount of change in power level; the Xe-135 concentration change is greater for a larger change in power level. When reactor power is decreased, the process is reversed.[1]

Because Sm-149 is not radioactive and is not removed by decay, it presents problems somewhat different from those encountered with Xe-135. The equilibrium concentration and (thus the poisoning effect) builds to an equilibrium value during reactor operation in about 500 hours, and since Sm-149 is stable, the concentration remains essentially constant during reactor operation

What are accumulating fission product poisons? There are numerous other fission products that, as a result of their concentration and thermal neutron absorption cross section, have a poisoning effect on reactor operation. Individually, they are of little consequence, but taken together they have a significant impact. These are often characterized as lumped fission product poisons and accumulate at an average rate of 50 barns per fission event in the reactor. The buildup of fission product poisons in the fuel eventually leads to loss of efficiency, and in some cases to instability. In practice, buildup of reactor poisons in nuclear fuel is what determines the lifetime of nuclear fuel in a reactor: long before all possible fissions have taken place, buildup of long-lived neutron-absorbing fission products damps out the chain reaction. This is the reason that nuclear reprocessing is a useful activity: solid spent nuclear fuel contains about 99% of the original fissionable material present in newly manufactured nuclear fuel. Chemical separation of the fission products restores the fuel so that it can be used again.

Other potential approaches to fission product removal include solid but porous fuel which allows escape of fission products and liquid or gaseous fuel (Molten salt reactor, Aqueous homogeneous reactor). These ease the problem of fission product accumulation in the fuel, but pose the additional problem of safely removing and storing the fission products.

Other fission products with relatively high absorption cross sections include Kr-83, Mo-95, Nd-143, and Pm-147. Above this mass, even many even-mass number isotopes have large absorption cross sections, allowing one nucleus to serially absorb multiple neutrons. Fission of heavier actinides produces more of the heavier fission products in the lanthanide range, so the total neutron absorption cross section of fission products is higher.

In a fast reactor the fission product poison situation may differ significantly because neutron absorption cross sections can differ for thermal neutrons and fast neutrons. In the RBEC-M Lead-Bismuth Cooled Fast Reactor, the fission products with neutron capture more than 5% of total fission products capture are, in order, Cs-133, Ru-101, Rh-103, Tc-99, Pd-105, Pd-107 in the core, with Sm-149 replacing Pd-107 for 6th place in the breeding blanket.

What are control poisons? Throughout the operation of a nuclear reactor, the amount of fuel contained in the core constantly decreases. If the reactor is to operate for a long period of time, fuel in excess of that need for exact criticality must be added when the reactor is built. The positive reactivity due to the excess fuel must be balanced with negative reactivity from neutron absorbing material. Movable control rods containing neutron absorbing material is one method, but control rods alone are unable to balance the excess reactivity depending on the reactor design.

What are non-burnable poisons? A non-burnable poison is one that maintains a constant negative reactivity worth over the life of the core. While no neutron poison is strictly non-burnable, certain materials can be treated as non-burnable poisons under certain conditions. E.g. Hafnium.

What is a decay poison? As well as neutron poisons in the reactor core, other materials in the reactor can decay into materials that act as neutron poisons.

What is the difference between burnable and soluble poisons? Burnable poisons are used to control large amounts of excess fuel without control rods, burnable poisons are loaded into the core. Burnable poisons are materials that have a high neutron absorption cross section that are converted into materials of relatively low absorption cross section as the result of neutron absorption. Due to the burn-up of the poison material, the negative reactivity of the burnable poison decreases over core life. Ideally, these poisons should decrease their negative reactivity at the same rate the fuel's excess positive reactivity is depleted. Fixed burnable poisons are generally used in the form of compounds of boron or gadolinium that are shaped into separate lattice pins or plates, or introduced as additives to the fuel. Since they can usually be distributed more uniformly than control rods, these poisons are less disruptive to the core's power distribution. Fixed burnable poisons may also be discretely loaded in specific locations in the core in order to shape or control flux profiles to prevent excessive flux and power peaking near certain regions of the reactor. Current practice however is to use fixed non-burnable poisons in this service.

Soluble poisons, also called chemical shim, produce spatially uniform neutron absorption when dissolved in the water coolant. The most common soluble poison in commercial pressurized water reactors (PWR) is boric acid, which is often referred to as soluble boron, or simply *solbor*. The boric acid in the coolant decreases the thermal utilization factor, causing a decrease in reactivity. By varying the concentration of boric acid in the coolant, a process referred to as boration and dilution, the reactivity of the core can be easily varied. If the boron concentration is increased, the coolant/moderator absorbs more neutrons, adding negative reactivity. If the boron concentration is reduced (dilution), positive reactivity is added. The changing of boron concentration in a PWR is a slow process and is used primarily to compensate for fuel burnout or poison buildup. The variation in boron concentration allows control rod use to be minimized, which results in a flatter flux profile over the core than can be produced by rod insertion. The flatter flux profile occurs because there are no regions of depressed flux like those that would be produced in the vicinity of inserted control rods. This system is not in widespread use because the chemicals make the moderator temperature reactivity coefficient less negative.

Soluble poisons are also used in emergency shutdown systems. During SCRAM the operators can inject solutions containing neutron poisons directly into the reactor coolant. Various solutions, including sodium polyborate and gadolinium nitrate ($Gd(NO_3)_3 \cdot x\ H_2O$), are used.

What is the purpose of using Boron as a neutron absorber? There is sufficient U-235 in a commercial power reactor to power the unit for 18 to 24 months before refueling.

Without additional absorbers present in the system, the keff would be greater than 1. It is impractical to provide the absorption needed using just control rods. Therefore, for long term control $k_{eff}=1$ is maintained by soluble boron in the moderator (boron dissolved in water) and/or burnable absorbers. Burnable absorbers contain a limited concentration of absorber atoms such that, as neutrons are absorbed, the effectiveness of absorption decreases and essentially burns out with time. Burnable absorbers may be in the fuel rods or in fixed absorber rods that are placed in fuel assembly guide tubes. The boron concentration in the coolant can be near 1800ppm at the beginning of a fuel cycle (BOC) and yet this concentration can decrease to near zero at the end of the fuel cycle (EOC) when a significant portion of the uranium is depleted and fission products have built up. The critical boron concentration is the boron level required to maintain steady-state reactor power levels. The burnable absorber limits the amount of soluble boron that is used at the beginning of the cycle.

What is control rod poison worth? The percentage (fraction) of the available neutron flux that the control element assembly is capable of absorbing, depending on relative magnitude flux at the rod tip and relative importance of these neutrons to the fission process. (SL-1)

What is Differential Boron Worth (DBW)? The reactivity change per ppm change in Boron concentration.

What is Inverse Boron Worth (IBW)? The amount of Boron (in ppm) required to make a percent change in reactivity.

What are the effects of fission product poisons (FFPs) on core reactivity? Fission product poisons (FFPs) are isotopes produced as fission fragments which have a large cross section for absorption. When present in the reactor core, they "rob" neutrons from the neutron life cycle. As FPP concentration increases, more neutrons are removed from the neutron life cycle causing K_{eff} to decrease, inserting negative reactivity to the core. To have a significant effect of core reactivity, a FPP must have: a large cross section and be in high abundance (high fission yield).

What are two major fission product poisons? Two major FPPs are Xenon (Xe) and Samarium (Sm).

What factors affect Xenon production? Xenon is produced by direct methods being produced in about .3% of all fissions and accounts for about 5% of all Xenon produced in the core. Xenon is also produced indirectly as a result of β- decay of I-135 and accounts for about 95% of all Xenon produced in the core.

How is Xenon removed as a poison? Xenon can burnout by neutron absorption into Xe-136 or it can β- decay into Cs-135.

How is Samarium produced and removed from the system? Samarium is produced by the β-decay of Promethium and is removed only by neutron absorption (burnout).

What concentration of boron is used in the reactor vessel? The concentration of boron is high when the reactor is being started up from an outage and throughout the next 18 to 24 months the concentration of boron is gradually diluted as fuel burn-up is calculated. By diluting the boron concentration in the reactor vessel core reactivity is maintained and controlled in a manner that protects the entire plant system. If the boron concentration wasn't diluted over time in reactor vessel, reactivity in the core would decrease to a point where the core would not be able to generate enough thermal energy to meet the steam temperature demand of the unit. Therefore the unit would have to be removed from service.

Why is boron injected into the reactor vessel? Boron is injected into the reactor vessel to control reactor core reactivity by absorbing neutrons. PWRs use boron, BWRs do not use boron.

Core Refueling

How does the core refueling process work? The unit is taken off-line and cooled down to a safe temperature. During the refueling process, approximately one third of the fuel assemblies are replaced and the remaining assemblies are reorganized within the reactor core. Nuclear Fuel Engineers will have a plan of action in place that tells them what fuel assemblies must be removed from the core, the re-organization of the remaining fuel assemblies within the core as well as where the new fuel assemblies are going to be placed. This process is all done underwater with special machinery that allows plant personnel to move and replace the nuclear fuel assemblies without increasing their risk of radiation exposure.

The fuel handling equipment cover three basic points of operation: main fuel handling from the reactor core, the fuel transfer (between the reactor core and spent fuel pool) and the spent fuel/new fuel handling area. The main fuel handling system, used to remove and install fuel in the reactor vessel, consists of a motorized bridge spanning the canal which rides on rails. A handling mast, mounted to and extending below the carriage, is positioned over the fuel assembly to be moved. The grapple (gripping device) is then lowered and engages the assembly. Once engaged, the grapple is raised and the fuel is drawn into the hollow mast. Sensors indicating load and fuel assembly position assist the operator. The bridge is then moved to the fuel transfer area, where the fuel is lowered into a vertical transfer basket and released from the grapple. Then, the basket is lowered to a horizontal position and is moved through the transfer tube into the spent fuel area. The fuel is then upended and moved to its new location within the spent fuel pool to continue cooling down for the next several years.

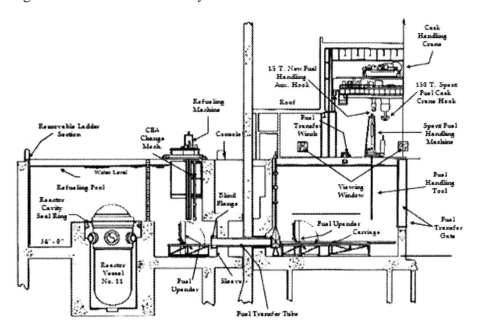

As a note, there are some type of nuclear reactors that can be refueled while remaining online. CANDU reactors are one type of reactor that can be refueled while online.

What is the purpose of the spent fuel handling machine? The refueling machine is a traveling bridge which spans the containment building refueling pool. Mounted on this bridge is a trolley that travels perpendicular to the bridge travel. These motions allow the rectilinear positioning of the grapple over the center of the fuel assembly to be installed or removed. The trolley contains and supports the mast, hoist box and grappling tool which are used for installing and removing the fuel within the reactor vessel. The hoist box, into which the fuel assembly is drawn prior to its movement in the refueling pool, also supports the underwater TV camera and spreader device. The underwater closed circuit television system allows the operator to view the remote grappling and un-grappling of the fuel assembly and the spreader device moves adjacent fuel assemblies radially away from the fuel assembly which is either being removed or installed.

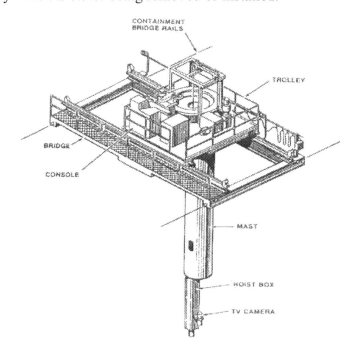

How is the spent fuel pool monitored? About one-fourth to one-third of the total fuel load of a reactor is removed from the core every 12 to 18 months and replaced with fresh fuel. Spent fuel rods generate intense heat and dangerous radiation that must be contained. Fuel is moved from the reactor and manipulated in the pool generally by automated handling systems, although some manual systems are still in use. The fuel bundles fresh from the core normally are segregated for several months for initial cooling before being sorted in to other parts of the pool to wait for final disposal. Metal racks keep the fuel in safe positions to avoid the possibility of a "criticality"— a nuclear chain reaction occurring. Water quality is tightly controlled to prevent the fuel or its cladding from degrading. Current regulations permit re-arranging of the spent rods so that maximum efficiency of storage can be achieved.

The maximum temperature of the spent fuel bundles decreases significantly between 2 and 4 years, and less from 4 to 6 years. The fuel pool water is continuously cooled to remove the heat produced by the spent fuel assemblies. Pumps circulate water from the spent fuel pool to heat exchangers then back to the spent fuel pool. Radiolysis, the dissociation of molecules by radiation, is of particular concern in wet storage, as water may be split by residual radiation and hydrogen gas may accumulate increasing the risk of explosions. For this reason the air in the room of the pools, as well as the water must permanently be monitored and treated.   Courtesy of Wikipedia

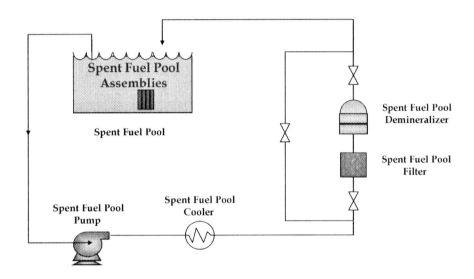

Spent Fuel Pool Cooling & Purification

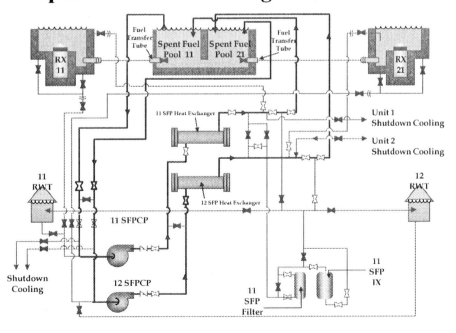

Where is the spent fuel sent after the spent fuel pool? Dry cask storage is a method of storing high-level radioactive waste, such as spent nuclear fuel that has already been

cooled in the spent fuel pool for at least one year. The fuel is surrounded by inert gas inside a large container. These casks are typically steel cylinders that are either welded or bolted closed. Ideally, the steel cylinder provides leak-tight containment of the spent fuel. Each cylinder is surrounded by additional steel, concrete, or other material to provide radiation shielding to workers and members of the public. Some of the cask designs can be used for both storage and transportation.

There are various dry storage cask system designs. With some designs, the steel cylinders containing the fuel are placed vertically in a concrete vault; other designs orient the cylinders horizontally. The concrete vaults provide the radiation shielding. Other cask designs orient the steel cylinder vertically on a concrete pad at a dry cask storage site and use both metal and concrete outer cylinders for radiation shielding.

What is the purpose of the spent fuel pool? Spent fuel pool (SFP) are storage pools for spent fuel from nuclear reactors. Typically 40 or more feet deep, with the bottom 14 feet equipped with storage racks designed to hold fuel assemblies removed from the reactor. These fuel pools are specially designed at the reactor in which the fuel was used and situated at the reactor site. In many countries, the fuel assemblies, after being in the reactor for 3 to 6 years, are stored underwater for 10 to 20 years before being sent for reprocessing or dry cask storage. The water cools the fuel and provides shielding from radiation.

While only about 8 feet of water is needed to keep radiation levels below acceptable levels, the extra depth provides a **safety margin** and allows fuel assemblies to be manipulated without special shielding to protect the operators.

About one-fourth to one-third of the total fuel load of a reactor is removed from the core every 18 to 24 months and replaced with fresh fuel. Spent fuel rods generate intense heat and dangerous radiation that must be contained. Fuel is moved from the reactor and manipulated in the pool generally by automated handling systems, although some manual systems are still in use. The fuel bundles fresh from the core normally are segregated for several months for initial cooling before being sorted in to other parts of the pool to wait for final disposal. Metal racks keep the fuel in safe positions to avoid the possibility of a "criticality"— a nuclear chain reaction occurring. Water quality is tightly controlled to prevent the fuel or its cladding from degrading. Current regulations permit re-arranging of the spent rods so that maximum efficiency of storage can be achieved.

The maximum temperature of the spent fuel bundles decreases significantly between 2 and 4 years, and less from 4 to 6 years. The fuel pool water is continuously cooled to remove the heat produced by the spent fuel assemblies. Pumps circulate water from the spent fuel pool to heat exchangers then back to the spent fuel pool. Radiolysis, the dissociation of molecules by radiation, is of particular concern in wet storage, as water may be split by residual radiation and hydrogen gas may accumulate increasing the risk of explosions. For this reason the air in the room of the pools, as well as the water must permanently be monitored and treated.

Without cooling, the fuel pool water will heat up and boil. If the water boils or drains away, the spent fuel assemblies will overheat and either melt or catch on fire. Fear has been expressed that sabotage, an accident, or an attack which partially or completely drains a plant's spent fuel pool or disables it's cooling, might be capable of causing a high-temperature fire that could release large quantities of radioactive material into the environment. Since there is no standard design, most SFPs are housed in far less robust structures than reactor containment vessels and moreover, an SFP often contains much more radioactive material than the reactor core.   Courtesy of Wikipedia

Spent Fuel Pool

What is the purpose of the High Integrity Container (HIC)? This container is thick walled polypropylene and designed to hold radioactive resin that can be buried at a specific site.

What about the storage of spent fuel? There are two acceptable storage methods for spent fuel after it is removed from the reactor core:

- Spent Fuel Pools - Currently, most spent nuclear fuel is safely stored in specially designed pools at individual reactor sites around the country.
- Dry Cask Storage - If pool capacity is reached, licensees may move toward use of above-ground dry storage casks.

How We Regulate

The NRC regulates spent fuel through a combination of regulatory requirements, licensing; safety oversight, including inspection, assessment of performance; and enforcement; operational experience evaluation; and regulatory support activities.
                                                        Courtesy of NRC

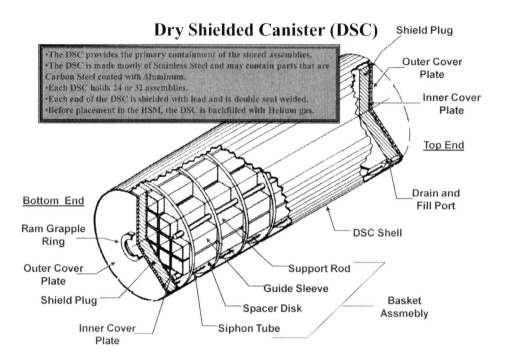

Dry Shielded Canister (DSC)

- The DSC provides the primary containment of the stored assemblies.
- The DSC is made mostly of Stainless Steel and may contain parts that are Carbon Steel coated with Aluminum.
- Each DSC holds 24 or 32 assemblies.
- Each end of the DSC is shielded with lead and is double seal welded.
- Before placement in the HSM, the DSC is backfilled with Helium gas.

Shield Plug
Outer Cover Plate
Inner Cover Plate
Top End
Drain and Fill Port
DSC Shell
Bottom End
Ram Grapple Ring
Outer Cover Plate
Shield Plug
Inner Cover Plate
Support Rod
Guide Sleeve
Spacer Disk
Siphon Tube
Basket Assmebly

What about the transportation of spent nuclear fuel? Spent nuclear fuel refers to uranium-bearing fuel elements that have been used at commercial nuclear reactors and that are no longer producing enough energy to sustain a nuclear reaction. Once the spent fuel is removed from the reactor the fission process has stopped, but the spent fuel assemblies still generate significant amounts of radiation and heat. Because of the residual hazard, spent fuel must be shipped in containers or casks that shield and contain the radioactivity and dissipate the heat.

Over the last 30 years, thousands of shipments of commercially generated spent nuclear fuel have been made throughout the United States without causing any radiological releases to the environment or harm to the public.

Most of these shipments occur between different reactors owned by the same utility to share storage space for spent fuel, or they may be shipped to a research facility to perform tests on the spent fuel itself. In the near future, because of a potential high-level waste repository being built, the number of these shipments by road and rail is expected to increase.

How We Regulate

The NRC regulates spent fuel transportation through a combination of safety and security requirements, certification of transportation casks, inspections, and a system of monitoring to ensure that requirements are being met.   Courtesy of the NRC

What about radioactive waste safety research?

Spent Fuel

NRC research is developing the technical basis to ensure the continued safe performance of long-term dry storage systems for spent nuclear fuel and high-level radioactive waste under extended service conditions (20 to 100 years) and the structural integrity of spent fuel transport casks during severe accidents.

Decommissioning and Waste Management

Research programs are developing tools for assessing the performance of decommissioning and waste management options. This research examines the potential mechanisms that could cause environmental movement of residual radioactivity remaining after license termination and provides computational methods to estimate the effects of such movement on the public.   Courtesy of the NRC

What is low-level waste? Low-level waste includes items that have become contaminated with radioactive material or have become radioactive through exposure to neutron radiation. This waste typically consists of contaminated protective shoe covers and clothing, wiping rags, mops, filters, reactor water treatment residues, equipment and tools, luminous dials, medical tubes, swabs, injection needles, syringes, and laboratory animal carcasses and tissues. The radioactivity can range from just above background levels found in nature to very highly radioactive in certain cases such as parts from inside the reactor vessel in a nuclear power plant. Low-level waste is typically stored on-site by licensees, either until it has decayed away and can be disposed of as ordinary trash, or until amounts are large enough for shipment to a low-level waste disposal site in containers approved by the Department of Transportation.   Courtesy of NRC

What is high-level waste? High-level radioactive wastes are the highly radioactive materials produced as a byproduct of the reactions that occur inside nuclear reactors. High-level wastes take one of two forms:

- Spent (*used*) reactor fuel when it is accepted for disposal
- Waste materials remaining after spent fuel is reprocessed

Spent nuclear fuel is used fuel from a reactor that is no longer efficient in creating electricity, because its fission process has slowed. However, it is still thermally hot, highly radioactive, and potentially harmful. Until a permanent disposal repository for spent nuclear fuel is built, licensees must safely store this fuel at their reactors.

Reprocessing extracts isotopes from spent fuel that can be used again as reactor fuel. Commercial reprocessing is currently not practiced in the United States, although it has been allowed in the past. However, significant quantities of high-level radioactive waste are produced by the defense reprocessing programs at Department of Energy (DOE) facilities, such as Hanford, Washington, and Savannah River, South Carolina, and by commercial reprocessing operations at West Valley, New York. These wastes, which are generally managed by DOE, are not regulated by NRC. However they must be included in any high-level radioactive waste disposal plans, along with all high-level waste from spent reactor fuel.

Because of their highly radioactive fission products, high-level waste and spent fuel must be handled and stored with care. Since the only way radioactive waste finally becomes harmless is through decay, which for high-level wastes can take hundreds of thousands of years, the wastes must be stored and finally disposed of in a way that provides adequate protection of the public for a very long time.   Courtesy of NRC

What about low-level waste disposal? Low-level waste disposal occurs at commercially operated low-level waste disposal facilities that must be licensed by either NRC or Agreement States. The facilities must be designed, constructed, and operated to meet safety standards. The operator of the facility must also extensively characterize the site on which the facility is located and analyze how the facility will perform for thousands of years into the future.

There are three existing low-level waste disposal facilities in the United States that accept various types of low-level waste. All are in Agreement States.

The Low-level Radioactive Waste Policy Amendments Act of 1985 gave the states responsibility for the disposal of their low-level radioactive waste. The Act encouraged the states to enter into compacts that would allow them to dispose of waste at a common disposal facility. Most states have entered into compacts; however, no new disposal facilities have been built since the Act was passed.

How We Regulate

NRC and the Agreement States regulate low-level waste disposal through a combination of regulatory requirements, licensing, and safety oversight. For more information see the following:

- Regulations, Guidance and Communications
- Licensing
- Oversight

See our Agreement States page for more information on the roles of NRC and the Agreement States in regulating low-level waste disposal and other activities associated with nuclear materials.   Courtesy of NRC

What about high-level waste disposal? On June 3, 2008, the U.S. Department of Energy (DOE) submitted a license application to the U.S. Nuclear Regulatory Commission (NRC), seeking authorization to construct a deep geologic repository for disposal of high-level radioactive waste at Yucca Mountain, Nevada. The NRC's review of that application will require evaluation of a wide range of technical and scientific issues. The

NRC will issue a construction authorization only if DOE can demonstrate that it can safely construct and operate the repository in compliance with the NRC's regulations. See What We Regulate and How We Regulate (on this page) for the latest news and information about the NRC's high-level waste disposal activities and the process the agency will use to decide whether to authorize DOE to construct a geologic repository at Yucca Mountain.

What We Regulate

United States policies governing the permanent disposal of HLW are defined by the Nuclear Waste Policy Act of 1982, as amended (NWPA). This Act specifies that HLW will be disposed of underground, in a deep geologic repository, and that Yucca Mountain, Nevada, will be the single candidate site for characterization as a potential geologic repository. Under the Act, the NRC is one of three Federal agencies with a role in the disposal of spent nuclear fuel, as well as the HLW from the Nation's nuclear weapons production activities:

- The U.S. Department of Energy (DOE) is responsible for designing, constructing, operating, and decommissioning a permanent disposal facility for HLW, under NRC licensing and regulation.

- The U.S. Environmental Protection Agency (EPA) is responsible for developing site-specific environmental standards for use in evaluating the safety of a geologic repository.

- The NRC is responsible for developing regulations to implement the EPA's safety standards, and for licensing and overseeing the construction and operation of the repository. In addition, the NRC will consider any future DOE applications for license amendments to permanently close the repository, dismantle surface facilities, remove controls to restrict access to the site, or undertake any other activities involving an un-reviewed safety question.

In accordance with its mission, the NRC focuses its regulatory actions on protecting the health and safety of the public and the environment both before and during the active life of the proposed Yucca Mountain repository, and after the facility has been decommissioned. The NRC staff accomplishes this mission by performing the following activities:

- Establish and enforce safety and security regulations
- Perform a comprehensive safety review of DOE's license application
- Perform an adoption review of DOE's environmental impact statement
- Conduct a public, formal adjudicatory hearing
- Decide whether to authorize or deny repository construction
- Decide whether to authorize or deny a license to receive and possess waste at Yucca Mountain (if DOE is authorized to construct the repository)
- Inspect and oversee any construction, waste emplacement, and/or repository closure activities at Yucca Mountain   Courtesy of NRC

Steam Generator

Why would a low water level in the steam generator trip the reactor? A low water level in the steam generator would trip the reactor because of a loss of heat sink. If there is not enough water to pull the heat away from the reactor coolant the reactor begins to overheat is no appropriate action is taken.

Why would low pressure in the steam generator trip the reactor? Low steam generator pressure trip protects the reactor against excessive heat removal from the steam generators and subsequent cool-down and reactivity addition. Low steam pressure would be caused by either an abnormally high steam demand or a line rupture which would also create a high steam demand.

What is the purpose of the feed ring in the steam generator? The feed ring in the steam generator is designed to distribute feedwater evenly throughout the steam generator which helps reduce thermal stress in the steam generator.

Where is the water level maintained in the steam generator? The feedwater level in the steam generator is maintained below the steam separators and above the U-tubes. Keeping it above the U-tubes ensures that the heat is continuously transferred from the primary side to the steam generator without over-heating. Keeping it below the steam separators ensures that there no water carryover to the turbine.

Where does the steam generator blow-down water from? The blow-down line can be valved in to remove solids from the water surface line in the upper half of the steam generator or it can valved in to remove solids from the bottom plate of the steam generator. Chemistry of the steam generator, the amount and types of chemicals used will determine which blow-down supply is to be used.

How does this build up affect steam generator performance? Buildup inside the steam generator reduces heat transfer between the primary and secondary system by insulating the tubes. This buildup also creates temperature differential stresses between the clean tubes and the dirty tubes. The more buildup a steam generator has an increase in reduced efficiency will be found.

How many types of steam generators are there? There are two basic fundamental designs for steam generators: Once-through and Recirculating type.

- The **once-through steam generator (OTSG)** is a vertical shell counter-flow straight-tube heat exchanger design which directly generates superheated steam as the feedwater flows through the steam generator in a single pass.

- **Recirculating steam generator (RSG),** only part of the feedwater is converted to steam as the water passes through the unit. After the steam is separated from the water, the steam is sent to the turbine for power generation while the water is returned to the tube bundle for additional steam generation.

The steam generators also come in three basic designs: A vertical U-tube arrangement, a horizontal arrangement, and a vertical once through arrangement.

What are the internal structures of a steam generator?

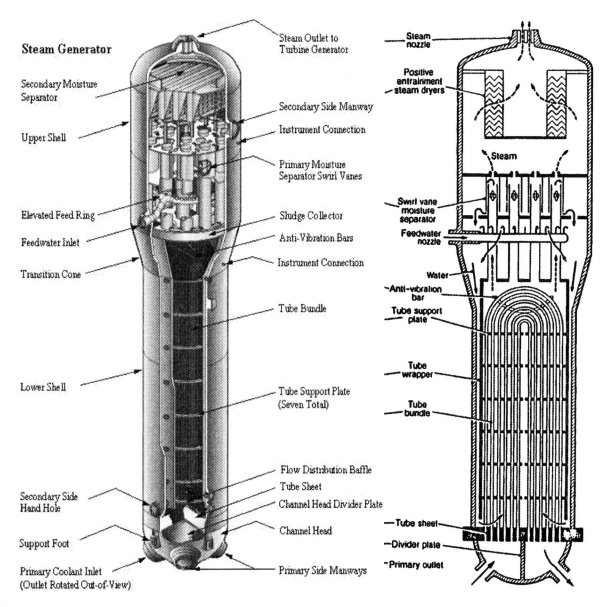

Steam Generator

- Secondary Moisture Separator
- Upper Shell
- Elevated Feed Ring
- Feedwater Inlet
- Transition Cone
- Lower Shell
- Secondary Side Hand Hole
- Support Foot
- Primary Coolant Inlet (Outlet Rotated Out-of-View)

- Steam Outlet to Turbine Generator
- Secondary Side Manway
- Instrument Connection
- Primary Moisture Separator Swirl Vanes
- Sludge Collector
- Anti-Vibration Bars
- Instrument Connection
- Tube Bundle
- Tube Support Plate (Seven Total)
- Flow Distribution Baffle
- Tube Sheet
- Channel Head Divider Plate
- Channel Head
- Primary Side Manways

- Steam nozzle
- Positive entrainment steam dryers
- Steam
- Swirl vane moisture separator
- Feedwater nozzle
- Water
- Anti-vibration bar
- Tube support plate
- Tube wrapper
- Tube bundle
- Tube sheet
- Divider plate
- Primary outlet

This drawing is only one of many designs for a PWR steam generator.

What is the purpose of the flow restrictors in the steam generator? The flow restrictors are designed to limit the flow rate from the steam generator in the event of a steam line rupture to the steam turbine.

What is the secondary system? Not all nuclear reactors have a secondary system. Pressurized water reactors use a secondary system to transfer heat from the reactor coolant (primary system) to the secondary coolant to use the heat to generate steam to power the turbine/generator.

What is the purpose of the egg crate? The purpose of the egg crate is to the tubes within the steam generator in place. These egg crates also allow a place for sludge buildup to take place. In some cases, the egg crates have also caused tube leaks to develop because of vibration taking place between the tubes and the egg crate.

What is the purpose of the steam generators? The purpose of the steam generators is to create steam to turn the turbine/generator by exchanging heat from the primary system to the secondary system. It also establishes an intermediate barrier between the radioactive fluid in the primary system (reactor coolant) and the secondary system (feedwater system). It also removes heat from the reactor for normal cool down. Aids in the natural circulation within the reactor core should the reactor coolant pumps not be operational by removing heat from the primary system and transferring the heat to the secondary system.

What is the secondary system? Not all nuclear reactors have a secondary system. Pressurized water reactors use a secondary system to transfer heat from the reactor coolant (primary system) to the secondary coolant to use the heat to generate steam to power the turbine/generator.

What does the steam generator tube temperature profile look like? A steam generator tube temperature profile at 100% power may have an inner tube temperature of 572^0F and an outer tube temperature of 525^0F. The tube wall reduces heat transfer by 47^0F.

How does feedwater flow through the steam generator? The reactor coolant enters one side of the lower head from the hot leg and flows up the U-tubes, down into the other side of the lower head and out through the cold leg nozzle into the suction of the reactor coolant pump. The feedwater enters midway up the shell and flows down through the annulus to the tube region. Feedwater then enters the tube region and begins to absorb heat from the tubes. The top of the U-tubes is always covered with a saturated steam-water mixture. As the feedwater pool continues to absorb heat, saturated steam is produced. This steam is forced through moisture separators at the top of the steam generator before t is sufficiently dried to exit into the main steam headers to the turbine. The moisture extracted from the steam is re-circulated to the feedwater pool surrounding the U-tubes.

What would happen if the feedwater entering the steam generator was not heated? If the feedwater was not heated before it entered the steam generator, severe thermal stress on the metal will occur possibly seriously damaging the steam generator internals.

What is the purpose of the moisture separator reheater? The moisture separator reheater (MSR) is designed to reheat steam that has passed through the high pressure section of the turbines in which the steam has lost pressure and temperature. The steam needs to be reheated in order to delay the steam from reaching saturation temperature as it is passing through the low pressure section of the turbines. By reheating the steam it helps to increase the efficiency of the plant heat rate.

How is the moisture separator reheater constructed? The tubes are supported throughout their length in both the evaporator and economize regions by a series of tube support plates. These plates are of an eggcrate design.

This design allows the maximum open flow area while providing sufficient structural support for the tubes to protect from mechanical or flow induced vibrations. By providing the maximum open flow area the steam/water mixture sees minimum resistance while passing through the bundle region. In addition, the localized flow blockage, stagnation and associated corrosion problems of the older perforated plate type tube supports are minimized.

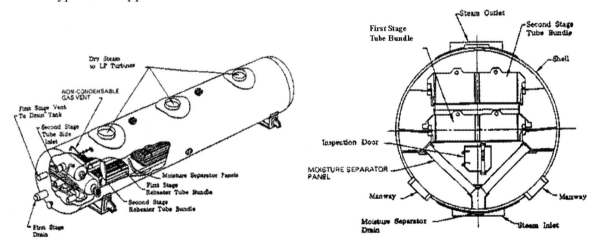

Pressurizer

What is the purpose of the pressurizer? The basic design of the pressurized water reactor includes a requirement that the water (reactor coolant or coolant) in the reactor coolant system not boil. Another way to put this is that the coolant must remain in the liquid state at all times, especially in the reactor vessel. To achieve this, the coolant in the reactor coolant system is maintained at a pressure sufficiently high that boiling does not occur at the coolant temperatures experienced while the plant is operating or in an analyzed transient. To pressurize the coolant system to a higher pressure than the boiling point of the coolant at operating temperatures, a separate pressurizing system is required. That is the function of the pressurizer.

In a pressurized water reactor plant, the pressurizer is basically a cylindrical tank with hemispherical ends, mounted with the long axis vertical and directly connected by a single run of piping to the reactor coolant system. It is located inside the reactor containment building. Although the water in the pressurizer is the same reactor coolant as in the rest of the reactor coolant system, it is basically stagnant, i.e. reactor coolant does not flow through the pressurizer continuously as it does in the other parts of the reactor coolant system.

Because of its innate incompressibility, water in a connected piping system adjusts equally to pressure changes anywhere in the connected system. The water in the system may not be at the same pressure at all points in the system due to differences in elevation but the pressure at all points responds equally to a pressure change in any one part of the system. From this phenomenon, it was recognized early on that the pressure in the entire reactor coolant system, including the reactor itself, could be controlled by controlling pressure in a small interconnected area of the system and this led to the design of the pressurizer. The pressurizer is small vessel compared to the other two major vessels of the reactor coolant system, the reactor vessel itself and the steam generator(s).

Pressure in the pressurizer is controlled by varying the temperature of the coolant in the pressurizer. Water pressure in a closed system tracks water temperature directly; as the temperature goes up, pressure goes up and vice versa. To increase the pressure in the reactor coolant system, large electric heaters in the pressurizer are turned on, raising the coolant temperature in the pressurizer and thereby raising the pressure. To decrease pressure in the reactor coolant system, sprays of relatively cool water are turned on inside the pressurizer, lowering the coolant temperature in the pressurizer and thereby lowering the pressure.

The pressurizer has two secondary functions. One is providing a place to monitor water level in the reactor coolant system. Since the reactor coolant system is completely flooded during normal operations, there is no point in monitoring coolant level in any of the other vessels. But early awareness of a reduction of coolant level (or a loss of coolant) is important to the safety of the reactor core. The pressurizer is deliberately located high in the reactor containment building such that, if the pressurizer has sufficient coolant in it, one can be reasonably certain that all the other vessels of the reactor coolant system (which are below it) are fully flooded with coolant. There is therefore, a coolant level monitoring system on the pressurizer and it is the one reactor coolant system vessel that is normally not completely full of coolant. The other secondary function is to provide a "cushion" for sudden pressure changes in the reactor coolant system. The upper portion of the pressurizer is specifically designed to NOT contain liquid coolant and a reading of full on the level instrumentation allows for that upper portion to not contain liquid coolant. Because the coolant in the pressurizer is quite hot during normal operations, the space above the liquid coolant is vaporized coolant (steam). This steam bubble provides a cushion for pressure changes in the reactor coolant system and the operators ensure that the pressurizer maintains this steam bubble at all times during operations. Allowing this steam bubble to disappear by filling the pressurizer to the top with liquid coolant is called letting the pressurizer "go hard" meaning there is no cushion and any sudden pressure change can provide a hammer effect to the entire reactor coolant system.

Part of the pressurizer system is an over-pressure relief system. In the event that pressurizer pressure exceeds a certain maximum, there is a relief valve called the Pilot Operated Relief Valve (PORV) on top of the pressurizer which opens to allow steam from the steam bubble to leave the pressurizer in order to reduce the pressure in the pressurizer. This steam is routed to a large tank (or tanks) in the reactor containment building where it is cooled back into liquid (condensed) and stored for later disposition. There is a finite volume to these tanks and if events deteriorate to the point where the tanks fill up, a secondary pressure relief device on the tank(s), often a rupture disc, allows the condensed reactor coolant to spill out onto the floor of the reactor containment building where it pools in sumps for later disposition.

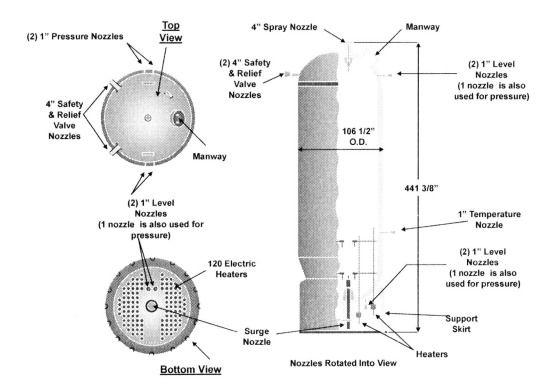

What supplies water to the pressurizer? The water level in the pressurizer is supplied by a primary source coming from the discharge piping of the reactor coolant pumps.

How is the pressurizer maintained? The pressurizer is controlled by the pressurizer pressure controllers by opening/closing spray valves to maintain level and turning on/off heaters to adjust temperature within the pressurizer. The pressurizer level controllers automatically adjust the CVCS letdown and charging pumps to maintain a level within the reactor.

How is the pressurizer protected? The pressurizer is protected by a safety valve and a relief valve. Both are needed in order to ensure that if pressure builds up the safety valve will be able to handle the steam bubble while the relief valve will be able to handle a high water level in the pressurizer if this condition occurs.

How does the pressurizer keep the PWR from boiling? The reactor coolant does not boil because it is maintained by the pressurizer, at a pressure which is greater than the saturation pressure for the existent reactor coolant temperature.

Where does the pressurizer power operated relief valves (PORV) relieve to? The PORV's when open relieve to a quench tank where the steam is condensed.

Where is the pressurizer located in the piping system? The pressurizer is located attached to the hot leg reactor coolant pump piping.

What does it mean to draw a bubble in the pressurizer? When the primary system is put back into service after an outage, etc. the pressurizer is full of water. The heaters in the pressurizer are turned on and allowed to bring the temperature of the primary coolant up to operating conditions. As the water begins to boil, the steam begins to collect at the top of the pressurizer. As more and more water is converted into steam, the water level slowly decreases until it is where the set point for its operation is reached.

What would happen if the pressurizer level was too high or too low? If the pressurizer level becomes too high the pressurizer heaters turn on to bring the water temperature above the boiling point in order to generate more steam. By generating more steam, the water level decreases to accommodate room for the steam. If the pressurizer level gets too low, the pressurizer heaters are turned off and/or the spray system inside the pressurizer opens up to add more water to the pressurizer to bring the water level back up.

What is the purpose of the steam bubble in the pressurizer? The pressurizer is like a huge shock absorber for the primary system. Since liquid is non-compressible it would become extremely hard if not impossible to control pressure within the system and it would not leave any room for fluctuation within the system. By having a bubble in the pressurizer, the steam bubble allows for better control of the pressure within the primary system. This gives the operating margin room to actually be able vary pressure within a given operating range, making the system much more user friendly.

How many heaters are in the pressurizer? Depending on the design of the unit a pressurizer may have a few heaters or more than a hundred depending on the amount of heat is needed to maintain a certain temperature/pressure balance.

What controls the pressurizer heaters? Primary system pressure controls the pressurizer heaters.

What is the purpose of the pressurizer heaters? Pressure is controlled in a pressurized water reactor to ensure that the core itself does not reach its boiling point in which the water will turn into steam and rapidly decrease the heat being transferred from the fuel to the moderator. By a combination of heaters and spray valves, pressure is controlled in the pressurizer vessel which is connected to the reactor plant. Because the pressurizer vessel and the reactor plant are connected, the pressure of the steam space pressurizes the entire reactor plant to ensure the pressure is above that which would allow boiling in the reactor core. The pressurizer vessel itself may be maintained much hotter than the rest of the reactor plant to ensure pressure control, because in the liquid throughout the reactor plant, pressure applied at any point has an effect on the entire system, whereas the heat transfer is limited by ambient and other losses.

Why would high pressure in the pressurizer trip the reactor? The high pressure in the pressurizer is designed to causes a reactor trip (which lowers reactor coolant pressure) and thus prevents excessive reactor coolant system blow-down due to the pressurizer safety valves.

Circulating Water System

The circulating water system is a fairly simple system. It is manly composed of several high capacity pumps, a cooling tower or basin (lake, river, etc.), a venting and priming system, a trash rack and/or traveling screen system and a chemical addition system.

Even though this system is not very complex when compared to other plant systems, it is extremely important. This system has one of the most important jobs in the plant cycle, to condense steam back into water inside the condenser. The condenser is a huge piece of equipment and is located directly under the Low Pressure Turbine. A condenser will have thousands of tubes running through them, which allows the circulating water to flow through. The steam leaving the low-pressure turbine section enters into the condenser and flows around the condenser tubes, which allows heat transfer to take place. This is a very simple yet very important function.

I will now take you through a simple flow pattern of a circulating water system that uses a lake for cooling water. I will start at the intake section of the lake. Lake water will flow through a trash rack and a traveling screen section to remove any debris in the water. Remember if the debris were allowed to flow through it would begin plugging up the tubes in the condenser. The traveling screen section is normally monitored by operating personnel and can automatically be set to operate in the event of a high differential pressure develop between the traveling screen section. The trash rack is monitored for big debris like tree branches, big fish, animals and basic trash coming off the lake. The chemical addition system allows chemicals to be added to the circulating water system so that biological growth can be minimized within the system. The next stop is the massive circulating water pumps. These pumps move massive amounts of water at a low pressure. The number of pumps is determined by the design of the plant and is operated when the unit is online. There is a venting and priming system that is normally operated when the condenser needs to be filled with water. This prevents air pockets from developing in the top of the condenser. Air pockets would greatly reduce the heat transfer in the condenser, which would cause the upper section of condenser tubes to fail. As water flows from the circulating water pumps, it flows through a massive discharge piping headed toward the plant. When the water enters the plant it will enter the condenser inlet water box and begin flowing through the condenser tubes toward the other side of the condenser. The water leaving the condenser tubes enters the condenser outlet water box and enters the discharge piping that leaves the condenser and then flows through the discharge piping back into the lake where the water is to be cooled naturally.

There are several different designs and sub-systems that are used with the circulating water system; it will change from plant to plant.

What is the purpose of the circulating water system? The circulating water system provides cooling water for the condenser and other heat exchangers through the use of pumps and piping. Without this system and its basic functions, the power plant is unable to operate.

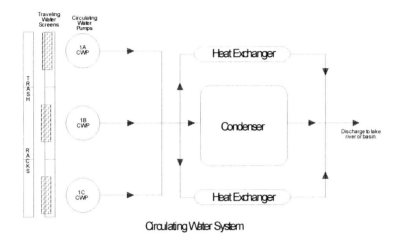

Circulating Water System

Circulating Water System (Elevation View)

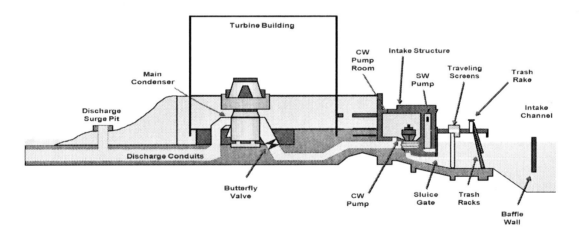

What is a trash rack? A trash rack is a large mesh screen that is located on the suction side of the circulating water pumps. It provides protection to the circulating water system and condenser and heat exchangers by stopping large debris like tree limbs or large fish from entering the system.

What is a cooling tower? A cooling tower is a device used to remove heat from the circulating water that is returning from the main condenser through the use of fans or a natural draft of air to cool the water.

What is the purpose of the traveling screen wash system? The traveling screen wash system is used to filter out debris like leaves, small fish, and other trash. These screens are rotated every so often out of the water so the screens can be cleaned by the cleaning system which sprays the inside of the screens to dislodge any fish or trash that has accumulated on them. This trash is blown off into a discharge basin where it is collected. This system may provide several other functions with the use of the screen wash pumps like irrigation, or to prime the fire system.

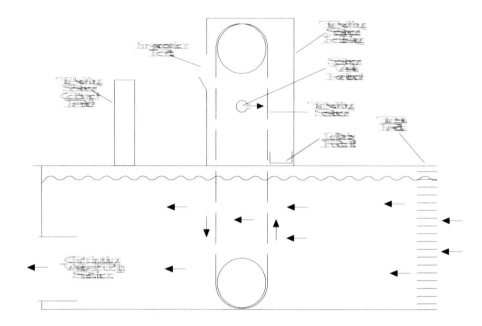

What are two types of cooling towers? The natural draft and mechanical draft are the two types of cooling towers.

Why must the circulating water system be primed before starting a circulating water pump? Priming the circulating water system is very important, if this was not done, a water hammer might occur in which piping, expansion joints, or even the condenser may become damaged. The pressure in the circulating water system is relatively low; the volume of water is incredibly high and powerful.

As water evaporates in a cooling tower, what happens to the concentration of impurities in the water? As the water falls through the cooling tower some of the water will evaporate, which will cause an increase in the concentration of impurities.

What two factors affect the type of cooling water system is used in a plant? One factor is the protection of the environment and the other is the source of water supply available.

What is the function of the circulating water pump? A circulating water pump is usually a big centrifugal pump capable of pumping a huge volume of water at a relatively low pressure. These pumps pump the cooling water through the condenser and other heat exchangers to remove heat from the plant systems.

What is the purpose of the vacuum priming system? The vacuum priming system is used to pull the circulating water up to the top of the condensers water box through the use of vacuum pumps. This process allows all of the condenser tubes to be used by allowing cooling water to flow through them. With this process, condenser vacuum and efficiency should be close to the condensers design specifications. If any air should enter the system, a set of float traps will remove the air from the system with minimum water loss.

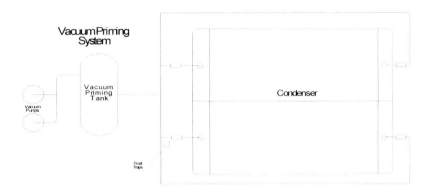

What is hydraulic gradient? Hydraulic gradient is a line of atmospheric pressure. In the circulating water system it will vary depending on the number of circulating water pumps operating and the lake level. Water at the elevations above this line will be under a vacuum and the water that is below this line will be under a positive pressure.

How can we control the buildup in the circulating water system? A buildup of growth in the circulating water system will reduce the amount of heat transfer in the condenser and other heat exchangers that use this water system. The buildup in the condenser tubes will act as an insulator and reduce the amount of heat transfer or in severe cases the tube can become plugged and possibly rupture. Chemicals are added to the water to reduce the buildup on the condenser tubes and allow for maximum heat exchange.

What would happen if the pH in the circulating water system should become too high or too low? If the pH of the water were high, scale would begin to form on the inside of the condenser tubes and eventually begin to cause pluggage problems. But if the pH was too low, the acidity of the water would eat away at the condenser metal causing them to eventually to become weak and possibly fail.

What are some disadvantages of a cooling tower? The disadvantages of a cooling tower are that 1. They require a strong (heavy) structure, 2. Metal structures are highly corrosive to the atmosphere, 3. It is difficult to make major repairs when necessary, and 4. Evaporation of water is high.

What is the purpose of the blowdown on a cooling tower? The cooling tower blowdown is used to remove impurities from the cooling water by removing water from the cooling tower while supplying fresh makeup water to the tower. This process helps to control the suspended solids in the cooling water.

On a mechanical draft cooling tower, what determines the amount of fans to be put into service (turned on)? The number of fans to be put into service is usually decided by the amount of condenser backpressure. Putting all of the fans in service will increase the evaporation rate and use excessive electrical power that is not really needed. So condenser backpressure is monitored continuously to determine when another fan is to be put into service.

What are some common functions that most cooling towers have?

1) Air circulation system
2) Water distribution system
3) Maximizing the surface area of the water
4) Collection and discharge basin
5) Minimize the water droplet carryover or water loss
6) A blowdown & water makeup system to minimize impurities and water chemistry.

How can the pH be corrected in a circulating water system? Correcting pH in a circulating water system can be controlled by the addition of chemicals, the addition of fresh water, and/or a blowdown system.

What are the advantages of using bromine instead of chlorine for cooling water treatment? 1) Bromine is more effective than chlorine as a biocide allowing reduction in doses. 2) Bromine undergoes a rapid decay process thus minimizing the impact on aquatic life.

What are some influences for the design of a plant's circulating water system? 1) environmental protection regulations, 2) the type of water supply available, 3) the cost, and 4) efficiency of the entire plant.

How does a mechanical draft cooling tower work? In this type of cooling tower, water from the condenser is pumped to the top of the cooling tower and distributed throughout the area of the tower. On the distribution deck there are holes that allow the water to free-fall through the tower. As these drops of water fall they collide with the splashboards layered across the tower from top to bottom. These boards break up the water into droplets, exposing more surface area thus cooling the water faster. A mechanical fan on top of the tower pulls air through the sides of the tower and upward through the fan. So as the water falls downward and the air flows upward a heat exchange is made. The cool water now falls into the bottom of the cooling tower called a basin. Here the cool water will again be pumped into the plants condenser and/or auxiliary heat exchangers.

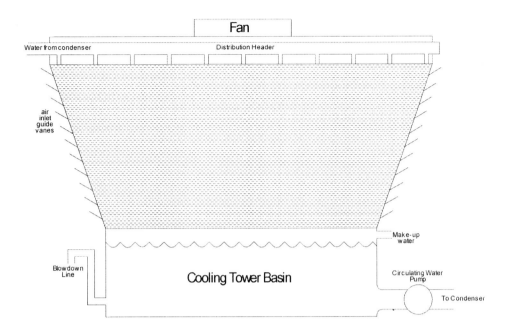

Where is the world's tallest natural draft cooling tower? The World's Tallest Natural Draft Cooling Tower is built, northwest of Cologne, Germany standing over 200 meters tall.

Where is the tallest natural draft cooling tower in the U.S.? Nine Mile Point nuclear plant in Oswego, New York has the tallest natural draft cooling tower in the U.S. at 532 ft. The cooling tower is visible from Chimney Bluffs in Sodus, New York…nearly 30 miles away.

How does a natural draft cooling tower work? The natural draft cooling tower works very similar to the mechanical draft cooling tower in that it works the same principles to cool the incoming water. The major difference is that there is no fan to pull the air through. The tower is designed to create a venture effect. As the hot air rises and flows through the narrow section of the tower, its velocity increases. This increase causes a low-pressure area to form at the bottom of the tower causing fresh air to enter the cooling tower. This fresh air in turn cools the hot water coming from the plant. Again as the water falls downward and the air upward, a heat exchange is made. The now cooler water falls to the basin and is ready to be used again in the plant.

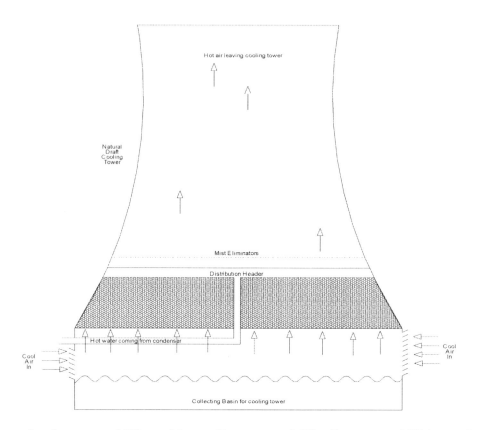

What are two basic types of fill used in cooling towers? The first type of fill is a splashboard used to break up the flowing water into droplets exposing more surface area. The second type of fill is film, water flows through a thin film over vertically oriented sheets of fill that is spaced for either horizontal or vertical air passage.

Which type of cooling tower fill is better? The film type of fill is better because it takes approx. 10 ft of splash fill to achieve the cooling effect of just 1 ft of film fill. Yet, the narrow passages between the film fill makes it more vulnerable to fouling.

What are sources that could be used to supply coolant to a nuclear plant? Four of these coolants are water based: a river, a lake, an ocean, or a reservoir (cooling tower) could be used to supply coolant. The fifth source is air and is used in air cooled condensers when water is not available.

The Condenser & Condensate System

The condenser as mentioned before is a massive piece of equipment. I will cover the different parts of the condenser. There are several sub-systems that work in conjunction with the condenser and will also be covered in this section. The design of the condenser has changed from the early experimental days when condensers were first used. I have seen a condenser system where the condenser was the size of a van to the size of a small house and the hotwell looked like a big piece of pipe at the bottom of the condenser where the condensed steam would collect to a hotwell that was as big as the condenser's floor and several feet deep. There are several different designs that are used and of course will change in size and design based on the how big the unit is and other factors involved.

There are several systems that are combined with the condenser housing like internal feedwater heaters, vacuum pumps or air ejector systems, hotwell level controls, corrosion coupons, conductivity meters, etc… On some condensers you might find one or several feedwater heaters located in the top portion of the condenser. This is to use some extraction steam from the turbine before it flows into the condenser. The condenser might use vacuum pumps or an air ejector system to remove air from the condenser. During unit startups the condenser must be under a vacuum before steam is allowed to flow through the turbine. If not, the steam would build-up inside the turbine casing creating a positive pressure, overheating the turbine and possibly allowing water induction to occur in the turbine blading. When the unit is online, the condensing steam mainly creates the vacuum in the condenser and the vacuum pumps or air ejector system is used to aid in the removing of non-condensable gases from the condenser. Together the condensing steam and equipment allow the vacuum to be maintained in the condenser.

Inside the condenser, there are structures that hold the condenser tubes in place so that they don't move or bend, yet allow for expansion and contraction to take place. These structures are called tube sheets and a condenser will have several of them. In the upper middle section of the condenser is the air exit lane where the vacuum pumps or air ejector system pull the non-condensable gases out of the condenser. This area is considered the lowest pressure area in the entire plant cycle. It is amazing to realize that the plant cycle may obtain pressures as high as several thousand psi and as low as 29 inches of vacuum (Hg).

There are several simple systems used with the condenser like the conductivity meters. These meters simply measure the conductivity of the water in the hotwell. Another system is the corrosion coupons, this system uses pieces of the same material in the condenser tubing and is exposed to the same conditions as the condenser tubes. Its purpose is to allow the lab personnel to determine if any material from the sample material is deteriorating. If the material is breaking down, lab personnel could determine what actions to take to reduce or eliminate the problem or even shutting the unit down. There are also many different instruments measuring different things like water flow through the condenser, differential pressures from the circulating water inlet and outlet flows, monitoring vacuum in the condenser, temperature differential from the inlet and outlet water flow and exhaust hood temperature at the turbine. You will see different types of controls that may be attached to the hotwell of the condenser allowing the condensate system to be controlled.

Again these systems may seem confusing at first but give it time. It will take some time to gain the experience and knowledge of these systems but when you know how the system operates you can be an extreme benefit to your company.

I will now cover the condensate system. This system is one of the first systems you will become very familiar with. It is a system that begins a process starting at the condenser's hotwell and ends at the deaerator. I will describe some equipment that might be used in this system; of course, plant design will determine what equipment to be used.

Starting at the condenser's hotwell, there will be a set of pumps called condensate pumps. These pumps pull water from the hotwell and add pressure to it. When the water is discharged from the pump it begins its journey toward the deaerator. The first piece of equipment we come to is the gland steam condenser. This device is a mini condenser that condenses steam coming from the turbine seal system, more on that later. The condensate flows through the gland steam condenser doing two things at once. 1) Condensing the gland steam coming from the turbine and 2) absorbing heat from the condensing steam. As the condensate water leaves the gland steam condenser it continues onward to the first of several heat exchangers called feedwater heaters. These heaters are bigger than the gland steam condenser yet the condensate water that is flowing through them does the same two functions as it did before. Once the condensate water is finally leaving the feedwater heaters it is on its final journey to the deaerator. Once the condensate water enters the deaerator it is exposed to extraction steam from the main turbine. This exposure allows the condensate water to be scrubbed by the extraction steam. This process aids in removing non-condensable gases from the condensate water. There is also a series of trays that allow the condensate to fall on and become broken up into droplets again this allows those non-condensable gases to become removed from the water. When the condensate water reaches the bottom of the deaerator, it is allowed to flow into the deaerator storage tank.

What is the purpose of the condenser? The main purpose of the condenser is to condense steam that has been exhausted from the turbine back into water (condensate).

How does a condenser work? A condenser is a heat exchanger that condenses steam that is exhausted from the turbine with the use of a circulating water system that is used on the tube side of the condenser to cool the exhausted steam. The circulating water enters the condenser through the inlet side water box and flows through the condenser tubes and leaves the condenser through the outlet side water box. As the water flows through the tubes of the condenser it cools the steam that is exhausted from the turbine allowing the steam to condense into water. As the steam is condensed, air and non-condensable gases are removed from the water as they fall (like raindrops) toward the bottom of the condenser, called the hotwell. The air and non-condensable gases are removed from the condenser by the air removal system.

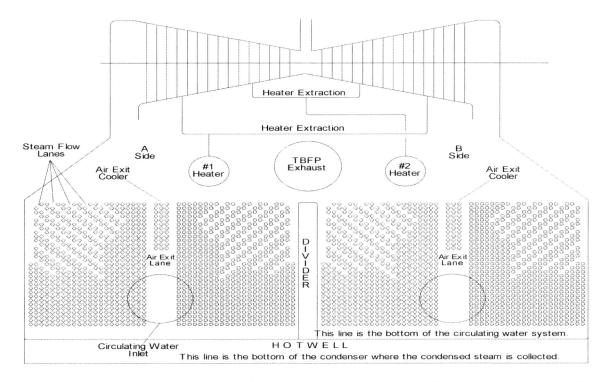

What are four functions of the condenser? Provides an area for the turbine's exhaust to flow into which maximizes the thermal efficiency of the unit. Serves as a reservoir by holding condensing steam to be used over again. It deaerates the condensing steam to minimize corrosion. Serves as a collection point for the steam cycle drains and plant auxiliaries.

What determines the size of the condenser? The size of the condenser to be used is determined by the base thermal load to be handled by the condenser and is set by the turbine at maximum load conditions. To handle transient thermal load conditions, which can exceed the maximum load conditions such as a turbine bypass system.

How is the vacuum efficiency of a condenser calculated? This formula is used to calculate condenser efficiency: vacuum efficiency = actual vacuum divided by ideal vacuum.

How is the efficiency of a condenser calculated? Condenser efficiency = rise of temperature of cooling water divided by temperature corresponding to the vacuum in the condenser minus the inlet temperature of the cooling water.

What graphs can be used to compare a condenser's actual performance with its design performance? The condenser performance curve will compare the condenser's actual and design performance.

What factors can affect a condenser's performance? Three basic factors the can have an effect on the performance of a condenser. One factor is the amount of vacuum in the condenser; the less vacuum in the condenser the less amount of work is done by the steam flowing through the turbine. The second factor is the circulating water temperature, the cooler the water is the better the heat transfer rate will be which in turn helps to maintain a high vacuum in the condenser. The third factor is the number of circulating water pumps running, the more pumps that are running the greater the amount of flow through the condenser the more heat is transferred from the steam to the water.

If a condensers actual backpressure is below that of design value, is the condenser operating more efficiently? If the condenser backpressure is increased, the heat rate will also increase causing the condenser and plant operation to become less efficient. When the condensers backpressure is higher than normal, the entire plant system must work harder to make up for the power lost in plant generation, thus causing undue stress. An increase or decrease in backpressure from the condenser's design point will be less efficient for the unit.

What are two types of coolant used for condensers? Two types of coolant used for condensers are the air-cooled condensers and the water-cooled condensers. The air-cooled condensers use air as the coolant. Refrigeration and air conditioning systems work in the same manner. There are plants that use air-cooled condensers because of the lack of water around them. Most power plants use the water type condensers, which use water as the coolant.

How does a high circulating water inlet temperature affect the vacuum in the condenser? The vacuum in the condenser will be affected by a high circulating water inlet temperature; the higher the circulating water inlet temperature the less heat is absorbed from the steam to the circulating water. This will slow the condensing process down causing the condenser backpressure to increase.

What is a direct contact condenser? A direct contact condenser is a condenser where two fluids come in direct contact with each other.

What is a surface condenser? A surface condenser is a condenser in which two fluids are separated by a solid surface, like tubes.

What is the air exit lane? The air exit lane is located in the center of the tube section of the condenser. This area is the lowest pressure region in the entire condenser. This area causes the steam air mixture exhausted from the turbine to flow around the tubes in the condenser.

What is a boundary layer? The boundary layer is a thin motionless layer of fluid that is created next to the surface of a stationary object in the flow path of a fluid.

How does a single pass condenser work? In a single pass condenser, the circulating water enters the inlet side water box of the condenser and flows through the tubes where it enters the outlet side water box of the condenser where it exits the plant system through the discharge piping.

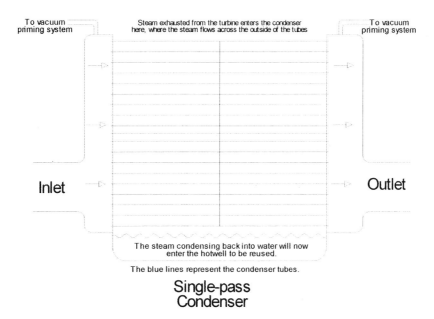

Single-pass
Condenser

How does a double pass condenser work? In a double pass condenser, the circulating water enters the inlet side water box of the condenser; from here the water flows through the tubes cooling the steam. After flowing through the tubes the circulating water enters the backside of the condenser where the water is directed to flow through the condenser again through another set of tubes. Once the water flows through the second set of tubes it enters the discharge piping of the condenser where it will exit the plant.

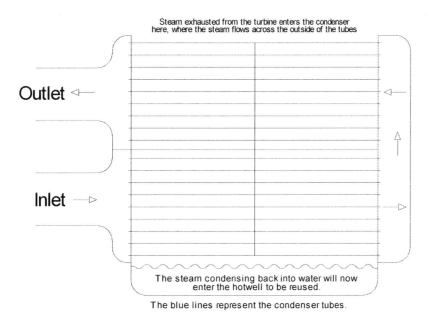

Double-pass
Condenser

What is the function of the air removal system? The air removal system removes air and non-condensable gases that have entered into the condenser by whatever means and aids in maintaining vacuum in the condenser. This system is also used to build-up or establishes vacuum in the condenser prior to the startup of the unit.

Where is the air cooler section of the condenser located? The air cooler section is a section of tubes located above the air exit lane in the condenser. It is this section where nearly all of the steam flowing through the air exit lane is condensed.

What does condensing mean? Condensing is a process where the state of water changes from a vapor (gas) to a liquid.

What is condensate? Condensate is a term used to represent the water that is used in a power plant system. Condensate is the water that appears as the steam from the turbine condenses in the condenser.

What are the steam inlet lanes? The steam inlet lanes are spaces or lanes located between the condenser tubes that guide or directs the steam flow around all of the condenser tubes so that even heat transfer occurs throughout the condenser.

What does it mean to break vacuum? To break vacuum means that the vacuum in the condenser is no longer required because of a unit trip or a normal shutdown. The vacuum is reduced by opening a vacuum breaker to allow air to enter the condenser until the condenser is at atmospheric pressure.

What is a vacuum breaker? A vacuum breaker is a valve that is connected to the condenser that can be opened to reduce or totally remove vacuum in the condenser or the valve can be closed to allow a vacuum to be established in the condenser with the use of the air removal system.

Why is the recirculation regulator located after the gland steam condenser? The recirculation regulator is located after the gland steam condenser in order to insure that there is enough condensate flow through the gland steam condenser should the recirculation regulator open.

What would happen if the hotwell level should become too low or too high? It is important that the hotwell level be maintained within its normal operating range. If the level was low in the hotwell it is possible that the condensate pumps could lose suction, which would disable their ability to pump. But if the level in the hotwell were high it would reduce the efficiency of the condenser by flooding the lower section of tubes with condensate; this causes the lower section of tubes to become useless for condensing steam.

What is the purpose of the hotwell? The hotwell is the area directly underneath the condenser tubes that collect and holds the water for use by the condensate pumps.

What is the function of the makeup water system? The makeup water system is composed of valves, piping, regulators, and many other devices. This system is controlled by the level of the hotwell. Should the level in the hotwell become low, a regulator valve would open to allow condensate from a storage tank to enter the condenser's hotwell to bring the level of the hotwell back to normal.

What is the purpose of the condensate makeup pump? The condensate makeup pump is used to provide a means of keeping water in the condensate system or to add condensate to other systems like a stator cooling water system or chemical system.

How is the hotwell level controlled? The hotwell level is controlled through the use of regulators. There are several regulated valves used to control the hotwell level. The makeup valve regulator is used to supply condensate to the hotwell to raise the hotwell level should it be low. A dump valve regulator is used to lower the level in the hotwell should it become too high.

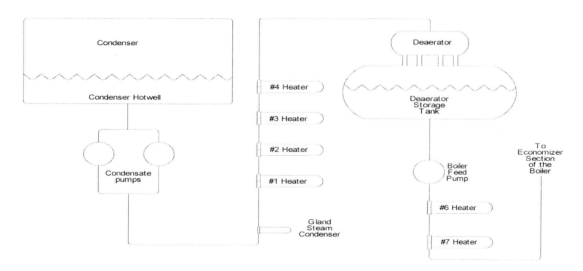

Are there any overrides to protect the condenser's hotwell? Depending on the plant design, one override could be a low level in the condensate storage tank whereas the vacuum make up regulator would close if opened so as not to allow any air to become sucked into the condenser, reducing the condensers vacuum.

Why is there a minimum flow requirement for the condensate pumps? A minimum flow requirement is required by the condensate pumps to assure that the gland steam condenser is kept cool. Without this requirement, the gland steam condenser may become over heated and possible tube failure could occur.

What is the purpose of a hogging pump? A hogging pump is used to reduce the pressure in the condenser to about 26 inches of vacuum. The hogging pump works similar to a vacuum pump but it increases the vacuum faster in the condenser than a vacuum pump. It is intended to pull vacuum during a unit startup and for only emergency purposes only. It is not used to maintain vacuum in the condenser.

How can high efficiency in a condenser be maintained? Here is a list of a few things that could help with the condenser efficiency: 1) a high difference between the circulating water entering and leaving the condenser is maintained. 2) Adequate circulating water flow through the condenser. 3) Condenser hotwell level should be normal. 4) Low air in leakage. 5) Maintain design vacuum and backpressure. 6) a low-pressure difference between the circulating water inlet and outlet of the condenser.

What are some ways to clean the condenser tubes without having to shut-down the unit? There are four general ways to clean the condenser tubes without having to shut-down the unit.

1) Use of abrasive sponge like balls	3) Use of chemicals
2) Backwash	4) Manually clean tubes – one section of the condenser at a time. (This would be rarely ever done because of the dangers involved.

What are some methods for finding a condenser tube leak? There are several methods that can be used for finding a condenser tube leak, here are just a few. 1) With the section of the condenser isolated and vacuum in the condenser, a thin plastic sheet can be used to cover the condenser tubes from the water box side of the condenser. If the condenser tube has a hole in it, the vacuum in the leaking tube will pull the plastic sheet into the tube. 2) With the unit off line, the condenser can be flooded with demineralized water until all of the condenser tubes are covered. The tube with the leak will allow condensate to run out of the condenser tube, indicating the leak. 3) With the section of the condenser isolated and with vacuum in the condenser, a lit candle can be used moving it from tube to tube. The tubing with the leak will try to suck in the flame of the candle. 4) With the unit on line and the use of helium and a helium leak detector a condenser tube leak can be found.

How is the heat transfer affected in the condenser when the air leakage is high? When the air leakage is high, this indicates that the heat transfer has been reduced because not all of the non-condensable gases are being removed from the condenser causing the gases to blanket or insulate the condenser tubes preventing or reducing heat transfer.

How is efficiency and heat rate affected when the condensers vacuum falls below its design value? When condenser vacuum decreases, the turbine becomes less efficient because of the reduction in steam flow through the turbine. This results in less electrical energy that is produced while the plant's heat rate increases and its efficiency decreases.

Does the condensate level in the hotwell provide any information about the condensers performance? An excessively high hotwell level could flood the lower condensing tubes and the deaeration trays thus affecting the capacity of the condenser and the deaeration process. A low hotwell level could cause the condensate pumps to lose suction and become damaged.

How does steam, create a vacuum? Remember that a single drop of water will expand 1500 times its size when turned to steam. Now just reverse the process. When steam enters the condenser and condenses back to water, it will shrink 1500 times smaller than it was as steam.

What can be used to determine the proper circulating water flow rate required to maintain the proper vacuum in the condenser as the load on the unit changes? A graph called the circulating water pump selection curve is used to allow the operational changes in the number of circulating water pumps in service so that the desired amount of cooling and condenser vacuum can be maintained as load changes. This graph also calculates the number of pumps in service with the inlet cooling water temperature.

If a condenser has a severe tube-fouling problem, will the circulating water outlet temperature be affected? Yes, the circulating water outlet temperature will be abnormally high and a big pressure difference between the condenser circulating water inlet and outlet will be present indicating tube pluggage.

How can tube-fouling problems be minimized? The use of chemicals in the circulating water system, backwashing the condenser tubes, using abrasive balls to flow through and scrub the tubes clean.

How can air leak into a condenser? Air can leak into a condenser through many ways like an expansion joint, a valve with bad packing, a condenser tube leak, steam seals malfunctioning, rupture diaphragms on the L.P. turbine cracked. There are just too many things that could cause this type of problem.

What is the purpose of a steam air ejector? A steam air ejector is a jet pump that uses high-pressure steam that flows through a nozzle to draw air out of the condenser in order to establish a vacuum in the condenser.

What methods are used to detect air leakage into the condenser? 1) Monitor the performance of the condenser air removal system. 2) With the unit off line, listen for the sound of air being pulled into the condenser. 3) Check for any changes in related parameters. 4) Using a gas like nitrogen, and the use of the condenser air removal system air leakage can be detected.

What would cause an insufficient circulating water flow through the condenser? Plugged tubes from debris in the circulating water, mechanical damage, or a severe biological buildup.

How is the circulating water flow measured through the condenser? By measuring the differential pressure from the inlet of the condenser to the outlet of the condenser it can be determined that enough or not enough flow is flowing through the condenser. If the differential is low, good flow of water is flowing through the condenser but if the differential increases, it could be pluggage in the tubes or scale buildup. Determining factors are also, unit load, # of circulating water pumps in service and vacuum/backpressure.

What is backpressure? Backpressure is the difference in pressure between the condenser's manometer and the barometer reading.

How does backpressure affect the performance of the unit? By decreasing the backpressure in the condenser by one inch of mercury can mean a tremendous amount of savings. Turbines have a designed set point for backpressure and any deviation from this point contributes to a higher rate and a loss of megawatts produced.

Why is the air off-take core located in the center of the condenser? The air off-take is located near the center of each tube bank so that the flow of steam is inward from the steam space around the banks toward the central core. This type of tube arrangement becomes progressively smaller as the flow of steam nears the center of the bundle. As the flow is reduced through condensation, the velocity remains constant.

How is a two halved condenser flexible in operation? One half of the condenser can be shut down while the other side is isolated for cleaning or inspection. With only one side of the condenser in operation, unit load is reduced to half load.

Can problems with the condenser vacuum cause a generating unit to shut down? If the condenser vacuum should drop below a designed set point, an automatic trip system will actuate and close all of the turbines steam valves. Closing off the turbine steam valves stops the flow of steam to the turbine in order to protect the turbine and condenser from a high-pressure buildup in the turbine low-pressure casing and condenser in which they are not designed to handle.

What is the purpose of the crossover valve in a condenser? On this type of condenser, a crossover valve is positioned between the two halves of the water box. With each circulating water pump having its own discharge piping up to the condenser, should one side of the condenser's water box need to be isolated, the cross-over valve can be closed allowing total isolation to the isolated half of the condenser. Not all condensers have a crossover valve. Condensers that do not have a crossover valve, probably have the circulating water pump discharge piping come together in one pipe to supply water to the entire condenser.

How does an air cooled condenser work? An air-cooled condenser works in the same manner, as a water-cooled condenser except the air is the coolant causing the steam to condense.

What are two types of air-cooled condensers? Two types of air-cooled condensers are the direct steam condenser and the indirect steam condenser w/coolant loops.

What are some advantages and disadvantages of an air-cooled condenser? There are four major advantages with using an air-cooled condenser.

1) Makeup water is not required for cooling
2) Maintenance is less expensive
3) Chemicals are not needed, no blow down required
4) No fogging, misting, or icing is encountered during the seasons

There is one major disadvantage with using an air-cooled condenser. The minimum temperature and corresponding steam condenser pressure are not as low as with a water type condenser. The result is a higher condenser backpressure (3.5 to 9), which lowers plant efficiency, and an increased heat rejection for the same power production.

What is the purpose of the condenser water box differential transmitter? The condenser water box differential transmitter is an indication of possible pluggage within the condenser tubes.

On a condenser vacuum pump, what is the difference between the hogging mode and the holding mode? In the hogging mode, the vacuum pump pulls air directly from the condenser through the pump and out to atmosphere. In the holding mode, the vacuum pump pulls air indirectly from the condenser through an ejector nozzle or ventura by pulling air from the separator tank through the ventura, back to the separator tank then out to atmosphere. The air is pulled from the condenser by the low-pressure area created at the ventura.

What are the effects of air leakage in the condenser? 1) an increase in back pressure 2) greater amount of cooling water is needed to condense steam 3) reduces the rate of steam condensation 4) exhaust hood temperature is increased.

How can the condenser performance be determined by the inlet and outlet temperature differences? A higher than normal temperature difference could indicate reduced flow through the condenser, possibly tube pluggage. A lower than normal temperature difference could indicate that scale has built up inside the condenser tubes, this would also indicate a higher than normal backpressure.

Condenser Air Removal Unit

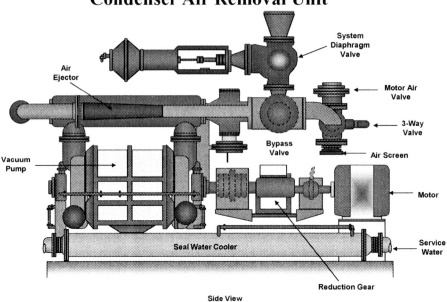

Side View

What is the difference between deaeration and deoxygenation? Deaeration is the removal of both O_2, CO_2 and other dissolved gases from the water while deoxygenation removes only O_2 from the water.

What are the factors that affect the deoxygenation reducing capabilities of hydrazine? 1) Temperature - hydrazine can remove dissolved oxygen at an appreciable rate only at water temperatures at or above 373^0K. 2) pH value - at a pH value below 7, hydrazine will fail to prevent corrosion and can even increase it.

What are two common causes of condenser tube fouling? 1) biological fouling and 2) scale fouling.

What is the purpose of an expansion plug? An expansion plug is used to isolate a single tube inside a heat exchanger that is damaged beyond repair and can no longer be used. The expansion plug is simple in design constructed with a bolt, a nut, expanding material, and two washers. When the tube is found and needs to be plugged, the expansion plug is pushed into place and then is screwed into place. As the screw is turned the washers push towards each other causing the expanding material to push outward against the tube. This outward pressure is what holds the plug in place.

What could cause a loss of condenser vacuum?

1. Increase in circulating water temperature
2. Biological fouling of the condenser tubes
3. Malfunction of a vacuum pump
4. Leaking miscellaneous drains (startup, shutdown, etc.)
5. Feedwater heater emergency drain malfunction
6. Low Pressure Turbine Rupture Disc
7. Steam seals malfunction (too high or too low)
8. Condenser vacuum breaker malfunction
9. Steam bypass valve open to isolated condenser waterbox
10. Low pressure turbine / condenser expansion joint malfunction
11. Steam traps malfunction
12. Extremely high level in the hotwell
13. Condenser vacuum priming system malfunction
14. Condenser vent / drain open
15. Reduced flow through the condenser (condenser discharge valve closed or broken).
16. Condenser hogging pump mis-aligned.
17. Restricted water flow through intake structure
18. Condenser instrumentation leaking (hotwell indication, pressure/temperature sensors, sight glasses, transmitters, etc.)
19. If more than one condenser is in operation check the equalizing lines top and bottom for leaks.
20. Any drain tanks draining into condenser…if they are drained the condenser will suck air.
21. Circulating water pump problems
22. Condenser structure damaged
23. Vacuum priming pump malfunction
24. Too few vacuum pumps in service

What is the purpose of a condensate booster pump? A plant might use a condensate booster pump to increase feedwater pressure instead of paying enormous amounts of money for a much bigger condensate pump. These decisions are based upon multiple variables like the number of feedwater heaters being used, the maximum flow rate required, and/or the availability of service and parts for the equipment…this is only a couple of variables mentioned.

Feedwater Heaters

The use of feedwater heaters in the plant cycle allows the efficiency of the unit to be increased by using extraction steam from the turbine to be feed into these heaters to preheat the condensate/feedwater heading toward the boiler. If feedwater heaters are not used, the efficiency as well as life expectancy of the equipment like the boiler would be reduced because the boiler would have to create that much more heat to compensate for the low temperature feedwater entering the boiler in order to ensure that the steam leaving the boiler would be superheated and not saturated.

There are two general types of feedwater heaters, the opened feedwater heater and the closed feedwater heater. The open type feedwater heater is when extraction steam and condensate/feedwater mix together, like in the deaerator. The closed type feedwater heater, the extraction steam and the condensate/feedwater remain separated by the tubing in the heater. The extraction steam flows around the tubes while the condensate/feedwater flows through the heater tubes.

There are several different types of feedwater heaters, again based on plant design and heater manufacture. I will describe for you a 3-zone feedwater heater. Basically there are three zones for this heater, they are: Desuperheating zone, Condensing zone and the Sub-cooling zone. On Fig. 4-1, you can see that the feedwater enters the bottom of the feedwater heater while the extraction steam enters the feedwater heater at the top. This is what we call counter flow, when one fluid is flowing in the opposite direction of the other fluid. As you can see there are several baffles placed inside the heater to 1) support the heater tubing and 2) direct the extraction steam flow. This design makes the heater more efficient, by allowing the condensate/feedwater to absorb as much of the heat from the steam as possible causing the steam to condense and the condensate/feedwater to increase in temperature. The extraction steam is bled from different sections of the turbine. Ex: High-pressure heaters might use Cold Reheat steam as an extraction while a Low-pressure heater might use an extraction coming from the last few stages of the Low-pressure turbine. Fig 4-2 will give you an idea on how extraction steam could be used in the heaters.

There are several devices that might come with the heater like a sight glass for the sub-cooling zone, an impingement plates, safety valves on water and steam side, heater drain entrance, shell side drains, tube side drains, normal and emergency heater drain system and a nitrogen addition valve (during shutdown). Again these devices might or might not come with the heater again depending on plant design and manufacture.

Each feedwater heater will have its own drain system. This system protects the heater from a low or high level in the sub-cooling section of the heater. This system can be setup in many different ways. One popular way is the cascading drain system. This system like Fig 4-3 can show you how the normal drains from the heater might flow and if an abnormal condition occurs how the alternate drain system might allow the drains to flow. Normally the normal drain setup is from a high-pressure heater to a lower pressure heater to another low- pressure heater and so on, until the drains enter into the condenser. You are probably asking yourself, "Why go through all that trouble to collect the drains from these heaters?" Remember, the water that we purified cost us money to create. Now if we just ran all the drains from the heaters to the sewer, we would be throwing a lot of money out the window and the efficiency of the plant would greatly decrease and a lot more water would have to be purified.

I will now go over a basic flow from the deaerator to the boiler entrance. The condensate/feedwater in the deaerator storage tank I will now refer to it as feedwater is fed into a boiler feed pump. The boiler feed pump increases the energy in the water (pressure) from just static head pressure to a few thousand psi. There is no specific number of heaters to flow through again depending on plant design. Normally there is 2 or 3 high-pressure feedwater heaters between the boiler feed pump and the boiler. The feedwater will flow through one heater at a time again gaining temperature along the way to the boiler while the extraction steam is reduced in temperature as it cascades toward the next heater. When the feedwater finally leaves the last high-pressure heater it will flow toward the boiler and enter into the economizer section of the boiler. This section of the boiler is the first heat exchanger in the boiler that the feedwater flows through.

Now you should have a general understanding of what is actually happening to the feedwater as it has traveled from the condensate and feedwater system to the boiler. These systems will take a while to learn and understand. But once you have achieved this knowledge you will be a much better troubleshooter when something goes wrong. I speak from experience of course.

What is the purpose of a feedwater heater? Feedwater heaters increase the efficiency of a power plant cycle by using steam that is extracted from the turbine to heat up the feedwater before it enters the boiler. For an increase of every 10 degrees F in feedwater temperature will save 1% in fuel consumption. If extraction steam was not used, it would be wasted heat spent to the condenser.

What determines how many heaters are used in a plant cycle? Power plants range from having five to ten feedwater heaters. The more heaters used means the plant's efficiency will increase. But a point is reached where only a small improvement in efficiency is gained by adding another heater. And more steam would be needed from the boiler for the turbine to produce the same amount of work and adding more heaters means a larger plant is needed to produce the same amount of electrical power.

What are two common types of feedwater heaters? A closed feedwater heater and an open feedwater heater are two common types of heaters. An open feedwater heater actually allows the extraction steam to come in contact with the feedwater. This process aids in the removing of non-condensable gases. A closed feedwater heater keeps the extraction steam and feedwater from mixing through the use of tubes. These tubes contain the feedwater directing it through the heater while allowing extraction steam to pass around the outside of the tubes.

How does a cascading drain system work? As extraction steam flows through a feedwater heater it condenses, the water will pass through a sub-cooled section of the heater and from here the drips from this heater will flow downstream to the next lower pressure heater. When the heater drips enter the heater, it will flash into steam and mix with the extraction steam and begins the process all over again.

Heater Drain System

How does the cascading drain system improve plant efficiency? The cascading drain system improves plant efficiency by providing the maximum amount of heat to be recovered from the extraction steam.

What is meant by air binding? Air binding is when air is trapped in either the waterside or shell side of a feedwater heater as a result of improper venting of the heater; this problem causes the terminal temperature difference to increase while the drain cooler approach temperatures remains normal.

What are three types of shell and tube heat exchangers that are named for the direction of fluid flow through them? One type is a counter flow heat exchanger where two fluids travel in the opposite direction along a tube. Another type is a parallel flow heat exchanger where both fluids are traveling in the same direction along the tube. The third type is a cross flow heat exchanger where both fluids are flowing at right angles to each other along a tube.

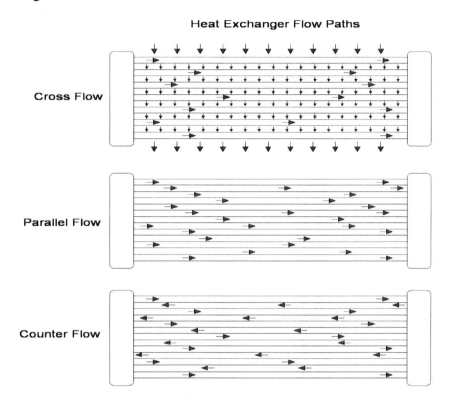

What is terminal temperature difference? The terminal temperature difference is the difference between the saturation temperature of the steam that's entering the heater and the temperature of the feedwater leaving the heater.

How does bypassing a high-pressure feedwater heater affect the unit? Bypassing a high-pressure heater will cause a decrease in unit efficiency due to the inadequate feedwater temperature entering the boiler. Since the temperature of the feedwater is lower, the boiler will have to be over fired to compensate for the temperature loss. Over firing in the boiler will lead to abnormally high furnace temperatures, which might cause tube failures and create more pollution than normal.

What is meant by drain cooler approach? The drain cooler approach is the difference between the temperature of the heater drips leaving the drain cooler section of the feedwater heater and the temperature of the feedwater entering the tubes from the inlet of the heater.

What is the purpose of the normal and emergency drain regulators on a heater? The normal and emergency drain regulators control the water level on the shell side of a feedwater heater. The normal drain regulator tries to maintain a normal level in the heater, but if for some reason the level is getting higher and higher, the emergency drain regulator will open to help bring the level back to a normal setting. When a heater reaches its high alarm, the drain cascading from another heater will close so as not to compound the problem in the troubled heater.

What is the purpose of the non-condensable gas vents? The non-condensable gas vents aid in the removal of non-condensable gases that enter a feedwater heater. If these non-condensable gases were allowed to accumulate in a heater they would blanket or insulate the tubes not allowing the proper heat transfer to take place.

Where do the feedwater heaters non-condensable gas vents go? The non-condensable gas vents on low-pressure feedwater heaters usually are sent to the condenser where these gases are removed from the system by an air ejector or vacuum pumps. On intermediate or high pressure feedwater heaters the non-condensable gases would usually go to the condenser.

What is meant by the tube side of a feedwater heater? The tube side of a heater is where the feedwater enters the heater and flows through the tubes to the outlet side of the heater.

What is meant by the shell side of a feedwater heater? The shell side of a heater is where the extraction steam and cascading drain system actually flows around the outside of the tubes in the heater.

What is the purpose of the shell side drains? The shell side drains on a heater allows the water remaining on the shell side of the heater to be drained when the unit is shut down or if the heater is isolated for repairs.

What is the purpose of the tube side drains? The tube side drains on a heater allows the feedwater in the heater to be drained. This is usually done when the heater is isolated for repairs.

What is the purpose of the extraction block valves? An extraction block valve is an isolation valve used to protect the turbine from an excessively high level in a feedwater heater from reaching the turbine blading and cause damage.

What is the purpose of the extraction non-return valve? The extraction non-return valve is a swing check valve that is weighted and air assisted to close to prevent the turbine extraction steam or water from a heater to flow back into the turbine causing the turbine to over speed and / or water induction into the turbine blading.

What are the three sections of a three sectional feedwater heater called? The three sections of a three sectional feedwater heater are the desuperheating section where the extraction steam begins to cool down, the condensing section is where the steam is condensing back into water, and the sub-cooling section where the water in the heater is cooled enough so that when it leaves the heater to cascade to another heater it have a big temperature differential.

Feedwater Heater

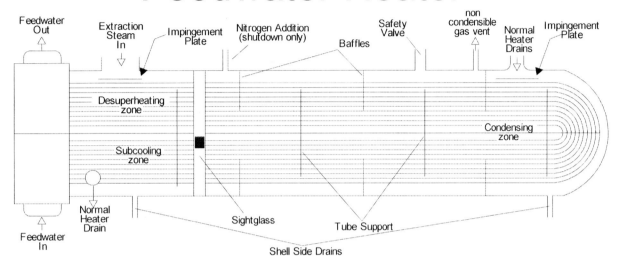

What is the purpose of the heater side extraction drain valves? The heater side extraction drain valves will be open when the extraction block valve closes. This valve is open to relieve any steam and/or water built up between the extraction non-return valve and the extraction block valve, which drains into the condenser. These valves may be manually operated or setup to automatically operate depending on plant design.

What is the purpose of the turbine side extraction drain valves? The turbine side extraction drain valves will be opened when the extraction block valve closes. This is opened to maintain a flow through the extraction piping in order to keep water from accumulating in a dead space when the extraction block valve is closed.

How should the performance of a feedwater heater be monitored? The performance of a heater is important because it can affect the rest of the plant cycle. The heater levels, feedwater inlet & outlet temperature, extraction steam temperature, and heater drains temperature should all be monitored for maximum performance.

What is a surface heat exchanger? A surface heat exchanger will transfer heat from one fluid through a surface to another fluid of a lower temperature. Example, feedwater heaters, condensers.

How does bypassing a low-pressure feedwater heater affect a unit's heat rate? A low-pressure feedwater heater will affect a unit's heat rate but not as much as an intermediate or high-pressure heater will. The further from the hotwell the heater is (in piping), the more of an affect it will have on a unit's heat rate.

What is the purpose of the impingement plate in a feedwater heater? The impingement plate in a heater is a plate located on the top part of the heater where the extraction steam enters the heater. This plate protects the heater tubes from being destroyed by the erosion process of the extraction steam flowing across the tubes.

Is flashing steam harmful? Flashing is a process where a rapid change occurs like when water turns into steam. This usually happens when a sudden drop in pressure occurs. A drop of water will expand approximately 1500 times its size into steam. So you can see the potential energy found in this condition.

What is the purpose of the flash pot? The flash pot is used as an indication that feedwater has flashed into steam in the TBFP circuit. The flash pot has a vent that allows the steam to be vented from the system to the atmosphere and allowing water to return back to the TBFP circuit. You might see something like this if you had a booster pump feeding the TBFP.

What is the difference between a low-pressure heater and a high-pressure heater? Low-pressure heaters are located between the condensers hotwell and the feedpump. A low-pressure heater operates at or below 900psi of pressure on the tube side. High-pressure heaters are located after the turbine boiler feed pump, where pressures can be anywhere from 900psi up to 5,000psi on the tube side.

What is sub-cooled water? Sub-cooled water is when the temperature of the heater drips are lowered below the saturation temperature or boiling temperature. Sub-cooling the water prevents the heater drips from flashing into steam when flowing through a drainpipe into the next heater.

What are tube sheets? Tube sheets can be found in heat exchangers like the condenser, feedwater heaters, or other pieces of equipment. Tube sheets are metal plates that are attached to the end of the tubes to support the tubes and separate the shell side fluid from mixing with the tube side fluid.

What do baffles do in a heat exchanger? Baffles are used to direct the flow of the shell side fluid allowing the maximum amount of heat to be exchanged between the tube side fluid and the shell side fluid.

What are several ways of classifying shell and tube type heat exchangers? The basic ways of classifying these types of heat exchangers. One way is the direction or flow path of the two fluids in the heat exchanger like parallel flow, counter flow or reverse flow. The second way is the number of times that the fluid will make within the heat exchanger like is it a single pass heat exchanger or does it pass through several times. The third way is the construction of the heat exchanger like the shell design, is it vertical or horizontal, is it straight tubes or u-tubes in the heat exchanger.

Why are horizontal feedwater heaters preferred over vertical feedwater heaters? Vertical feedwater heaters minimize floor space yet are higher in maintenance cost and in operational problems. Horizontal feedwater heaters have a greater liquid (drain) volume that is available for level control during plant transients.

What is the function of the extraction steam system? The extraction steam system is composed of valves and piping to direct and control the flow of extracted steam from the turbine to the feedwater heaters. This design will increase the units' efficiency while maintaining a designed heat rate.

Where would straight tube heat exchangers be used? Straight tube heat exchangers are easier to clean so they are commonly used in heat exchangers where the tubes are susceptible to buildup of rust, scale, and sludge.

Where would U-tube type heat exchangers be used? The U-tube type heat exchangers are more commonly used where tubes are not susceptible to a lot of contaminant buildup; this is because the U-tubes are much harder to clean than straight tubes. Another advantage of U-tubes is that they allow for expansion and contraction to occur in systems where temperatures change frequently.

What are three advantages of using extraction steam in feedwater heaters? One advantage is that extraction steam reduces the thermal stress on the boiler, by preheating the feedwater. The second advantage is the extraction steam system improves the overall efficiency of the plant. And third of all, this type of system allows for the use of smaller turbines.

What would happen if the feedwater entering the steam generator was not heated? If the feedwater was not heated before it entered the steam generator, severe thermal stress on the metal will occur possibly seriously damaging the steam generator internals.

How does a feedwater heater that is located in the condenser work? The basic difference between a feedwater heater in a condenser and one that is located outside of a condenser is that the heater in the condenser receives its' extraction steam almost directly from the turbine. The drains from the heater can be routed to the condenser or a heater drains cooler or even a heater drains receiver. Plant design will vary.

What four factors influence the amount of heat transferred in a cooler? The temperature differential between the two fluids, the thickness of the tubing or material, the flow rate between the two fluids, and the fluid levels within the cooler.

What is a water hammer? A water hammer is a condition that occurs when condensate slams into the wall of a pipe. This condition can happen in any system, but it is more common in steam piping.

What are two symptoms of a water hammer? The symptoms of a water hammer are vibration and noise. It is called a water hammer because that's what it sounds like someone hitting the corner of a pipe with a sledgehammer.

How can water hammers be prevented? Water hammers can be prevented by the use of steam traps in steam lines, opening or closing valves slowly, or reduce the flow of a fluid when starting a pump.

When isolating a feedwater heater from service, what are some precautions that should be remembered? When isolating a feedwater heater from service, first do not interrupt the flow of feedwater to the boiler. Secondly, do not stop the feedwater flow to the heater until the extraction steam has been removed and isolated from the heater. Finally, release pressure on the heater slowly, boiling water can cause serious damage if a hand or face is close by.

What are three factors that can affect the operating heat rate of a plant? One factor is the temperature of the superheated steam entering the turbine. The second factor is the condenser vacuum. And the third factor is the temperature of the feedwater in the system.

If a heater is isolated and the safety valve lifts, what would this indicate? An improperly isolated heater or a tube leak could cause a safety valve to lift. Extreme caution should be taken whenever isolating a feedwater heater.

What are two concepts to remember about heat transfer in a heat exchanger? First, heat can only be transferred from one substance to another as long as there is a temperature difference between them. Secondly, a higher temperature fluid will always transfer heat to a lower temperature fluid.

What are two ways heat is transferred in a heat exchanger? One way heat is transferred is by conduction, where heat is transferred into a solid. The second way is convection, where heat is transferred into and through a fluid.

Can the amount of heat transferred be affected by a change of one or both flow rates? The amount of heat transferred can be affected by changing the rate of flow of either one or both of the fluids. Being able to control the flow rate of one fluid will vary the amount of heat transfer, which in turn controls the temperature of the other fluid.

Where is the highest pressure in the primary system? The highest pressure in the primary system is at the discharge of the reactor coolant pumps. From this point forward the pressure gradually drops until it which the suction of the reactor coolant pump, this is where the lowest pressure is in the primary system.

Where is the highest pressure in the secondary system? The highest pressure in the secondary system is at the discharge of the steam generator feed pump, from this point on the pressure will continue to drop until it enters the condensate system again. The lowest pressure in the secondary system would be in the condenser.

What can cause tube failures in feedwater heaters? Here is a list of a few factors that could cause tube failures. (oxygen pitting, weak welds, tube vibration, improper plugging, tube joint failure, improper pH, cavitation type erosion, steam impingement erosion, poor maintenance practices, poor operating practices, laminated corrosion, stress corrosion cracking, inlet tube end corrosion/erosion, and de-alloying.)

What is the purpose of the orifice in the feedwater heater non-condensable gas vent? The orifice reduces the flow and pressure of the steam/air mixture flowing through the vent line.

What are two functions of a feedwater heater? The first function of a feedwater heater is to use extraction steam bled from the turbine to bring the feedwater temperature up, minimizing thermal differential within the boiler. The second function is for the feedwater heater to condense the extraction steam so that it can be reused within the system while adding in the removal of non-condensable gases.

What determines the type of extraction steam to be used on a feedwater heater? The type of extraction steam to be used on a feedwater heater is determined by the position of the heater in the feedwater process (where the heater is located going to the boiler). Low-pressure heaters use low-pressure extraction because the position of the heaters will allow the heat from the extraction steam to slowly warm up the water. Intermediate and high pressure heaters use higher pressure extraction steam to continue heating the feedwater in the heaters and not allowing the feedwater to flash into steam inside the heater.

What determines how many zones are needed in a closed feedwater heater? The number of zones in a closed feedwater heater is dependent on if the extraction steam entering the heater is superheated or saturated. The desuperheating zone is used usually when superheated extraction steam is to enter the heater.

In what part of the closed feedwater heater, does most of the heat transfer take place? Approximately 60% to 80% of the total heat transfer takes place within the condensing zone of a feedwater heater.

What determines the type of extraction steam to be used on a feedwater heater? Depending on the type of feedwater heater (High, intermediate or low pressure) will determine what part of the turbine the extraction steam will come from. A low pressure heater located in a condenser would receive extraction steam coming off of the last few stages of the low pressure turbine. A low-pressure heater located outside of a condenser may use extraction coming from the middle to the beginning stages of the low-pressure turbine. An intermediate heater will normally receive extraction steam from an intermediate pressure turbine while a high pressure heater can receive its' extraction steam from the cold reheat line leaving the high pressure turbine. This is just an example; every power plant is different in some way depending on plant design, number of feedwater heaters, etc.

How does feedwater temperature affect boiler fuel consumption? For every 10^0F increase in feedwater can produce approx. 1% savings in fuel consumption. The higher the feedwater temperature, the lower the heat rate and the greater the plant efficiency.

Turbine

The turbine is another piece of equipment that will make you wonder, how. I mean, this equipment is not as big as the boiler but when you imagine a metal shaft with blading rotating at 1800 rpm (30 times in one second) it is hard to believe. Then when you learn that the turbine itself weights many tons it becomes even harder to believe. The turbine is an incredible machine and fairly efficient at converting thermal energy into mechanical energy. When you finally understand how the turbine works, the tolerances, the design and its limits, you will be amazed how this all comes together. There are many sub-systems that work with the turbine in order to ensure that it operates properly and efficiently. As for right now, we will focus on just the turbine.

The turbine can be classified in many ways depending on certain factors like, types of blading used (impulse or reaction), how many stages the turbine has (single or multiple), steam flow into the turbine, condensing or non-condensing, extraction or non-extraction and the amount of steam pressure in the turbine (high, intermediate or low pressure). In the example below I will be describing a tandem compound turbine that uses a combination of impulse and reaction blading, multiple stages, tangential steam flow arrangement, condensing and extraction setup. Please don't let all this mumbo gumbo bother you. Like before it will take time to get use to the terminology and functions.

Our turbine with the basic turbine sections labeled and arrows to help show you how the steam flows through the turbine and finally ending up into the condenser. On the far left side you see where the main steam enters the high-pressure turbine section. The first section the steam flows through is called the nozzle block. The nozzle block or first stage is divided into sections to allow the main steam to flow evenly through this section of the turbine. Up until now you realize that steam is flowing from the boiler and now beginning its way into the turbine. But I didn't mention on what controls the steam flow to the turbine. There are valves that control the main steam to the turbine. There are two sets for the high-pressure section; they are the main steam stop valves and the control valves. Between these two sets of valves is an area called the steam chest. The steam chest allows the steam to slowdown and become smoother so it is easier to control into the turbine. The main steam stop valves are normally wide open while the control valves open or close enough to maintain the turbine speed. If the turbine speed was to slow the control valves would open more to allow more steam in thus increasing the speed of the turbine and should the turbine be going too fast, the control valves would close back to reduce the amount of steam going to the turbine thus slowing the turbine down.

Now, each control valve allows steam to flow through a section of the nozzle block (first stage). If the turbine is in full arc admission, it means that all of the control valves are wide open and the main steam stop valves are used to control the flow of steam through the turbine meaning that all the sections of the nozzle block have steam flowing through them. If the turbine is in partial arc admission it means that the control valves are controlling the steam flow through the turbine and that only certain sections of the nozzle block has steam flowing through them. That is enough for now. Back to the flow through the turbine.

We see the main steam entering the high-pressure section turbine and flowing toward the left through the turbine until it leaves the high-pressure section of the turbine as Cold Reheat steam. The cold reheat steam goes back to the boiler to be reheated and then is brought back as hot reheat steam to the turbine. The hot reheat steam is now going to flow through the Intermediate Pressure turbine section and the steam flow through this section is going to the right. The reason why the HP and IP turbines steam flow are different is because the flow of steam through them creates thrust on the turbine rotor, and in order to balance the thrust you can setup an opposing turbine to counteract the thrust. This arrangement greatly decreases the thrust on the turbine. Again I didn't mention this until now but we have another set of valves that work with the turbine. These valves are called the reheat stop valves and the interceptor valves. These valves are normally wide-open and only throttle or close in the event of a turbine over speed condition. Back to the basics. When the steam is finished in the Intermediate pressure turbine and leaves, the steam will continue to flow through a pipe called the Crossover. The Crossover is just a pipe that allows the steam to flow from one turbine section to another. As the steam flows through the crossover it will now enter the Low-Pressure turbine and with this arrangement, the steam will be divided in two allowing half the steam to flow toward the right while the other half to flow toward the left. Again this design helps minimize the turbine thrust. As the steam flows through the low-pressure turbine section it will come to the end of the turbine and enter into the condenser section to be condensed back into water.

What is the purpose of the turbine? The purpose of the turbine is to convert energy stored in the steam or water (hydro-plant) into mechanical energy in order to turn (rotate) the generator to make electricity.

How is extraction steam beneficial to the plant cycle? The use of extraction steam increases plant efficiency by preheating feedwater before it enters the boiler. If extraction steam were not used to heat feedwater the steam that would be wasted in the condenser and the boiler would have to be fired harder in order to raise the temperature of the feedwater.

What is the purpose for the stationary blades? The stationary blades direct or guide the steam into the rotating blades on the rotor.

What is the purpose for the rotating blades? The rotating blades use the energy from steam (impulse or reaction) to move the turbines rotor.

What is considered a stage of a turbine? A set of nozzles (blades) and rotor blades.

What is extraction steam? Extraction steam is steam that is extracted or bled from between the turbine blading to be utilized somewhere else in the plant cycle mostly for heating feedwater.

Heater Extraction System

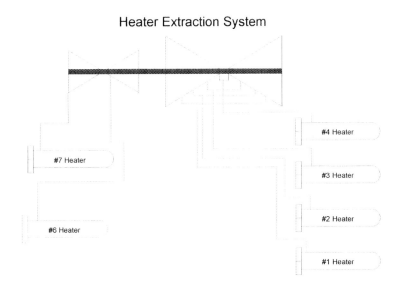

When does a turbine operate most efficiently? The turbine will operate most efficiently when all of the turbine design conditions have been met. Like throttle pressure, main steam and reheat temperatures, back pressure and vacuum.

What is after-seat pressure? After-seat pressure is the pressure of steam between the main stop valves and the control valves.

What function does the turning gear oil pump provide? The turning gear oil pump provides lubrication to the turbine bearings and the turning gear itself.

What is the purpose of the auxiliary oil pump? The auxiliary oil pump provides hydraulic pressure and lubricating oil during startups and shutdowns, since the main oil pump cannot operate effectively below the turbine rated speed.

What are the turbine blades made of? Turbine blading is usually made up of stainless steels, super alloys, titanium, and ferritic steels. The type of material used for blading depends on where the blading will be located (high pressure, intermediate pressure, low pressure, or last stage blading).

What function does the booster oil pump provide? The booster pump is powered by the discharge of the main lube oil pump. Its function is to supply oil to the suction of the main lube oil pump and to the turbine bearings.

How are the turbine blades put together? 1) Single root tree, 2) Double root tree, 3) Side entry root, and a variety of pinned types

What steam drives the steam driven oil pumps? Main steam is used to drive the steam driven oil pumps.

What is the function of the turbine casing? The turbine casing contains all of the turbine internal parts, keeping them in place and containing the steam within the turbine.

What is the turbine oil reservoir? The turbine oil reservoir contains all the oil pumps and oil supply essential for turbine lubrication and hydraulic control.

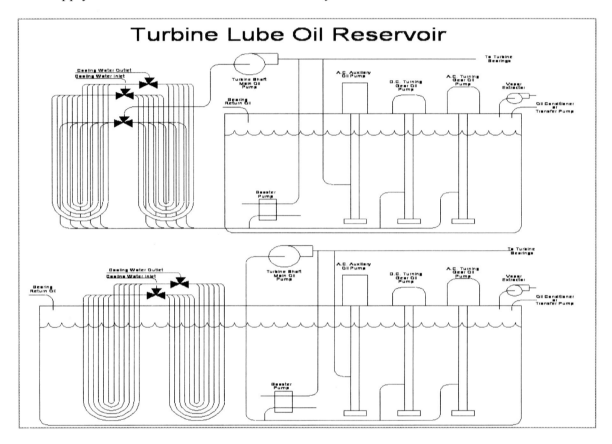

What part of the turbine does the cold reheat line come off of? The cold reheat line comes off of the end of the turbines high-pressure section.

What stage of the turbine is the control stage? The control stage is the first stage of the turbine, consisting of a nozzle block and a row of rotating blades.

What are control valves? Control valves are hydraulic actuated valves that control the flow of steam to the turbine, in which controls the speed of the turbine.

What are governor valves? Governor valves are the same thing as control valves. The difference between these valves is their maker. General Electric uses the term control valves while Westinghouse uses the term governor valves.

What are interceptor valves? Interceptor valves are hydraulic actuated valves that stop the flow of reheat steam to the I.P. turbine in the event of turbine over speed. Normally, these valves are wide open.

What are main stop valves? Main stop valves are hydraulic actuated valves that are located upstream of the control valves and are either open or closed. These valves are open during startup and closed when the turbine is tripped or the event of turbine over speed.

What are reheat stop valves? Reheat stop valves are hydraulic actuated valves that are located upstream of the interceptor valves and are either open or closed. These valves are open during startup and closed when the turbine is tripped or the event of turbine over speed.

What is a variable speed set point? A variable speed set point is electronic logic that is used to vary the speed of the turbine during pre-warming. The speed of the turbine will vary approximately ±100 rpm from the pre-warming set point. This varying of the speed of the turbine allows the turbine to be pre-warmed without sitting at a critical speed where vibration would cause a problem. Not all turbines use a variable speed set point.

What is a ventilator valve? A ventilator valve is a valve used to bleed steam from the main steam leads during an over speed condition. This valve is normally closed during normal operation. The source of this steam is cold reheat steam forcing the steam back through the H.P. turbine. This steam flow is used to cool down the H.P. turbine.

What is a steam chest? The steam chest is an area designed to reduce the turbulence of the steam before the steam enters the turbine control valves.

Where is the steam chest located? The steam chest located between the turbine main stop valves and the control valves.

What are nozzles? Nozzles are fixed turbine blades, they guide the steam into the path of the rotating blades and are used to expand the high-pressure steam to extract its energy and direct the resulting steam jets toward the rotating blades.

Where is the nozzles block located? The nozzle block is the first stage blading the H.P. section. It is the point where steam first enters the turbine.

What is critical speed? Critical speed is a speed where the turbine rotor will increase in vibration substantially above normal because of resonance.

What does cross-compound mean? Cross compound means that different turbine sections are on different shafts.

What does tandem compound mean? Tandem compound means that different turbine sections are on a common shaft.

What is a cylinder? A cylinder is the turbine rotor assembly and casing as a unit.

What are diaphragms? Diaphragms are rows of fixed blading that consists of an inner ring, which surrounds the shaft, and an outer ring, which attaches the rows to the turbine casing.

What is differential expansion? Differential expansion is the difference between the rate at which a turbines rotor and turbines shell expand of contract in response to temperature changes.

What is a disc type rotor? A disc type rotor is a turbine rotor in which the moving blades are mounted on discs, called wheels that are raised up from the shaft.

What does double casing mean? Double casing is a turbine casing consisting of two casings: an inner casing and an outer casing.

What is a drum type rotor? A drum type rotor is a turbine rotor that consists of an enlarged portion of the shaft, called the drum. The turbines moving blades are mounted directly onto the drum.

What does double flow mean? Double flow means a steam flow arrangement in which steam enters the center of the turbine cylinder and flows in two opposite directions.

What does eccentricity mean? Eccentricity is a measurement used to determine the straightness of a shaft.

What types of control systems are used to control the turbine speed? There are two general types of turbine speed control systems, Electro-hydraulic Control Systems (EHC) and the Mechanical Hydraulic Control Systems (MHC). The MHC systems are used on the older type turbine-generator systems while the EHC systems are used on most of the turbine-generator systems presently in use.

What does EHC stand for? EHC is the acronym meaning Electro-hydraulic Control.

What does MHC stand for? MHC is the acronym meaning Mechanical Hydraulic Control.

How does the Electro-Hydraulic Control system work? The EHC system has the same basic principles of operation as the MHC system yet there are a few differences. The motive fluid to actuate the hydraulic actuators is supplied from an independent source (not turbine oil) that is usually synthetic and fire resistant that can be used under high pressures. Instead of using a flyweight governor like the MHC system, the EHC system uses an electronic governor to monitor and maintain (control) the speed of the turbine. The hydraulic pressure in the EHC system can fluctuate between 1300 to 2000 psi as compared to the MHC system whose hydraulic pressure ranges from 150 - 300 psi.

How does the Mechanical Hydraulic Control system work? The mechanical hydraulic control system (MHC) is a turbine control system that uses turbine oil as the motive fluid used to actuate the turbine valves. The hydraulic pressure in the MHC system is lower than the hydraulic pressure used in the EHC system. This system consists of a flyweight governor, a hydraulic actuator, a servo valve, and connecting linkage. The governor consists of a shaft that is connected to the turbine shaft through the use of gears. Two brackets are held together by a spring and are attached to the governor shaft by pivot point of the brackets. Centrifugal force from the turbines rotor rotating causes the weights on the governor to move outward when the turbine speed is increased and move inward when the turbine speed is decreased. The position of the governor determines the amount of oil that enters the servo valve and then flows into the hydraulic actuators.

What do flyweights do for turbine control? Flyweights are found on units that use mechanical hydraulic control systems. These flyweights consist of a set of weighted arms that are connected together by a spring that is attached to a pivot rod that controls oil flow to the turbine valves. The weights work off of centrifugal force from the speed of the turbine. If the speed of the turbine increases, the weights will spin faster causing them to move outward, thus reducing the amount of oil supplied to the turbine valves, slowing the turbine down. And should the turbine slow down to much, the flyweights will come closer together and allow more oil to enter the turbine valves, causing them to open and again increase the speed of the turbine.

What is the purpose of the mid-span packing? The mid-span packing is a seal located between the H.P. and I.P. sections of the turbine. It prevents steam from leaking or bleeding between the two sections during normal operations.

Mid-Span Packing

Where is the exhaust hood located on the turbine? The exhaust hood is located on the outer rim of the L.P. section of the turbine, before the steam enters into the condenser.

What is the purpose of the exhaust hood? The exhaust hood is to direct or guide steam leaving the turbine and entering the condenser.

What is the function of the exhaust hood spray system? The exhaust hood spray system is used to minimize exhaust hood expansion, which will reduce thermal stress on the turbine casing. The system consists of a pump(s) that supplies water to a ring shaped header that is mounted to the outer ring of the turbine L.P. turbine casing. Normal operation is to control the temperature of the exhaust hood by spraying water into the L.P. turbine steam exhaust as it enters the condenser, reason being that there is not enough steam flow through the turbine to cool the exhaust hood so during startups and shutdowns this system is used. Ex. In automatic the system would begin spraying the exhaust hood at 600rpm on the turbine and would continue to do so up to 10% turbine load. This system is only monitoring the temperature of the exhaust hood and functions when needed (startup & shutdown). If the temperature of the exhaust hood rise to high and cannot be controlled, the unit should be removed from service to prevent increasing the damage to the L.P. turbine section.

What is full arc admission? Full arc admission is a method of rolling the turbine in which steam is distributed through all of the control valves so that each section of the nozzle block and the turbine is heated evenly. Full arc admission is in service during startups and initial loading.

What is partial arc admission? Partial arc admission is a condition that occurs in a certain part of the turbine startup. It is a state in which steam is admitted through only certain parts of the nozzle block at a time, as a result of the sequential opening and closing of the control valves. Partial arc admission is in service during normal operation.

What is the purpose for the steam seal system? There are two reason for the steam seal system, prevent air in leakage into the system and limit the turbine steam losses where the rotor shaft penetrates the turbine casing.

What does the governor control switch do? A governor control switch is a switch that controls the turbine control valves, and regulates load changes.

What does it mean to soak a turbine? Soaking the turbine means to allow the turbine metal both stationary and rotating to heat up evenly so that differential expansion can be minimized and thermal stress reduced.

Where does the hot reheat enter the turbine? Hot reheat enters the turbine through the reheat stop and interceptor valves and into the I.P. section of the turbine.

What is impulse (pressure) blading? Impulse blading is moving blading that has the same space between the blade inlet and the blade outlet. The energy transferred in impulse blading results in a decrease in steam velocity. This type of blading uses the actual pressure of the steam to drive the turbine blading.

What is an impulse (pressure) turbine? An impulse turbine is a turbine that uses the force of steam to move the turbine blading.

How does the gland steam seal supply system work? The gland steam seal supply system is a turbine support system designed to prevent air from being sucked in and steam from blowing out of the turbine casing by using steam under pressure that is admitted to the steam seals along the turbine shaft. The steam supply will normally come from the unit itself when online. Main steam is used when the unit is at low load and the cold reheat steam is used when the unit is at a higher low. Sometimes the unit if shutdown may be able to use steam from another unit that is online through the auxiliary steam piping. The steam pressure to the seals is maintained to hold a constant pressure on the seals at all times. As the steam enters the steam seal it begins to flow through a very tight path like a maze which causes the steam to spread through the entire seal trying the point of lowest pressure. This point of lowest pressure is right before the seal is exposed to the atmosphere. The gland steam exhauster creates a vacuum in the gland steam condenser which pulls the steam through the seal while keeping the steam from blowing out of the end of the seal to atmosphere. If that were to happen someone standing close by might get burned. As the steam flows into the gland steam condenser the steam condenses and drains into the main condenser to be used again. The gland steam condenser is cooled by the incoming water from the condensate pump. The heat removed from the condensing steam is transferred to the condensate water allowing it to utilize the heat for the plant cycle which increases efficiency and prevents this extra heat from being wasted.

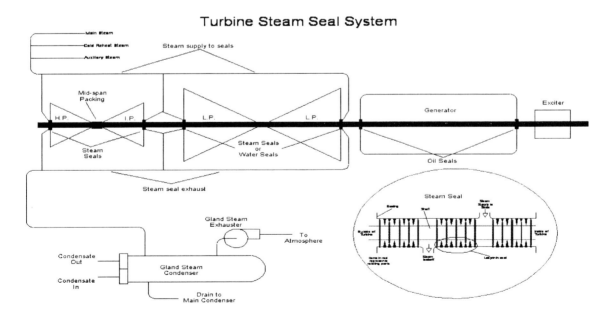

Turbine Steam Seal System

What is the purpose of the gland steam exhauster? The gland steam exhauster is a fan that is used to draw air/steam from the turbine steam seals to the gland steam condenser.

What is reaction (expansion) blading? Reaction blading is moving blading with the space between the blades is smaller at the exhaust end than at the inlet end of the blading. The energy transferred in reaction blading results in a decrease in steam pressure and velocity. This type of blading uses the expansion force of the steam to drive the turbine blading.

What is a reaction (expansion) turbine? A reaction turbine is a turbine that uses the expansion of steam to move the turbine blading.

What is the purpose for a turbine lube oil trip? The turbine lube oil trip is a device that protects the turbine should oil pressure drop to low to protect the turbine from damage.

Where does the main steam enter the turbine? Main steam enters the turbine through the first stage (nozzle block) in the H.P. section of the turbine.

What should happen if the main turbine oil pump malfunctions? Should the main oil pump malfunction, there are back up oil pumps to take over and protect the turbine from any damage.

What is the purpose for a condenser low vacuum trip? The purpose for the condenser low vacuum trip is to protect the turbine and condenser against positive pressures, which might cause severe damage to both and excessive heat buildup.

What is the purpose of the internal bypass valve? The internal bypass valve is used to reduce the steam pressure differential between the main stop valves and the steam chest and can be used to prewar the turbine.

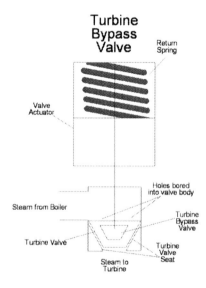

Do all the turbine valves have an internal bypass valve? No, just on specific turbine models.

What is the function of the main turbine oil pump? The function of the main oil pump is to supply hydraulic pressure and lubrication to the vital turbine systems when on-line.

Manual trip lever

&

Overspeed sensor

Main turbine oil pump

When does the main turbine oil pump operate? The main oil pump is fully operational when the turbine is a rated speed. The main oil pump becomes operational around 90% rated speed.

What is a multi-cylinder turbine? A multi-cylinder turbine is a turbine consisting of two or more cylinders.

What does the over speed mechanism do for the turbine? The over speed mechanism is a device that is used to actuate the turbine trip system should the turbine exceed its maximum limit in speed.

What is the purpose of pre-warming a turbine? Pre-warming the turbine allows the metal temperature throughout the turbine to expand slowly and evenly. 1) reduces thermal stress, 2) minimize the depletion of the turbine rotor cyclic life, 3) reduces the chance of a catastrophic rotor failure, 4) minimizes compression and tension stresses on the turbine rotor, 5) minimizes the chance of a turbine rub.

What is the rotor of a turbine? The rotor of the turbine is all the rotating part of the turbine.

What causes a rotor long condition? A rotor long condition is a condition that occurs when the rotor of a turbine expands faster than the turbine's shell.

What causes a rotor short condition? A rotor short condition is a condition that occurs when the rotor of a turbine contracts faster than the turbine's shell.

How does a rupture disc protect the turbine? A rupture disc is designed to remove a positive pressure build up inside the L.P. section of the turbine by rapidly venting off unwanted steam in this section and to prevent any damage to the condenser or turbine.

Where are the rupture discs located on the turbine? Rupture disc are located on the upper end of the L.P. section of the turbine. This prevents anyone from being exposed to the venting of the steam.

What is the rupture discs made of? Rupture disc are made of either lead or thin copper.

How can a leak in a rupture disc be found? A leak in a rupture disc can be found by pouring water on top of the disc while there is a vacuum in the condenser. If there is a leak in the rupture disc, the water will be pulled into the L.P. turbine and condenser. There are several other methods that could also be used.

What does it mean to seal up? To seal up means to seal all of the <u>turbines</u> points where the shaft goes through the casing with steam, using the steam seal system.

When can the turbine be sealed up? The requirements to seal the turbine up can be found by taking the superheat temperature (final) and subtracting it from the drum middle center metal temperature with at least 25^0F difference (superheat temperature above drum metal temperature) and 200 to 300 lbs of steam pressure. During an overnight cycle, the drum pressure should be above requirement so the 25^0 difference in temperature is the only thing left. During a cold start, the temperature requirement will be achieved before the steam pressure in order to seal up.

Where would you find shrouding on the turbine? Shrouding can be found at the end of all the rotating blades on the rotor, except for the very few last stages of the L.P. section.

What is the purpose of the turbine shrouding? The turbine shrouding keeps the steam from passing around the turbine blade by closing the clearance between the rotating and stationary sections of the turbine. The shrouding also helps to dampen turbine blade vibration.

What is a single casing turbine? A single casing turbine is a turbine that has all of the turbine stages in one casing.

What is a cross-compound turbine? A cross-compound turbine is a turbine where two or more sections of the turbine has their own casing and is on a different shaft (rotor).

How could steam to metal temperature mismatch, cause a problem? If the steam entering the turbine is too hot, the turbine rotor will grow to quickly causing a long rotor condition. If the condition exists where steam entering the turbine is to cool, chilling of the turbine blading will occur possibly causing thermal stress and possibly a rotor short condition.

What is a tandem compound turbine? A tandem compound turbine is a turbine where two or more sections of the turbine have their own casing yet remain attached on a common shaft (rotor).

Turbine Configuration

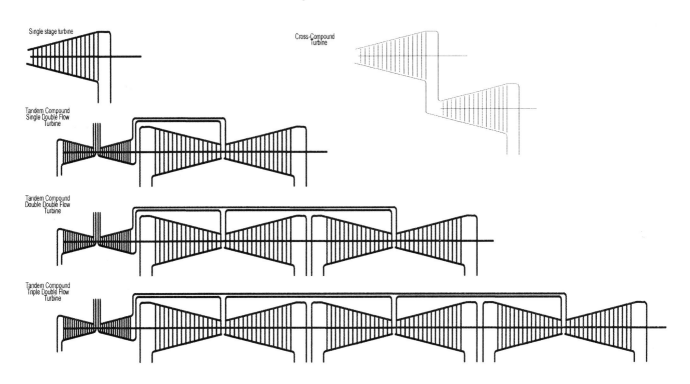

What is a single flow-condensing turbine? A single flow-condensing turbine is a turbine where all of its exhaust steam passes through one exhaust opening into the condenser.

What is a double flow-condensing turbine? A double flow-condensing turbine passes the exhaust steam from the turbine into two directions (opposed flow) and enters the condenser through two openings. A plant that has three L.P. turbines can have as many as six different flows into a condenser.

What is a tenon? A tenon is a raised knob on the end of a moving blade that is peened over shrouding to rivet the shrouding to the blading.

What is throttle pressure? Throttle pressure is the steam pressure between the boiler and the turbine main stop valves.

How does a thrust-bearing trip protect the turbine? The thrust-bearing on the turbine protects the turbine from excessive thrust in one direction or the other by tripping the turbine before the turbine thrust causes the different stationary and rotating parts of the turbine to touch each other.

When do you transfer from full arc to partial arc on a turbine? The transfer from full arc to partial arc differs per unit. On certain units that have a main steam stop bypass valve, once the bypass valve is open all the way, the pressure difference between the valves is balanced allowing a transfer from main steam stops to control valves to take place. On other units, load is the factor on when to transfer from full arc to partial. When a certain load is reached, the transfer begins changing the governor valve position. On this unit all of the governor valves are in the same position, each sharing in the steam flow to the turbine. This transfer will take several minutes to arrange the governor valves in a sequence of opening. As certain governor valves close off others will open up to keep the turbine at its rated speed.

When do you transfer from partial arc to full arc on a turbine? There are two basic reasons for transferring from partial arc to full arc on a turbine, 1) transfers automatically or manually after a turbine trip, or 2) at low loads, helps to keep the turbine hot.

What is the transfer point of a turbine? There is a point on certain turbines were the transfer from throttle valves to governor valves is changed during startup only. The purpose of the transfer point is to allow uniform heating, optimum differential expansion and greatly reduce thermal stress at the steam chests when the turbine speed is control is transferred from the turbine throttle valves to the governor valves. After the transfer, the throttle valves are wide open and all of the governor valves are throttled evenly.

What protective devices will protect the turbine? The turbine is protected by several major devices that will trip the turbine in the event of an emergency. 1) Vacuum low trip- will trip turbine should the vacuum reach a low set point like 22" Hg. 2) Low oil trip - will trip turbine should low oil pressure occur. Without adequate oil pressure, the turbine bearings are not protected and will be damaged. 3) Thrust bearing trip - should the thrust of the turbine increase to a dangerous level, the thrust bearing trip will trip the turbine protecting it from stationary and rotating parts of the turbine from meeting. 4) Over speed trip - will trip the turbine should the speed of the turbine become uncontrollable. The over speed trip should trip the turbine at a set point speed to keep the turbine from centrifugally tearing itself apart. 5) Bearing High Vibration.

What does the turning gear do for the turbine? The turning gear rotates the turbine slowly so that the shaft of the turbine will remain straight.

How long must the turbine be on turning gear? The turbine can remain on gear for an indifferent period of time. Should the turbine need to be shut down for repairs, the turbine metal temperature will determine the amount of time the turbine must remain on gear. If the metal temperature is too high the babbitt on the bearings will begin to melt, destroying the bearing surface and render the turbine useless.

What causes a rotor to bow or sag? The turbine LP section when shutdown will want to bow because of the high temperatures in the center of the turbine section but once the turbine cools down the sheer weight of the turbine causes the turbine to sag in the center while the bearings support the outer ends. This is why is it important to put the turbine on turning gear.

What is a wheel on a turbine? A wheel on a turbine is a row of rotating turbine blades.

Will a turbine trip on high vibration? A turbine can be setup to trip on high vibration. During startups and shutdowns there will be times where the bearing vibration will increase until it passes through a critical speed area and when shutting down the turbine the same can occur. Normally, high vibration with the unit online is left for the operator to decide whether or not the turbine should be removed from service.

What is peening? Peening is a process in which shrouding is attached to a row of rotating blades by hammering tenons on the blades through holes in the shrouding to rivet the shrouding to the blading.

What controls the speed of the turbine? The speed of the turbine is controlled by the flyweight governor (MHCS) or electronically by the (EHCS); the turbine is also controlled by the electrical grid when the unit is online.

Why does steam flow from the boiler to the turbine? Steam flows from the boiler to the turbine because of a difference in pressure. As long as there is a difference in pressure there will be a flow to the point of least resistance.

Why are seals installed where the turbine shaft passes through the stationary blading? To keep steam from bypassing the turbine blading by leaking across the shaft of the turbine. Without these seals, the efficiency of the turbine would be reduced.

How does the pressure inside the turbine vary with load? The higher the load, the more steam is needed to turn the turbine in which increases pressure in the turbine. At low loads, most of the LP section is under a vacuum, but as the load on the unit increases, more and more of the turbine becomes pressurized until at full load 95% of the LP section will be under positive pressure while the very last stages of the turbine will be under a vacuum.

What is the purpose of the gland steam seal system? The purpose of the gland steam seal system is to seal the ends of the turbine where the shaft goes through the turbine casing with steam, to prevent air from entering the turbine and from steam leaking out of the turbine.

What are 4 types of steam seals used on a turbine? The four different steam seals used on a turbine are <u>end seals or shaft seals</u> - to reduce steam leakage from where the shaft penetrates the turbine casing. <u>N2 packing</u> is also used to reduce steam leakage between two turbine sections. <u>Retractable packing</u> is used on shaft seals to decrease the chance of radial rubbing on these seals. <u>Radial spill strips</u> are used to prevent steam from bypassing the rotating blades of the turbine.

Steam Seal

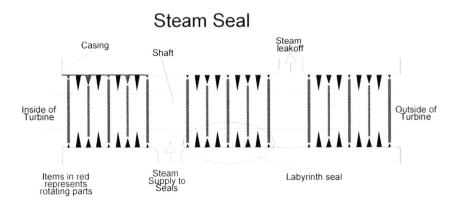

What is the nozzle block? The nozzle block is the first row of fixed blading in the H.P. section of the turbine. These blades are separated into several sections so that each section can be supplied with steam from their own individual control (governor) valve which in turn feeds one arc of the nozzle block. (The picture below only shows one half of the nozzle block.)

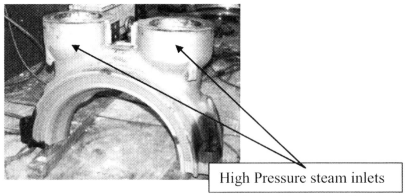

High Pressure steam inlets

What are some basic components of a turbine lube oil system? The turbine lube oil system consists of several oil pumps, a booster pump, a vapor extractor, oil conditioning system, and oil coolers.

What is the purpose of the reservoir in the lube oil system? The turbine oil reservoir holds all the oil needed to safely operate the turbine at any load and to store oil when shutdown.

What is the purpose of the turbine lube oil coolers? The turbine lube oil coolers, cools the oil with the help of a cooling water system, before returning back to the turbine bearings and auxiliary systems.

What are some common methods for filtering oil? The use of screens, bags, skimmers, centrifuge, or an oil conditioner can be used to filter oil.

What is the function of the vapor extractor on the turbine lube oil reservoir? The vapor extractor is used to remove any vapors that are in the reservoir. This removal of vapors aids in preventing an explosion and helps to pull the bearing oil through the return lines back to the bearing oil reservoir.

If oil has a milky color, what would this indicate? If oil has a milky color, this would probably indicate that water has entered into the oil system. Also if the oil looks like it has metal in it, it is probably babbitt from the bearing.

Is the turbine lube oil system a closed system? The turbine lube oil system is a closed system, it is self-contained.

What is an oil whip? An oil whip is a phenomenon that happens when oil is to cool and thick. When this occurs, an oil wedge is formed beneath the rotor and forces the rotor to lift up. As the rotor moves upward, the oil wedge moves from underneath the rotor, causing the rotor to fall hitting the bearing and creating severe turbine vibrations.

How can an oil whip be prevented? An oil whip could be prevented by keeping the oil temperature at or above 100^0F.

What are two main functions of the turbine valves? Two main functions of the turbine valves are to control and maintain turbine speed, and to isolate the turbine from the boiler.

What kind of oil is used in a mechanical-hydraulic control system? The turbine lube oil system is the basic fluid used for the mechanical hydraulic control system.

What kind of oil is used in an electro-hydraulic control system? What is usually used is a synthetic type oil that is resistant to heat and is non-flammable.

What is a servo valve? A servo valve controls the amount of oil going to the turbine valve actuators.

How does a servo valve work? A servo valve is a valve that is designed to control the flow of oil into and out of the turbine valve hydraulic actuators. It is operated by the flyweight governor through the connecting linkage to determine if oil is added to the hydraulic actuators or if oil is removed from the actuators.

What two ways is the position of the servo valve affected? The servo valve can be affected by the action of the governor or by the action of the hydraulic actuator.

What does the reset feature do? The reset feature allows turbine speed adjustments to be made in small increments; prevents hunting.

What does a trip valve do? A trip valve is designed to remove the hydraulic oil from the turbine valve actuators so the turbine valves can be closed quickly when the turbine is tripped.

What does hunting mean? Hunting is a term used meaning the fluctuating speed of the turbine, when the speed is not constant.

What are the turbine supervisory instruments? Instruments that are designed to give specific information about the turbine and its conditions.

What are five conditions that the turbine supervisory instrumentation is monitoring? The five conditions that are monitored are: Vibration/bearing conditions, Expansion/contraction, Eccentricity, Turbine speed/valve position, and steam pressures/temperatures.

How can a drop of water damage turbine blading? A drop of water hitting the turbine blading while the turbine is a rated speed is like, shooting a house fan with a gun. But with water it also causes thermal shock to the turbine blading and other turbine components while it makes little dings and cracks in the blading causing changes in the blade dynamics thus reducing turbine efficiency.

What is the purpose of the turbine drains? The purpose of the turbine drains is to ensure that all the water has been removed from the piping and casing of the turbine.

Where are these turbine drains located? These drains are located underneath the turbine, usually at the lowest point of the turbine or piping.

What are three phases of turbine operation? The three phases of turbine operation are startup, normal operation, and shutdown.

If the turbine should over speed, what will the turbine valves do? Should the turbine begin to over speed, the control valves will close and if the turbine's speed still continues to increases the interceptor valves will close, and finally to prevent any further over speed of the turbine the main stop valves and the reheat stop valves will close.

What is the purpose of the air pilot trip valve? The air pilot trip valve is designed to vent off the air pressure going to the feedwater heater extraction non-return valves causing them to close during a turbine trip to prevent extraction steam from backing up and re-entering the turbine, causing the turbine to possibly over speed or water induction.

What is the purpose of the gland water tank? Gland water tanks are usually used in older units to supply gland water to the low-pressure seals on the turbine.

What are two types of extraction non-return valves? The two types of extraction non-return valves are the free swing non-return valves and the air assisted non-return valves.

What are two types of rotor assemblies? The two types of rotor assemblies are the drum type rotor and the disc type rotor.

What is the purpose of the grounding brushes on the turbine rotor? The turbine grounding brushes are used to keep shaft voltages from building up and damaging the journals and bearings of the turbine. There are three types of turbine shaft voltage: 1) Electrostatic voltage, 2) Capacitive coupled voltage, and 3) Electromagnetic loop voltage.

The electrostatic voltage is a D.C. voltage that builds up between the turbine low-pressure stationary and rotating parts due to a charge separation when turbine blading is revolving at high speeds through steam. D.C. voltage levels can reach levels as high as 150 volts. Without the use of the turbine ground strap, the charge would buildup and discharge through the bearing causing damage to the journals and bearings or possibly plant personnel.

As with capacitive coupled voltage, this is an A.C. voltage on the generator/turbine shaft due to the exciter system supplying D.C. current to the generator field. If not probably grounded the voltage will discharge through the turbines bearings and journals.

And finally electromagnetic loop voltage is an A.C. voltage that is present in all generators because of magnetic dissymmetry. This type of voltage will vary between 1 and 30 volts A.C. The same type of damage can occur if the generator/turbine shaft is not properly grounded. Without this grounding device, voltage may appear along the entire length of the generator/turbine shaft.

What are two types of shaft grounding brushes? The two types of shaft grounding brushes are 1) an actual shaft grounding brush, and 2) spring loaded silver coated shoes. The silver coated shoes where used on the older units, while the shaft grounding brush has replaced them on the newer units.

What are two advantages of using the shaft grounding brushes over the spring-loaded silver coated shoes? The first advantage is that there is no adjustment to be made, as shoes wear, spring tension must be adjusted. And secondly, the shaft grounding brush eliminates the possibility of damage to the turbine rotor, should the shoes wear excessively.

What are two major functions of the turbine shell and casing?
1) Keeps steam in and air out. 2) Supports the stationary internals and holds these parts in alignment with the turbine rotor.

What is a turbine standard? A turbine standard is what supports the turbine shell and/or the turbine bearings.

Where is the front standard located? The front standard is usually located at the front of the turbine, the farthest from the generator.

What does the front standard consists of? A <u>typical</u> front standard could consist of low speed switches, over speed trip mechanism, manual mechanical trip, master trip solenoid, emergency governor, emergency trip solenoid valve, shell expansion detector, thrust bearing wear detector, extraction relay dump valve, eccentricity detector, speed sensor, and #1 bearing.

Turbine speed sensors

Why is the front standard greased? The front standard is greased to allow the standard to slide on foundation plates when the turbine is heated and expands or when the turbine cools and contracts.

What would happen if the front standard becomes cocked? Should the front standard ever become cocked, rotor misalignment and/or vibration could result. If thrust bearing is in the front standard, serious problems could occur. The thrust bearing could wear down faster and possibly fail due to abnormal thrust being applied from the rotor to the thrust bearing. With this condition, the front standard should be corrected before continuing with the unit startup. Differential expansion and rotor position could help indicate this.

What are some different turbine designs? There are three basic turbine designs, 1) Single casing, 2) Cross-compound, and 3) Tandem compound.

What is the difference between simple extraction and controlled extraction? In a simple extraction turbine, there is no effort made to control the pressure or the amount of extracted steam flowing from the turbine so the extraction pressure and flow will vary with the load of the turbine / generator. And with a controlled extraction turbine, regulating valves are used to control the pressure and/or flow of the extraction coming from the turbine.

What is a non-extraction turbine? A non-extraction turbine is a turbine where all of the steam flowing through the turbine is exhausted into a common atmosphere.

What is a non-condensing turbine? A non-condensing turbine is a turbine where once the steam has left the turbine, it is exhausted and is not reused.

In what part of the turbine does the greatest amount of steam expansion occur? The greatest amount of steam expansion occurs in the low-pressure section of the turbine.

What is the purpose of tie wires on the low-pressure turbine? The tie wires on the low-pressure turbine are used to reduce vibration stress in the turbine by joining blades together allowing them to vibrate as a group instead of each blade producing vibration.

What is the capacity of the internal bypass valve on a main stop valve? The capacity of the internal bypass valve is approximately 25% to 35% rated steam flow.

How does the bar lift method control the control valves? With the bar lift method the valve stems are opened when the bar is raised. Depending on the length of the individual valve stem will determine the sequence that the valves are opened.

What is the purpose of the heating steam block valve? The heating steam block valve is a turbine valve used to bleed off steam from the first H.P. packing leak off. This valve is closed during pre-warming and opened during normal operation.

What is the purpose of the packing blow-down valve? The packing blow-down valve is a turbine valve that is used to bleed leakage steam from the mid-span packing during pre-warming and following a loss of load. This valve is closed during normal operation.

What is the purpose of the diaphragm leakage dump valve? The dump valve is closed during normal turbine operation. Should the turbine over speed trip mechanism actuate, the throttle or stop valves, control or governor valves and interceptor valves will close simultaneously. As the interceptor valve closes, the air pilot valve closes the connection to atmosphere and with the dump valve off of its seat; the leak off chamber of the labyrinth seals is connected to the condenser, thus preventing steam leakage through the seals from further over speeding the turbine.

Why must the turbine shaft be sealed when in operation? To keep steam from leaking out of the turbine and air from being sucked in.

Where does the steam come from that supplies the gland steam seal supply system? On General Electric units, main steam is the only supply and an unloading valve that bleeds off seal pressure as the unit increases load. On Westinghouse units there are two steam supplies, main steam and cold reheat while a spillover valve has the same function as the unloading valve on the G.E. unit.

How does a water seal work? A water seal consists of an impeller that is attached to the turbine rotor and the turbine casing contains the labyrinth seals. The impeller works like a centrifugal pump that increases the pressure of the water where it pumps it through the seal itself while the labyrinth seal limits the leakage from the chamber through which the impeller runs. The water used for these seal normally comes from a gland water tank

How does a pressure packing work? A pressure packing can seal against either positive pressure or vacuum. Need more!

How does a vacuum seal work? A vacuum seal is used in low-pressure turbines. The vacuum seal has only three rings of packing forming two chambers, the inner one is connected to the steam seal header and the outer one is connected to the steam packing exhauster header.

What is the purpose of the gland steam condenser desuperheater spray? The purpose of the desuperheater spray is to condense steam in the event that the gland steam condenser is not working properly. If there is no water flowing through the gland steam condenser, the steam would not condense, thus no vacuum would exist so the turbine seals would begin to steam outward into the atmosphere.

Where does the supply come from for the gland steam condenser desuperheater spray? The condensate system normally supplies the desuperheater spray to the gland steam condenser.

How much pressure is applied to the turbine steam seals? About 2 to 3 pounds of steam is supplied to the turbine seals. Yet, as load increases the amount of pressure will also increase to ensure that the steam seals contains the steam within the turbine.

What are some protective devices on the gland steam seal system? There are safety valves located on the steam seal header to prevent any overpressure in the system.

What is the purpose of the low-pressure gland desuperheater? The low-pressure gland desuperheater uses spray water to cool the gland steam to avoid sending super-hot steam to the vacuum seals of the L.P. turbine, which operates at a relatively low temperature. Hot sealing steam would cause thermal stress to the turbine seals.

What are some protective devices on the boiler feed pump turbine? The turbine boiler feed pump should have similar protective devices like the main turbine. Such as, low vacuum trip, low lube oil pressure trip, thrust bearing trip, and an over speed trip.

Turbine Boiler Feed Pump (Turbine)

What are some devices used to purify turbine lube oil? A centrifuge or an oil conditioner can be used to purify the turbine lube oil.

What is the purpose of the turbine lube oil system? The purpose of the turbine lube oil system is to supply the turbine with cool oil to all the turbine bearings and other auxiliary systems through the use of pumps, valves, and regulators.

How does a balanced design valve actuator work? A balanced design valve actuator is designed to allow fluid flow to push up against the valve seat while at the same time the fluid flow is pushing down against the other part of the valve seat, allowing the valve to actuate with ease because of the reduced difference in pressure across the valve.

What are two functions of the gland steam condenser? The two functions of the gland steam condenser are maintain a constant vacuum for the steam seal exhaust header by using a gland steam exhauster and to condense steam from the steam seals so that it can be reused within the plant cycle while also preheating the condensate.

What is a major disadvantage of using water seals? A major disadvantage for water seals is that for water seals to be effective, the main turbine must operate at a speed of 1500rpm or more. The water seals are not very effective at a vacuum between 10 and 15 inches with the main turbine on turning gear.

Why should the turbine steam sealing system remain in service until condenser vacuum is at atmospheric pressure? The reason for leaving steam seals in service while the condenser is under vacuum is because if the steam seals where removed from service the condenser vacuum would pull cool outside air across the hot steam seals causing them to warp because of the severe thermal stress.

What is the function of the turbine boiler feed pump seal water pumps? The turbine boiler feed pump seal water pumps provide water seal for the turbine feed pump and/or booster pump allowing feedwater to remain inside the pump casing and from boiling out of the pumps seals. These seal water pumps can provide water for other sub-systems in the plant, such as the exhaust hood sprays, gland steam desuperheater spray, etc.

How is the turbine oil booster pump controlled? Three valves control the turbine oil booster pump. The booster baffler valve is the inlet valve that controls the flow of oil from the main shaft pump to the oil turbine. The bypass baffler valve is the control valve that bypasses oil from the main pump around the oil turbine and on to the bearing header. The bearing header relief valve is the third control valve in the system; it is used to remove excessive amounts of pressure in the system.

What device automatically engages the turning gear? The low or zero speed switch is what allows the turning gear to engage to the turbine.

What are two functions of the turbines speed changer? The two functions of the turbines speed changer is to adjust the speed of the turbine when the unit is offline and secondly it is used to adjust the load on the generator without affecting the speed of the turbine, when the turbine is online.

What are two large subsystems of the EHC system? Two large subsystems associated with the EHC system are the hydraulic system and the electronic system.

What are two types of shaft vibration probes? Shaft riding and non-contacting vibration probes are two types used on turbines.

Where is the rotors' vibration probes located at? The vibration probes are located on the rotor bearing housing, where the vibration probe rides along the shaft of the rotor.

How does the shaft-riding vibration probe work? The shaft-riding probe has a teflon tip that rides on the shaft of the turbine and transmits motion from the shaft to a pickup head that converts the vibration into an electrical signal.

How does the non-contacting vibration probe work? The non-contacting vibration probe has an electronic sensor that measures the rotor's shaft movement. The probe does not touch the rotor, yet is extremely close about 0.040 of an inch from the rotor.

What is turbine phase angle? Turbine phase angle is the angle occurring between the peak of the eccentric disc versus the position of a machined slot on the same disc.

Where is the eccentricity detector located? The eccentricity detector is located within the front standard of the turbine.

How can water induction be detected in the turbine? Water induction can be detected within the turbine through the use of thermocouples located in the turbine casing to monitor the turbine metal temperature. A temperature on the bottom of the turbine that is much lower than the temperature on the top of the turbine would indicate the presents of water within the turbine. Possible vibration can be caused from excess water entering the turbine.

Where is the turbine over-speed mechanism located? The turbine over-speed mechanism is usually located in the front standard of the turbine.

What does rated speed mean? Rated speed means that the turbine will <u>normally</u> operate at this speed when in service.

What instrument gives an indication of axial clearance between the rotating and stationary blading of the turbine? The differential expansion recorder provides the information of the clearance between the rotating and stationary blading.

Where is the greatest variation of steam temperature within the turbine? The greatest variation of steam temperature occurs within the high-pressure area of the turbine. Main steam 1005^0F to cold reheat @ 400^0F.

Where is the greatest variation of steam pressure within the turbine? The greatest variation of steam pressure occurs within the intermediate and low-pressure turbine. 60psi to – 28.5 Hg

When the oil conditioner/centrifuge is in service, where does it discharge to? The discharge of the oil conditioner or centrifuge is normally discharged to the turbine oil reservoir.

When can the turbine over speed test be performed? The actual turbine over speed test is usually performed when the unit is coming down for overhaul or coming up from overhaul. On certain units, oil over speed test can be performed while the unit is on line.

When should the turbine valves be tested? If the unit does not cycle on/off for long periods of time it would be wise to test the valves. For units that cycle on/off every few days a turbine valve test could possibly be done once a month, just to check logic and valve operation.

What is the purpose of the turbine valve test? The purpose of testing the turbine valves is to insure that each turbine valve is opening and closing properly while being able to maintain the load of the turbine. The valve test will vary with different types of units.

What is the purpose of the throttle pressure controller? The throttle pressure controller is used to protect the turbine from an excessive decrease in throttle pressure. When the throttle pressure reaches a set point and falling, the TPC will begin to close the control valves on the turbine. These control valves will close to a certain point and stop so that there is enough steam flow through the turbine to keep the turbine from overheating.

What is the purpose of the Initial pressure limiter? The initial pressure limiter works similar to the throttle pressure controller it is used to protect the turbine from an excessive decrease in throttle pressure when the steam pressure of the boiler falls within a certain amount of time. Psi/min lost ramp rate.

What is the purpose of the valve position limiter? The valve position limiter is used to protect the turbine from over speeding by limiting the movement of the control valves.

On a unit startup, which turbine valves will open when the turbine is latched? When a turbine is latched up the reheat and interceptor valves will open along with the control valves. The main steam stop valves will remain closed until the turbine is ready to be rolled.

What is the purpose of the turbine valve tracking meters? The turbine valve tracking meters are used to give an indication of the position of each of the turbine valves.

When should the boiler feed pump turbine be on turning gear? The boiler feed pump turbine should be on turning gear whenever off-line. Being on turning gear will ensure that the pump will be ready for service when needed and that the rotor will remain straight.

Does the boiler feed pump turbine have a critical speed? All turbines have some type of critical speed. The boiler feed pump turbine is usually brought up to speed quickly thus passing through the critical speed areas without much of a problem.

What is the prime mover? The term prime mover is used to describe a device that converts energy whether thermal or hydraulic, into mechanical power (the main turbine).

What is a rub? A rub is when stationary turbine parts begin to make contact against the moving turbine parts, due to temperature distortion of turbine metal. This sound, sounds like a rub.

What could cause a rub? A turbine rub can be caused by rotor bowing, differential expansion, water induction, and turbine rotor misalignment or unbalanced.

What are the steam seals made of? The high-pressure steam seals are usually made of steel and ascoloy or have steel teeth inserts. The low-pressure steam seals are usually made of lead bronze.

Why are shaft seals segmented? Shaft seals are segmented to provide for expansion and contraction of the turbine.

What are six problems that could cause turbine vibration? Turbine vibration could be caused from turbine unbalance, generator rotor thermal sensitivity, bearing misalignment, an oil whip, rubbing, or a cracked rotor.

How long after the turbine has comes off-line can the turning gear be removed? Depending on the turbine metal temperature will determine how long the turbine has to be on turning gear and bearing metal temperature.

What is the purpose of the load limiter? The load limiter is used to protect the amount of load the turbine/generator can safely handle. This limiter will keep the generator from over producing its normal capability or falling below a stable limit on load.

What could happen if the turbine is rolled with a higher than normal eccentricity? If the turbine is rolled with a higher than normal eccentricity, excessive vibration and radial rub damage could occur with the possibility of permanently bowing the turbine rotor.

What is the purpose of the low-speed switch? A low-speed switch is a device that is used to indicate low shaft speed of the turbine. When the turbine shaft speed reaches zero, the switch will automatically engage the turbine turning gear.

Why should the main steam and reheat steam temperature remain within a certain temperature difference? Should a temperature difference become too large, severe thermal stress will be placed on the turbine casing and rotor of the turbine.

What are two types of compound turbines? Two types of compound turbines are the tandem compound turbine and the cross-compound turbine.

What are four limitations for starting most turbines?

1) Thermal stress   3) Rotor and shell differential expansion
2) Vibration   4) Eccentricity

What is a servomotor? A servomotor is a hydraulic powered actuator that opens the turbine valve by moving the valve stem against the valve spring force. Hydraulic fluid flow to the servomotor is controlled by the servo valve.

How fast does the turbine rotate on turning gear? The main turbine can rotate several rpm (under 30 rpm) while on turning gear.

How much can a turbine rotor expand? On average, a turbine rotor can expand up to 1 inch.

How much of the turbines' steam is exhausted into the condenser? Approximately 2/3rds of the turbine steam is exhausted into the condenser. The other 1/3 is used as extraction steam.

What is the primary purpose of the reheat stop and interceptor valves? The primary purpose of the reheat stop and interceptor valves is to protect the turbine from an over speed condition.

What is the difference between single valve control and sequential valve control? Single valve control is where only one valve or a group of valves working simultaneously as one valve to admit steam to the turbine. With sequential valve control, there are several valves that open and close depending on the sequence that the valve operates to admit steam to the turbine.

What is meant by turbine manual on the turbine control system? Turbine manual on the turbine control system allows the operator to have full control of the turbine. The operator can open or close turbine valves to maintain, increase or decrease turbine speed or to increase or decrease generator load.

Why are some low-pressure turbines designed with steam flow in opposite directions? There are two reasons for designing a low-pressure turbine with the steam flowing in opposite directions, 1) it reduces the diameter of the LP section, 2) it reduces axial thrust on the turbine thrust bearing.

How can the percentage of turbine slip be figured? The percentage of turbine slip is figured by taking the synchronous slip - actual rotor rpm divided by synchronous rpm multiplying that by 100.

What is the purpose of the turbine bypass system? The turbine bypass system allows operators to build up heat needed to generate steam before it is allowed to enter the steam turbine. This system also allows the operators to keep steam flow moving in the steam system in the event of a turbine trip in order to help cool down the reactor and the steam generator.

What are three advantages of using a turbine bypass system?

1) Unlimited matching capabilities for turbine temperatures.
2) Cooling protection at any load.
3) The reactor can operate while the unit is being started up / shutdown.

This allows the operator the ability to maintain positive control of the steam generator pressure and RCS temperature during normal heat up / cool down.

How are steam turbines classified? The turbine can be classified through several ways, the types of blading used (impulse, reaction, combination), how many stages are there in the turbine (single or multiple), the direction of steam flow (axial, radial, tangential, or helical), the amount of steam pressure at the turbine (high, low, or intermediate pressure), condensing or non-condensing, extraction or non-extraction.

What is turbine blade efficiency? Turbine blade efficiency is a ratio that is used to measure the work done from the blades to the energy supplied to the turbine blades.

How do turbine deposits affect the turbine? Turbine deposits affect the turbine in three ways, 1) Economically - reducing turbine output thus decreasing turbine efficiency, 2) Affecting reliability of the turbine - thrust on the turbine shaft, higher bending stresses on turbine blading, higher vibrations, and valve leakage due to deposits, 3) Corrosion - fatigue, pitting, and stress.

Low Pressure Turbine Section

What are two types of turbine deposits? Water-soluble and water insoluble are the two types of deposits found on turbine blading. Water-soluble deposits normally found on the High and Intermediate pressure turbine blading and consist of NaCl, Na_2SO_4, NaOH, Na_3PO_4 and water insoluble deposits are normally found on the Intermediate and Low pressure turbine blading which consist of SiO_2.

Where do most turbine rotor failures occur? Most turbine rotor failures occur in large low-pressure rotors.

How are these turbine deposits removed from the turbine blading? Water-soluble deposits can be washed off with either condensate or wet steam while insoluble deposits have to be removed mechanically by taking apart the turbine.

What could cause turbine bearings to fail? There are several things that could cause bearing failure, improper lubrication, misalignment, unbalance, bad bearing fit, excessive vibration, excessive heat and excessive rotor thrust.

How can you tell if a turbine bearing has a problem? If bearing metal temperature begins to change rapidly or erratically, could indicate a wiped or scored bearing. A good indication is when the turbine is tripped, as the speed of the turbine decreases the bad bearing will be unable to hold the hydrodynamic oil wedge like a normal bearing, this will cause the bad bearing to make metal to metal contact causing a rapid temperature rise. On certain units, the bearing oil return line may have an inspection port to view the oil leaving the bearing. Should the oil leaving the bearing look silvery in color is a possible indication of the babbitt being wiped off the bearing.

What are 3 forces that affect the turbine blading when the rotor is rotating? Centrifugal, torque, and vibratory force are the three forces that strain turbine blading when the turbine is rotating.

How can the position of the turbine rotor be checked manually? On certain turbines, an access hole is provided around the front standard in which a standard depth micrometer can be inserted to determine the axial location of the turbine rotor in relation to the cylinder.

The Generator

The generator is probably one of the hardest pieces of equipment to understand. Since there is hardly anything that you can actually see happening other than the rotor spinning. Yet the generator is one of the most efficient pieces of equipment found in a power plant. Almost completely converting all its mechanical energy into electrical energy. As with the turbine, the generator has several different sub-systems that operate with the generator. A generator falls into a category all its own. It is basically constructed in the same manner.

In Fig. 7-1 shows a typical generator with support systems included. These support systems include the hydrogen cooling system, hydrogen & carbon dioxide gas addition/venting system; seal oil system, stator cooling water system, gas dryer system, liquid detector system, exciter system and the core monitor system.

Have you ever heard of the expression, "It takes money to make money." Well the same goes for the generator but something like this, "It takes electricity to make electricity." In order for the generator to convert mechanical energy when it is rotating into electrical energy an exciter is used to generate DC voltage in the generator's rotor. There are several different types of exciters so I will not go into detail at this time. The exciter charges the generator's rotor creating a magnetic field. Now that the magnetic field is created and the generator's rotor is spinning up to speed we can increase the magnetic field so a voltage will be induced in the generator's stator windings, which in turn creates electrical power.

Pretty simple right...we've only scratched the surface.

But hang in there.

Just remember, the more voltage the exciter puts on the generator's rotor, the stronger the magnetic field will become, the more electrical power is produced and vice versa. But there is a limit.

I will go into the support systems for the generator because they are equally important. The hydrogen in the generator is used to cool the generator. That is its only purpose is to remove heat away from the generator. The hydrogen gas is circulated by a fan mounted on the shaft of the generator. A hydrogen gas dryer takes the moisture out of the gas and returns the gas back to the generator. The hydrogen/carbon dioxide/air addition and venting system is used to purge the generator in a safe and efficient manner. It is also used to maintain the hydrogen purity and pressure in the generator. The seal oil system is used to contain the hydrogen from leaking around the area where the generator's shaft penetrates the generator housing. On some generators there is a stator cooling system that is used to keep the stator-winding cool. And liquid detectors are used to determine if a liquid is leaking into the generator casing such as seal oil or stator cooling water. The core monitor basically determines the amount of particulates within the hydrogen gas in the generator. This would be the first indication that the insulation in the generator windings is beginning to break down. Of course the weakest part of the generator is the insulation.

What is a generator? A generator is a machine that converts mechanical energy into electrical energy using electrostatic or electromagnetic process to produce either direct current or alternating current. Generators can operate through the use of many different sources to drive them such as steam turbines, wind turbines, a hydro-turbines (water), and gas turbines.

What is a cycle? A cycle is when the generator's rotor makes one complete revolution, usually measured as cycles per second or as hertz.

What is frequency? Frequency is determined by the number of cycles completed per second by a given A.C. voltage. This term is expressed in hertz.

What is hertz? A hertz is a unit of frequency that is equal to one cycle per second.

What is an exciter? An exciter is a device that is used to produce and supply direct current to the rotor windings of a generator. Without this device a magnetic field would not be present within the generator and AC power exist.

How many types of exciters are there? There are eight different types of exciters. They are the brushless exciter (rotating rectifier exciter), static exciter, AC exciter (Alterrex, Althyrex, Generex), Pilot exciter, Flash exciter (flash circuit), DC exciter, and a portable exciter.

How does a D.C. exciter work? In a D.C. exciter, there is a main exciter and a pilot exciter. A magnetic field is produced in the pilot exciter's stator by a permanent magnet. The D.C. produced by the pilot exciter flows through the stator windings of the main exciter, creating a magnetic field. As the D.C. exciter rotates, voltage is induced into the exciter's rotor windings. This induced voltage creates an alternating current that is converted into D.C. by the commentator within the exciter. Then the D.C. from the exciter flows into the generator rotor windings, creating a magnetic field.

How does an AC exciter work? An AC exciter works in a similar manner to the DC exciter except the AC exciter uses a bank of rectifiers instead of a commentator. The magnetic field is created normally from the output of the main generator. During startup a separate AC source is used to create the magnetic field and when online the main generator takes over in supplying AC power to the exciter.

What is a pilot exciter? A pilot exciter is a generator that supplies current for the electromagnet in the stator of a D.C. exciter.

What is a flash exciter? A flash exciter or flash circuit is a circuit that is used on self-excited generators. In order to establish voltage in the generator for the exciter to work, the flash circuit supplies DC power usually from plant batteries briefly to the generator field to build up voltage. As the voltage increases enough to become self-sustaining, the external source of power (batteries) is disconnected.

How does a brushless exciter work? A brushless exciter is an exciter that does not have brushes or slip rings. In this exciter there are two rectifier banks used, one is a rotating rectifier bank and the second is a stationary rectifier bank. The current that creates the magnetic field within the stator is from a permanent magnet generator usually attached to the main generator. The permanent magnet generator and a pilot exciter are similar except the permanent magnet in a permanent magnet generator is the rotor, rather than the stator. The AC produced by the voltage in the permanent magnet generator stator windings, flows to the stationary rectifier bank where it is changed into DC. The DC then flows through the stator windings of the main exciter to create the magnetic field that induces voltage in the main exciter rotor. The induced voltage creates an alternating current that is converted into DC by the rotating rectifier bank. The DC is then used to create a magnetic field around the rotor of the main generator. A rotating rectifier exciter is a brushless exciter.

How does a static exciter work? A static exciter is unique because it has no moving parts. The current that produces the magnetic field around the main generator rotor comes from the voltage that is induced in the main generator stator. During generator startup, an outside AC source is needed like an AC exciter. The current from the outside source will flow through a transformer/rectifier combination. The transformer reduces the voltage while the rectifier converts the AC into DC so it can be used in the exciter. The DC then flows through the brushes and slip rings into the main generator's rotor windings, where the magnetic field is created around the main generator's rotor and induces voltage in the generator's stator windings. Once the generator is online, AC from the generator's output is fed into the static exciter making it self-exciting.

How does a generex exciter work? A variation of the static type and shaft driven exciter systems. This system uses two power sources. The first power source for excitation comes from three armature bars called "P-bars" which are installed on the top of the main stator bars. These three armature bars provide a three-phase power source that is rectified and is distributed into the main field through the collectors. The second power source comes from transformers located on top of the generator and receive power from the generator terminals. This type of excitation system is similar to the Althyrex excitation system and is used in an electrical system where stability is a concern.

What are two advantages that electromagnets have over permanent magnets?
Electromagnets are much stronger than permanent magnets and the magnetic strength can be controlled by regulating the current flowing through the magnet.

What are poles of a generator? The poles of the generator are the electrical circuits that are formed by the field winding on the rotor of a synchronous generator. This happens when direct current (DC) is passed through the field winding, the poles become the magnetic north and south poles. The north pole is positive while the south pole becomes negative.

What is the purpose of generator brushes and slip rings? The generator brushes and slip rings are attached to the shaft of the generator. The brushes are usually made of carbon while the slip rings are made of copper or steel. The slip rings are insulated from the generator shaft to prevent current from flowing through the shaft. Current flows from a D.C. power source through the brushes and the slip rings into the generator rotor a back to the D.C. power source.

How does the exciter output affect the load of the generator? Adjusting the strength of the magnetic field around the generator rotor regulates the output of the generator. If the load of the generator is to be increased, the exciter's output must also be increased and if the generator load is to be decreased the exciter's output must also be decreased. These functions are normally controlled automatically, yet can also be controlled manually when necessary.

If a generator is not at 60 hertz, how long can the generator remain online? Electrical power in the United States is maintained at 60 hertz. If for some reason 60 hertz is not maintained, damage to equipment will result with a short time period. It depends on the duration and how far away from 60 hertz the frequency is. The further away the frequency is from 60 hertz the less time the generator can stay on line.

Are there any other gases that could be used instead of hydrogen, to cool the generator?
Yes, air and helium are two other gases that could be used instead of hydrogen. The reason why air is not used is because the generator has to be built much bigger in order to remove enough heat from the generator using air, not cost affective. The reason why helium is not used is because of the cost and scarcity of this gas, again not cost affective. Hydrogen is used as a generator coolant because of its unlimited supply and relatively low cost.

What is the difference between the generator's stator and the rotor? The generator's stator is the stationary part that contains all the stator windings. The rotor itself rotates within the center of the stator windings, it rotates.

Why must the generator be kept within a certain temperature range? Should the generator become too hot the insulation on the windings would begin to break down and could possibly lead to a ground fault in the generator. Should the generator be kept to cool, the windings and other components within the generator would begin to produce condensation because of the temperature difference, which could lead to a series of problems.

Where are the liquid detectors located? The liquid detectors are located on the belly of the generator and possibly the exciter depending on the type of generator used.

What type of water system is used to cool the hydrogen in the generator? A closed cycle cooling system is generally used to cool the hydrogen gas in the generator.

When would the hydrogen vents on the generator be opened? These vents would be opened if the generator was to be purged or if there was a threat of a possible explosion might occur such as a fire around the generator.

What is the explosive range of hydrogen? The explosive range of hydrogen and the mixture of air is approximately 5% through 75% by volume. The maximum intensity is the halfway point between these limits.

What are lockout relays? A lockout relay is a relay that can be setup to perform many functions, usually designed to trip (open) a breaker and prevent the breaker from closing until the lockout relay is reset by plant personnel.

What is a switchyard? A switchyard is an area that receives power from the power plants generator and distributes it into the electrical grid for consumer use. This area contains many circuit breakers and protective devices to prevent any damage to the plant's equipment and/or the electrical grid should a malfunction occur.

What is a capability curve? A capability curve is a chart that illustrates a generator's ability to produce power safely and efficiently depending on specific operating conditions such as hydrogen pressure in the generator, VARs and power factor.

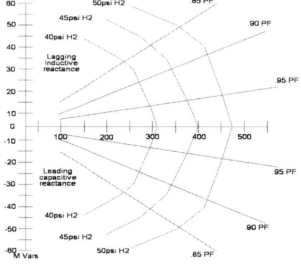

What is a circuit? A circuit is a path through which electric current flows. A circuit can be made simple like a battery attached with wires to a light bulb or more complex like those found inside a computer.

What is a circuit breaker? A circuit breaker is a device that is used in a circuit to control the amount of current flowing through a circuit by opening the circuit in the event of too much current flowing through the circuit causing an overload.

What does motorizing the generator mean? Motorizing the generator is a very unwanted condition. Instead of the generator putting out power to the grid, the grid puts electrical power to the generator turning the generator into a powerful motor. This condition is extremely harmful to the generator and the turbine. Motorizing needs to be prevented to prevent overheating of the low-pressure turbine and/or high-pressure turbine blading.

What are three types of generator motorizing?

1) Motoring the generator will occur when the steam flow to the turbine is reduced such that it develops less than no-load losses while the generator is still on line. Assuming excitation is sufficient, the generator will operate as a synchronous motor driving the turbine. The generator will not be harmed by synchronous motoring, but, if it occurs as a result of failure to complete a sequential trip, protection for the fault originating that trip is lost. In addition, the turbine can be harmed through overheating during synchronous motoring.

2) If the field excitation is lost, along with steam flow, the generator will run as an induction motor, driving the turbine. In addition to possible harm to the turbine, this will produce slip frequency currents in the rotor and could cause it to overheat if continued long enough.

3) The third type of motoring occurs when the generator is accidentally energized when at low speed or at a standstill. The rotor will accelerate and operate as an induction motor.

Damage to the rotor is imminent.

What is electricity? Electricity is the flow of electrons through a conductor, which can produce heat, magnetism, light, and even chemical changes. Electricity can be created by chemical changes, friction, or induction. There are two types of electric current, one is alternating current (A.C.) and the other is direct current (D.C.).

Is the generator's field A.C. or D.C.? The generator's field is D.C.

What is station service? Station service is an electrical system that supplies power to a power plant's equipment in order for the plant to remain in operation when the unit is shutdown. When the unit is running, power from the generator is used to power the plant equipment. This picture is of a station service board; it is where the Control Room Operator changes over power from the generator to the grid and vise versa.

Typical Station Service Board

When the generator's synchrometer is at top dead center, are the lights "on" or "off"? When the synchrometer is on, as the indicating needle moves away from top dead center, the lights will get brighter until it reaches 180^0 out of phase. At this point the lights are the brightest. As the needle moves closer to top dead center, the lights will become dimmer and dimmer until the needle reaches top dead center. At top dead center, the lights will be off.

What is a blackout? A blackout is a total loss of power to the electrical grid that supplies power to the city.

What is a brownout? A brownout is when partial loss of power to the electrical grid. This will happen when the demand for electricity is not met and power is supplied to only certain parts of the electrical grid.

What is a breaker and a half switchyard? This is a switchyard that contains one and a-half circuit breakers for each transmission line in the system.

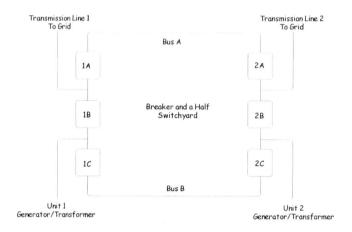

What is a double breaker switchyard? This is a switchyard that contains two circuit breakers for each transmission line in the system.

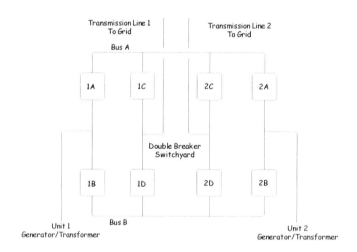

What is a delta connection? A delta connection is an electrical connection that is used in a three-phase circuit and is created by connecting each of the phases of a component at three different locations.

What is the difference between alternating current and direct current? Alternating current is electrical power that changes directions or alternates its flow path. Alternating current carries a sine wave. With direct current, electrical power flows only in one direction it does not alternate, and there is no sine wave with direct current.

What is a wye connection? A wye connection is an electrical connection that is made when joining one side of all three phases to a single point.

What is a synchroscope? A synchroscope is a device that resembles a clock with only one hand and is used to indicate the difference in frequency between the generator and the power grid. This will help the operator to be able to make adjustments to alleviate the difference in frequency between the generator and the power grid. Only used during the unit startup.

What are the two lights on the synchroscope used for? The two lights on the synchroscope are used for matching the running voltage with the incoming voltage. One light represents running voltage while the other light represents incoming voltage.

What two indications does the synchroscope provide? Phase angle matches generator frequency and system frequency.

What is a disconnect switch? A disconnect switch is a device that can interrupt the power supply to a piece of equipment, not allowing it to start. This switch is usually used in conjunction with a circuit breaker and is normally operated manually.

What is a transformer? A transformer is a device used for changing the voltage of A.C. electricity by electromagnetic induction. A transformer is made up of two coils of insulated wire that are wrapped around a metal core. One of these coils is supplied with electrical power, while the other supplies the changed output voltage to the grid. Two basic types of transformers are step-up and step-down transformers. A step-up transformer will increase the voltage that is coming into the transformer for transmission. A step-down transformer will decrease the voltage that is coming into the transformer for local distribution. When voltage increases, current decreases and when voltage decreases, current increases.

What are the parts of a transformer? The three main parts of a transformer are the core, the windings, the cooling system, the casing, the inert gas injection system, fire suppression system, relay protection system, the alarm system, and the insulation.

What is a distribution transformer? A distribution transformer transfers power from a bus to the distribution lines. This type of transformer is located in the substation.

How does a transformer stay cool? Depending on the type of transformer used will determine the way it is cooled. A transformer at a power plant would possibly have cooling fans and cooling pumps to keep cool. Other transformers may have a natural circulation system to keep them cool. Some transformers contain oil in them; this oil is used as a cooling medium.

Why is nitrogen injected into a transformer? Nitrogen is injected into the transformer to help prevent oxidation which reduces corrosion and to maintain an inert gas atmosphere in the space above the oil to keep any combustibles from exploding in an oxygen rich environment.

What is the difference between a wet type transformer and a dry type transformer? A wet type transformer contains a fluid like oil to help it stay cool where as a dry type transformer contains no fluid within it.

What are three types of dry type transformers? The three types of dry type transformers are the ventilated - used in general areas, the non-ventilated - used in dusty, dirty, chemical environments, and the epoxy filled - used for hazardous environments.

What are some advantages of using dry type transformers over wet type transformers? The dry type transformer has several beneficial features like they're more fire resistant, they're lighter, installation is much easier, much more versatile, and they do not require any special catch basin or vaults to contain any oil spill.

What does a differential relay do? A differential relay is a protective device used to compare the power generated by the generator with the power leaving the transformer. If the difference between the two is to great, the relay will open the circuit breakers.

What is a fuse? A fuse is an electrical protective device consisting of a piece of wire that will melt and interrupt the flow of current in a circuit when an excessive amount of current flows through the circuit.

What does an over current relay do? An over current relay is an electrical protective device that will interrupt a circuit when the relay detects an excessive amount of current flowing through the circuit. Unlike a fuse, when the over current relay trips it does not have to be replaced, just reset once the problem is fixed.

What is an under voltage relay? An under voltage relay is an electrical protective device used to open a circuit should the voltage in that circuit fall below an unsafe value. When voltage drops, current flow through the circuit will increase.

What is a sudden pressure relay? A sudden pressure relay is an electrical protective device used on transformers to detect a sudden pressure increase from internal faults within the transformer. If detected the relay will remove the generator and transformer form service.

What causes a transformer to hum? As power flows through a transformer a resonance is created, this is why the transformer hums.

What are effective values? Effective value is a measurement of the amount of work that can be accomplished by an A.C. circuit in relation to an equivalent D.C. value.

What does kilo stand for? A kilo is a unit of measure for reactive power that is equal to 1,000 vars.

What is a VAR? A VAR is a unit of measure for reactive power. Var means Volt Ampere Reactive.

What are megavars? A megavar is a unit of measure for reactive power that is equal to 1,000 kilovars.

What is load? Load is a term used to describe the amount of electricity being generated by a generator for usage on the electrical grid.

What is load rejection? Load rejection is a reduction in the amount of electricity used on the grid. This condition usually happens suddenly.

What is line current? Line current is the amount of current that is flowing through an electrical circuit.

What is line voltage? Line voltage is the amount of voltage that is present in an electrical circuit.

What is peak-to-peak value? A peak-to-peak value is an amount of voltage or current that indicates the distance between both the positive and negative peaks of a sine wave.

What is peak value? Peak value is an amount of voltage or current that is at its peak value on a sine wave. This can be either a positive or a negative value.

What is the difference between current and voltage? Current is the flow of electricity in a conductor whereas voltage is the pressure that causes the current to flow through a conductor. Without voltage there is no current flow.

What is a ripple? A ripple is when the output voltage of a D.C. generator is drifting or is not holding a constant output voltage.

What is the purpose of the substation? A substation is a facility that receives electrical power from high-powered transmission lines and distributes this electrical power for local consumption after the voltage has been stepped down through the use of transformers and circuit breakers.

What is the difference in rpm between a 4 and 2 pole generator? Depending on what country the generator is being used will determine the frequency the generator must operate at. Here in the U.S. a 2-pole generator will operate at 3600rpm to maintain the 60-hertz (cycles) needed to maintain the frequency at which the electrical grid is maintained at. A 4-pole generator would operate at 1800rpm to maintain the 60-hertz needed because it has twice the poles that the two-pole generator has, therefore it can operate at half the rpm to maintain the same frequency.

What is a volt? A volt is a unit of measure used to measure voltage. The term volt came from an Italian physicist, Alessandro Volta, 1745-1827. The term voltage is a term used to express more than one volt. 1 amp of AC is equal to .707 amps of DC.

What are Kilo Volt Amps? KVA are units that are used for the apparent power output of an A.C. generator.

What is an amp? An amp is a unit of measure used to measure electric current in a conductor.

What is a volt-ampere? A volt-ampere is a unit of measure used to measure apparent power.

What is a watt? A watt is a unit of measure used to measure electrical power. This unit of power is equal to the rate of work represented by a current of one ampere under a pressure of one volt. The term watt was named after a Scottish engineer, James Watt, 1736-1819. The term wattage is used to express more than one watt.

What is a milliamp? A milliamp is 1/1,000 of an amp.

What is phase current? Phase current is the amount of current that flows through a single phase of an electrical connection.

What is phase voltage? Phase voltage is the amount of voltage that is in a single phase of an electrical connection.

What is the purpose of a cuno filter? The purpose of the cuno filter is to filter out any impurities within the seal oil before the oil is supplied to the generator seals.

How does a cuno filter work? A cuno filter is a specially designed filter that allows the trapped debris within the filter to be removed through the use of a handle located at the top of the filter. The handle is connected to the filter elements through the shaft. The filter elements consist of spacers, cleaning blades or disk. When the handle is turned, the cleaning blades inside the filter elements will scrape the debris off, allowing the debris to fall to the bottom of the filter. Periodic inspection of the filter can be made from a drain plug on the bottom of the filter. These types of filters normally provide trouble free operation for many years.

How does the power demand change in a 24-hour period? Between the hours of 5 and 9pm the demand for power is at its highest point and after midnight the demand for power is at its lowest point within a 24-hour period.

What is the purpose of the seal oil system? The generator seal oil system provides a means of sealing the ends of the generator where the rotor passes through. If these ends were not sealed, hydrogen gas would escape and the cooling of the generator would continuously be a problem. The seal oil system provides a seal around each end of the generator using clean oil that has been filtered and degasified then pumped up to the generator seals.

The Seal Oil

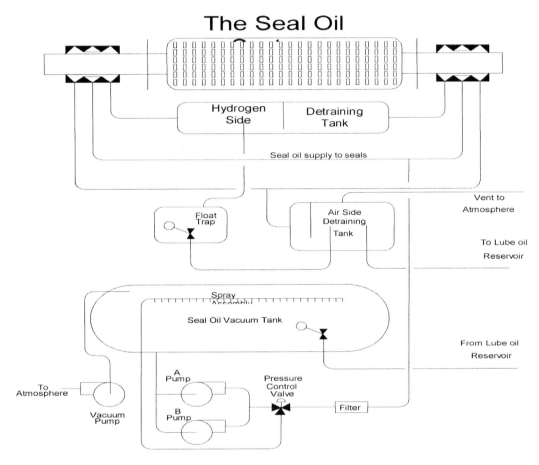

What is the purpose of the loop seal on the seal oil system? The purpose of the loop seal on the seal oil system is to keep the generator hydrogen from escaping into the turbine oil reservoir in the event the generator oil seals should fail. Works in the same principle as a sink trap.

What is the purpose of the seal oil detraining tank? The detraining tank allows gas bubbles that may be trapped in the oil to float to the surface, where they are removed by a vacuum pump. If these gases where allowed to accumulate in the oil there would be an increased chance of an explosion due to mixing of hydrogen and air in the oil.

When is the generator's field breaker opened and closed? The generator's field breaker is closed when the generator is ready to build up a magnetic field, during a startup and is opened when the generator is tripped to remove the magnetic field within the generator.

How is the terminal voltage of a generator controlled? The terminal voltage of the generator is controlled by the voltage regulator. Ex. As generator load increases, the generator output voltage will decrease causing the voltage regulator to bring the voltage back up to normal and vice versa.

What is terminal voltage? Terminal voltage is the generator's rated voltage; it is the amount of voltage that the generator is expected to produce during normal operation.

What does the number "86" represent on electrical trip devices? The number "86" represents lockout relays from the American Standard Device Function Number system. Look under IEEE.

What would happen to the station service, if the unit should suddenly trip? Station service should change over (auto) to allow power to flow back into the plant through the station service transformer should the unit trip because on an 86 lockout or the operator has tripped the turbine manually.

What is the purpose of the generator ground detector? The generator ground detector is used to detect voltage buildup on the generator stator neutral wire. This device is used to protect the generator from a potentially catastrophic event by removing the unit from service immediately.

What is an ampere? An ampere is a unit used to measure electric current. It is when a constant current that if maintained in two straight parallel conductors of infinite length and negligible cross section placed one meter apart in a vacuum, will produce between these conductors a force that is equal to 2×10^{-7} Newton per meter of length. One ampere of current is equivalent to 6.24×10^{18} (6.24 billion-billion) electrons passing through a cross section of a conductor per second. This unit of measure is named after French physicist Andre Marie Ampere (1775-1836). (Random House)

What is a kilowatt-hour? A kilowatt-hour is a unit of energy that is equal to the energy expended when power of one kilowatt is used for one hour.

What are the three types of buses? The three types of buses are the non-segregated, the segregated, and the isolated buses. In the non-segregated bus, all three phases are contained within one bus. With the segregated bus, all three phases are contained within one bus, but each phase is separated by partitions. And last, the isolated bus, all three phases are separated and contained within the own isolated bus.

What is the purpose of the isolated phase buses? The isolated phase buses connect the generator to the transformers. In a three-phase generator, each phase is insulated in it's own phase bus and remains centered within the tube through the use of ceramic insulators that hold the phase in place, not allowing it to move.

How is the isolated phase buses cooled? The isolated phase buses are cooled by air that is filtered and cooled (by cooling water) and blown through the phase bus system.

Phase Bus Cooling System

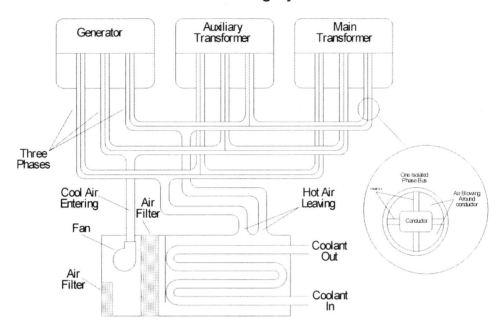

If the generator is at rated speed and is ready to be put online, is it possible to change over station service before putting the unit online? If control logic allows it, it is possible, because when the generator field breaker is closed, current is put on the main and auxiliary transformer. This is not recommended because of the unstableness of the generator load at such low loads.

What is the difference between active and reactive power? Active power is the electrical energy that actually turns on lights, motors, and other devices. Reactive power is the energy that supports the magnetic and electric fields necessary to operate equipment in the electrical system. Reactive power represents electrical power that is dissipated in the various components of the circuit, which does not do useful work.

What does it mean to synchronize the generator? Synchronizing the generator means that the rotating magnetic field of the generator and the rotating magnetic field of the electrical grid is being synchronized.

What is torque angle? Torque angle is the angle by which a rotating magnetic field of a synchronous machine lags or leads the rotating magnetic field of the system to which it connects. A motor has a negative torque angle while a generator has a positive torque angle.

How does a transformer work? A transformer has two coils of wire that are not physically connected and are placed close together on a common iron core. As the alternating current passes through one of the coils (primary winding), a changing magnetic field is created in the core. The building and collapsing of the electromagnetic field in the core caused by the alternating current flowing through the winding induces an electromotive force (emf) or voltage in the other winding (secondary winding). By varying the number of turns in each winding, the voltage that is transformed through the magnetic link can be changed. Example: If the primary windings had 10 turns and the secondary windings only had 5 turns, the voltage would be reduced to half but the amperage would double. If the turns on the windings were reversed, the secondary voltage would be doubled and the amperage would be halved.

What is phase angle? Phase angle is an angle one waveform will lead or lag another waveform. A leading phase angle would indicate that voltage is leading the current or in a lagging phase angle, voltage is lagging behind the current. Circuits where the reactance is mostly capacitive would have a leading phase angle and a circuit with inductive impedance would have a lagging phase angle. If thee impedance of the circuit is purely resistive, then the voltage and the current will be in phase and the phase angle will be zero.

What is transformer excitation? Transformer excitation is when the transformer has no load on the secondary winding (unloaded); the transformer will still draw current to magnetize the core in the transformer. This current used to magnetize the core is the excitation current.

What is the difference between a power transformer and an Instrument transformer? A power transformer is used to step-up or step-down electrical power voltages for transmission and distribution of electrical power. An instrument transformer is a transformer that steps-down the system voltages and currents so that it can power instruments within the plant or electrical system.

What is an Auto-transformer? An auto-transformer is a transformer that contains a single coil. This coil acts as a primary winding with a portion of the coil also being used as a secondary winding.

What are three levels of cooling for a transformer? There are three symbols to represent the level of cooling a transformer will receive, OA - means self-cooled due to natural circulation of oil and air, FOA – means a first stage of forced oil and air cooling using pumps and fans, FOA – means a second stage of forced oil and air cooling using pumps and fans.

What is in-rush current? In-rush current is the sudden rush of current when a transformer or motor is first energized. The peak magnitudes of the in-rush current may last only a few cycles but can reach levels 8-10 times the device's full load current.

What is complex power? Complex power is the sum of active and reactive power. Complex power is the total power that the transmission system is carrying.

What is a base load unit? A base load unit is a generating unit that continues to generate electricity all the time.

What is a cycling unit? A cycling unit is a unit that is started up and shutdown depending on the electrical requirement of the power grid. These units are used to supply electricity only when needed.

What is a peaking unit? A peaking unit is a unit that is only operated during peak load demands of the electrical grid.

What is meant by peak load? Peak load can be used to determine the maximum amount of power output of a generator or the total amount of electricity that is required for the power grid through the use of several power generating stations.

How efficient are transformers? Transformers are very efficient devices; their efficiency is approximately 97% to 99%.

How can a magnetic field be created? A magnetic field can be created by either a permanent magnet or by an electric current flowing through a conductor.

How much is a mega? A mega is a prefix used that represents one million.

What is a megawatt? A megawatt is equal to 1,000,000 watts or 1,341 horsepower. It is estimated that one-megawatt can supply enough electricity to power 1,000 average homes per day.

How many kilowatts are in a megawatt? There are 1,000 kilowatts of every megawatt.

What is a megawatt hour? A megawatt hour is 1,000,000 watts of power times 60 minutes.

What is a megawatt ampere? A megawatt ampere is a unit of measure for apparent power that is equal to 1,000 kilowatts.

What is a breaker failure relay? A breaker failure relay is a relay that will open a breaker automatically, should the main breaker malfunction and not open when it should. Used to protect the generator from becoming motorized. Should one or two poles of a generator line breaker fail to open, the result can be a single-phase load on the generator and negative sequence currents on the rotor.

What is power factor? Power factor is a relationship between apparent power and true power; true power is divided by apparent power. If the voltage and current are exactly in phase, then apparent power will equal true power. Whenever voltage and current are not in phase, then apparent power must be multiplied by some factor to account for the phase difference in order to get the true power in watts. This is called the power factor. If the voltage and current are in phase, the power factor will be equal to 1, and if they are not in phase the power factor will have a value less than one.

What limits the generators power factor? The stator core temperature mainly limits the power factor.

What limits the generators power factor from rated to 100%? The stator winding temperature limits the power factor from rated to 100%.

What limits the generators power factor from zero to rated? The rotor winding temperature limits the power factor from zero to rated.

What is unity power factor? Unity power factor is when true power and apparent power are equal; this would indicate that the system is operating at a power factor of 1.

What is true power? True power is the amount of power that is actually doing the work in an AC circuit.

What is apparent power? Apparent power is the amount of power that a generator would produce if inductive reactance or capacitive reactance were not present. Measured in volt-amperes, can be calculated by multiplying the voltage in an A.C. circuit by the current in the circuit.

What does lead and lag mean? Lead is a term used when comparing voltage and current waves. The wave that is heading positive and crosses zero first is the leading wave. In a leading load, the current wave leads the voltage wave. The term lag is used also when comparing voltage and current waves. The wave that is heading positive and crosses zero last is the lagging wave. In lagging load, the current wave lags behind the voltage wave.

What determines the speed that a generator will normally operate at? The number of poles on the generator determines a generator's speed that it normally operates at. A generator with two poles must rotate at 3600 rpm in order to maintain 60 hertz (cycles per second). A generator with four poles will rotate at 1800 rpm in order to maintain 60 hertz. This difference in rpm between a two pole and four pole generator is because a four pole generator will cut through twice as many lines of magnetic flux as a two pole generator and only needs to rotate half as fast as a two pole generator. Some generators rotate even slower than 1800 rpm yet they contain many more poles on the generator.

What keeps the turbine rotor from being electrically charged from the generator? The slip rings are insulated from the generator shaft, which keeps current from flowing through the shaft. The turbine also has grounding brushes that keep the generator/turbine from building up a static charge possibly causing damage to the turbine bearings and/or personnel.

What is a reverse power relay? A reverse power relay is used to protect the generator from becoming motorized.

What is field saturation? Field saturation is a point where a magnetic field reaches a certain strength; any further increase in the current that created the magnetic field will not result in any further increase in the strength of the magnetic field. At this point the magnetic field is saturated.

What is collector flashover? Collector flashover is a progressive loss of contact between the ring and brushes causing the current to be transferred by arching across the gap until the gap becomes too large for the arc to be sustained. The voltage regulator will call for higher exciter output to keep the generator terminal voltage constant.

What determines the number of generator brushes? The number of generator brushes is determined on the generator field current.

What is a synch-check relay? A synch-check relay is a protective device used to keep a circuit breaker from closing unless the frequency difference, voltage magnitude difference, or voltage difference across an open circuit breaker is within acceptable limits.

What are taps? Taps are fixed electrical contacts that are located on different positions on a transformer's winding. These taps are adjusted to change the voltage ratio of a transformer.

What is voltage ratio? Voltage ratio is the ratio between the primary and secondary voltages of a transformer. The transformer's voltage ratio is related to the transformer's turn ratio.

What is turn ratio? Turn ratio is the ratio of the number of turns in the primary winding of a transformer to the number of turns in the secondary winding. Ex: 10 turns to 5 turns would equal a 2 to 1 ratio.

What does the term "under excited," mean for a generator? A generator is said to be under excited when the excitation is less than what is needed to support the magnetic field around the generator at a given load. A generator that is under excited absorbs reactive power from the electrical grid system. This is where the terms "leading" and "bucking" are used in reference to an under excited generator. When a generator is overexcited, it is supplying reactive power to the system. An overexcited generator may be referred to as boosting, lagging, or pushing generator.

What elements are needed to induce voltage? There are three major elements that are needed to induce voltage; they are a conductor, a magnetic field, and relative motion between the conductor and the magnetic field. The conductor is like copper wire that is wrapped around a metal ring and a metal bar that is wrapped with copper wire and the ends of the wire are attached to a D.C. source in order to provide the magnetic field. As the D.C. power flows through the conductor around the metal bar, a magnetic field is produced. This will form an electromagnet. The electromagnet will have a north and south pole like a permanent magnet. Another element that is needed to induce voltage is relative motion between the magnetic field and the conductor; by rotating the electromagnet, the magnetic field will cut across the conductor. Every time one of the poles of the electromagnet cuts across the conductor, voltage is induced. As one of the poles in the electromagnet move away from the conductor, the voltage will decrease until it reaches zero.

What is the purpose of the stator core? The stator core supports and holds the stator windings in position around the rotor.

Why are three phase generators used? One main reason for using a three-phase generator is that no more than one of the three output voltages is at zero at any time. Meaning that the combined voltage will never equal zero, and power is always continuous.

What are three factors that affect the output of a generator? The three factors that affect the output of a generator are, the speed at which the rotor turns, the number of conductors, and the strength of the magnetic field.

What is a kilowatt? A kilowatt is a unit of measure for true power. It is the amount of true power that a generator is capable of producing at a particular power factor. A kilowatt is equal to 1,000 watts of power.

What is meant by KVA? KVA is the value that indicates the apparent power output of the generator. KVA is the acronym for kilovolt ampere. KVA can be found on the nameplate of the generator. Apparent power is calculated by multiplying the generator's rated voltage times the generator's stator amperage. If the generator is a three-phase generator multiply the last answer by 1.73 to give the rated KVA of the generator. The 1.73 comes from one of the characteristics of a three phase wye-connected generator in that the total voltage output of all three-generator phases is 1.73 times the voltage in each phase. KVA is equal to 1,000-volt amperes.

When a transformer steps up power, will the amps and voltage increase, or will just one of them increase? An AC transformer is used to raise or lower voltage. When the voltage is increased, the current will decrease and when the voltage is decreased, the current will increase.

What is capacitance? Capacitance is a physical property of an AC circuit that opposes changes in voltage. The effects of capacitance are opposite to the effects of inductance.

What is capacitive reactance? Capacitive reactance is the effect on an AC circuit. Capacitive reactance is usually created by a device called the capacitor within an AC circuit.

What is inductance? Inductance is a physical property of an AC circuit that opposes change in current flow.

What is inductive reactance? Inductive reactance is the effect of inductance on an AC circuit.

How is inductive reactance measured? Inductive reactance is measured in ohms.

Is there more inductance in a coil of wire or in a straight wire? If voltage were applied to a coil of wire, a magnetic field would begin to build and cut through each turn of the coil. Each turn in the coil would produce a magnetic field that would cut through each adjacent turn of the coil of wire. The result would be more inductive reactance than would be present in a straight wire.

What is self-induction? When there is a conductor, a magnetic field, and relative motion between them, a voltage is induced in the conductor. Since there is already a voltage applied to the wire, the voltage that is induced by the expanding magnetic field is called a second voltage. The second voltage is opposite to the existing applied voltage. Self-induction is the process of inducing a second voltage. Self-induction can be referred to as counter electromotive force or CEMF, because it opposes the applied voltage and limits the current flow.

What is electromagnetic induction? Electromagnetic induction is when electricity is produced by moving a conductor through a magnetic field.

What determines the strength of electromagnetic induction? Electromagnetic induction strength depends on the strength of the magnetic field and the rate of speed at which the conductor cuts through the magnetic lines of flux.

What is excitation voltage? Excitation voltage is the voltage that is used to create a magnetic field within the generator.

What is the purpose of the core monitor? The core monitor is used to determine if insulation breakdown inside the generator is occurring. The monitor samples the hydrogen gas within the generator to monitor the amount of particulate matter is present within the gas.

What is the difference between mhos and ohms? With ohms it is the measure of resistance while mhos is the measure of conductivity. They are just opposite of each other.

What is stator-cooling runback? Stator cooling runback is an automatic load rejection system that monitors the stator temperature and adjust load on the generator if the temperature exceeds a preset value over a given amount of time.

What is the purpose of the hydrogen gas dryer? The purpose of the hydrogen gas dryer is to remove any moisture from within the hydrogen gas circulating inside the generator through the use of a moisture absorbing material called desiccant. Eventually the desiccant will become saturated and will need to be dried out and again put back in service. If the hydrogen gas dryer was not used, moisture would build up within the generator and cause corrosion, which could lead to an electrical fault.

What is the purpose of maintaining the temperature of the stator cooling water? If the cooling water flowing through the stator were not maintained within a certain temperature the stator would overheat or would begin to form condensation on the windings by being to cool. Plus undue thermal stress would cause a breakdown in generator winding insulation.

The Stator System

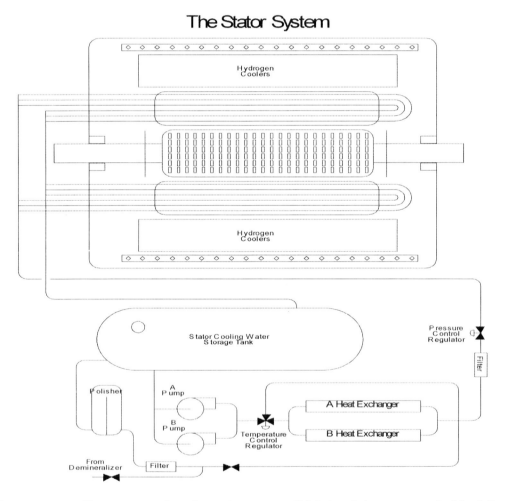

How is the stator cooling system chemistry important? Maintaining a neutral pH while maintaining a certain concentration of oxygen in the water. If the oxygen levels get too low, then there is excessive corrosion.

What is EMF? EMF is an acronym for electromotive force. It is the difference of potential produced by electrical energy, which can be used to drive currents through external circuits.

What is an amplidyne? An amplidyne is a device used to automatically control the generator field voltage.

What is a rheostat? A rheostat is a device that controls current flow by increasing or decreasing the amount of resistance within a circuit.

What is load shedding? Load shedding is the reduction of consumers of electricity within a power system in response to load demands in excess of generating capacity. This happens during the peak of summer when the electrical demand is at its highest or during a brownout or blackout. This process can also happen should a power plant go in the dark. All non-essential systems will shut-down so that enough power can remain for essential systems through the use of a back-up power systems.

What is a sine wave? A sine wave is a wave that is an electromagnetic or sound wave that represents an oscillating signal and can be graphically represented by a sine curve. A sine wave is used to represent the flow of electric current.

How does hydrogen density affect the generator fan pressure? As hydrogen density increases so does the generator fan pressure. An increase in hydrogen density occurs when the generator insulation begins to breakdown and particles become suspended within the gas making it denser. Problems with a seal oil system may cause other contaminants to enter the hydrogen gas also causing the gas to become denser. This is an indication that the purity of the hydrogen gas within the generator may need to be purged.

What is the purpose of generator purging? There are two reasons for purging the generator, 1) to remove the hydrogen within the generator using CO_2, then compressed air in order to enter the generator or storage of the generator 2) to increase the hydrogen purity in the generator.

Why must the generator frame be a strong and rigid structure? There are five major reasons for the generator frame to be strong and rigid in structure.

1) Must be able to support the weight of the entire generator.
2) Must be able to withstand the powerful torque of the rotor.
3) Must be able to withstand shocks that may come from electrical faults and surges.
4) Must be able to hold hydrogen gas pressure @ a constant pressure.
5) Must be able to withstand the force of a hydrogen explosion.

What is the biggest enemy of electrical devices? The biggest enemy of most electrical devices is heat.

How can heat be reduced in electrical devices? One way of reducing heat in electrical equipment is by increasing the voltage at which the device operates. If the voltage is doubled, the same amount of power can be produced with half the current. But if the voltage is too high, the insulating materials will begin to breakdown. Several other ways to keep devices from heat is to keep them cool, keep them properly lubricated, if equipment is being used for long periods of time, give it a rest, and last but not least, if equipment is running without being used, shut it off.

Why is most power generated as three phase power, rather than single-phase power? First, three phase generator has 50% greater capacity, than a single-phase generator of the same physical size. Secondly, single-phase power is easily available from a three-phase system by merely tapping from any two of the power leads. And finally, the cost of transmission is less, for the same voltage and current, in a three-phase system than it is for a single-phase system.

What are three conditions that each generator must satisfy in order to operate in an electrical system? 1) The operating frequency must be exactly the same as that of the system. 2) The generator must be exactly in phase with the system. 3) The generator voltage must be exactly the same as the system.

Can a generator be at the same frequency as the system, yet be "out of phase"? Yes! Example, one soldier may be marching in exact step with the other soldiers but will be putting his left foot down while the rest of the soldiers put their right foot down. He's out of phase. (phase angle) Don't ever do this!

How can an out of phase generator be corrected? An out of phase generator can be corrected by either increasing or decreasing the speed of the generator long enough to make the two synchronize.

What is the worst position on the synchroscope to be in? When the position on the synchroscope is pointing straight down (6 o'clock) the phase angle difference is 180^0 out of phase from the system. This is the worst position on the synchroscope when synchronizing the generator. Even though turbine/generators are designed to withstand these rare occurrences without catastrophic results, provided the stator current does not exceed the three-phase short circuit value, they can result in damage, such as slipped couplings, high vibration, loosened stator windings, and fatigue damage to the shaft and/or other mechanical parts.

What could happen should the main generator breaker be closed with the generator and system out of phase or have a large difference in frequency? Very large currents will flow into the generator stator windings causing damaging heat and tremendous forces will result as the system literally "pulls" the generator into synchronism. There are protective devices that should keep this type of condition from happening.

Where is the incoming and running voltage measured? The running voltage is usually measured on the bus side of the breaker while the incoming voltage is usually measured on the high side of the transformer or at the transformer side of the breaker.

What does it mean by slipping a pole? The generator voltage is reduced by reducing the current flowing through the generator field in normal operation. Should the DC current through the generator field fall too far, the magnetic force which holds the generator in synchronism with the electrical power system becomes weak. This could result in the generator "slipping a pole" which could result in very high currents being induced in the generator, resulting in damage from overheating. This may also cause very high mechanical forces, which can also be damaging.

What is voltage droop? Voltage droop is similar to speed droop in the turbine governor. It provides a means of reactive load sharing between generators, which are operating in parallel in a system with too little reactance between them.

What two functions does impedance compensation perform?
1) Voltage droop
2) Maintains voltage some point other than generator terminals at some constant voltage. Necessary when there is a substantial voltage droop in the transmission lines.

What is unit efficiency? Unit efficiency can be found by multiplying the three main systems of the unit. Unit efficiency = boiler efficiency x turbine efficiency x generator efficiency.

What does IMP IN and IMP OUT mean? IMP IN and IMP OUT are control functions on an EHC control system. IMP IN (load %)- if linear load response to reference change is desired, the system should be transferred to IMP IN mode. This gives the system a linear response to changes in load demand. IMP OUT (valve %)- the speed error signal assists in maintaining system frequency (turbine speed).

What is plant load factor? The plant load factor is a ratio of the total number of kWh supplied by a generator to the total number of kWh, which would have been supplied if the generator were operating continuously at full load.

What is the difference between Class A and Class B oils that are used in electrical devices? Certain devices and equipment like transformers or circuit breakers use oil such as mineral oil for cooling. The difference between these two classes of oil is Class A oils can be used in temperatures up to 80^0C and contain 0.1% maximum sludge while the Class B oils can be used in temperatures up to 75^0C and contain 0.8% maximum sludge.

What is the difference in windings between a step-up and a step-down transformer? In a step-up transformer the secondary windings have more turns than the primary windings;

voltage goes from low to high. In a step-down transformer the secondary windings have less turns than the primary windings; voltage goes from high to low.

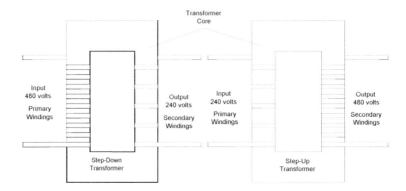

What is phase difference? Phase difference is the phase angle by which one waveform leads or lags another waveform of the same frequency.

What is inhibited oil? Inhibited oils are oils that are used in transformers and other electrical devices that contains an antioxidant that is used to delay the forming of sludge and acid formation.

What is an over-compound generator? An over-compound generator is a DC generator containing a series of windings that are specifically designed so that the voltage rises as the load increases.

What would cause the generator breakers not to open after a turbine trip? 1) A turbine steam valve didn't make a closed limit switch. 2) Breakers failed to open, generator motoring.

What formula is used to convert electrical energy into heat energy? The formula to convert electrical energy into heat energy is Watts = current^2 x resistance. $W = I^2 \times R$

What are windage losses? Windage losses is the motion of a rotor through the gas in the generator creates heat from friction just as a meteor burns up when it enters the Earth's atmosphere due to air friction.

What is the purpose of an open air-cooling system? This type of system is used on older and smaller generators. It uses fans attached to the rotor to pull cool outside air into the generator and is forced through the generator to remove heat and finally exits through openings in the generator's stator frame.

What are collector rings? Collector rings are smooth rings at the end of the shaft connected to the rotor windings.

What are three circuits that are needed for AC generator operation?

1) A magnetic field surrounds an electrical conductor that is to carry a current.
2) An electrical voltage will be induced in the electrical conductor when a magnetic field cuts across it.
3) Where the magnetic lines of flux flow through the air and steel.

What is the collector end of the generator? The collector end of the generator is where the exciter sends the current into the generator.

What are three major reasons for cooling a generator with hydrogen? The cooling properties of hydrogen are better than any other gas, including air. Hydrogen absorbs and removes heat better than any other gas. A pound of H2 can absorb fourteen times more heat than a pound of air at the same temperature. Also H2 can absorb heat from a solid surface 50% more than air. The density of H2 is proportional of its ability in absorbing heat, therefore the higher the pressure; the more heat can be removed from the generator. The windage losses are much less than that of air-cooling. Since air is 14 times more dense than H2 at a given pressure and temperature, the windage losses are greatly reduced with the use of H2 in the generator and less power is needed to drive the fan on the generator's rotor. Maintenance cost is greatly reduced when using H2 to cool a generator because the generator will be much more cleaner and drier than an air-cooled generator.

What is the purpose of the de-excitation circuit? The de-excitation circuit is designed to be used as a back up to the generator's exciter field breaker. When the exciter field breaker operates properly, the de-excitation circuit is not needed. But should the exciter field breaker circuit fail to open the exciter field breaker the de-excitation becomes operable and opens the exciter field breaker circuit by creating a short-circuiting effect. An additional feature of the de-excitation circuit is that exciter voltage buildup will be prevented in the event should the exciter field breaker be closed by mistake, while rolling the turbine up to speed.

What are some factors that determine what kind of trip is used to trip the turbine/generator? There are several different factors that determine the type of trip that will be used to trip the turbine/generator or just the generator alone. Here is a list of nine different factors: 1) severity of the fault to the generator, 2) probability of the fault spreading, 3) the amount of over speed resulting, 4) the probability of a high over speed, 5) the importance of removing excitation from the generator, 6) the need for maintaining auxiliary power, 7) the need for shutting down the unit, 8) the time required to resynchronize and 9) the total effect on the power system.

What is the difference between a simultaneous trip and a sequential trip? A simultaneous trip will trip the turbine valves closed, open the generator breakers, and remove excitation simultaneously, as with a lockout relay. A simultaneous trip is acceptable for all generator faults, and generally provides the highest degree of protection for the turbine/generator although it does permit a small over speed and there is a slight chance of a high over speed occurring. As with a sequential trip, it trips the turbine first. When the turbine steam inlet valves indicate that they are closed the recommended reverse power relay operates (3 second delay), the generator breakers are tripped. The opening of the generator breakers will then trip the excitation. This trip should prevent any turbine over speed. This type of trip has a disadvantage that certain multiple limit switch failures, or a reverse power relay failure, would prevent completing the trip.

What are over-volts per hertz? Over-volts per hertz is per unit voltage divided by per unit frequency, called volts/hertz, is a readily measurable quantity that is proportional to the flux in the generator and step-up transformer cores. Excessive flux can cause serious over-heating of metallic parts and, in extreme cases, localized rapid melting of the generator core laminations. This problem can be caused by generator regulator failure, load rejection while under control of the dc regulator, or excessive excitation while the generator is off line. It can also result from decreasing turbine speed while the ac regulator attempts to maintain the rated stator voltage.

How can the loss of excitation to the generator be harmful? The loss of excitation to the generator results in the loss of synchronism to the system and operation of the generator as an induction motor. This will result in the flow of slip frequency currents in the rotor as well as severe torque oscillations in the rotor shaft. The turbine/generator rotor is not designed to sustain such currents nor is the shaft designed to withstand the alternating torque. The result can be rotor overheating, coupling slippage, and even rotor failure. The length of time before serious damage occurs depends on the generator load at the time of the incident, slip frequency, and whether the field winding is open circuited or shorted, and maybe a matter of seconds. A loss of excitation normally indicates a problem with the generator excitation system. Because of the vars absorbed to make-up for the low excitation, some systems cannot tolerate the continued operation of a generator without excitation. If the generator is not disconnected immediately when it loses excitation, widespread instability may quickly develop, possibly causing a major system upset. A loss of excitation will result in a change in reactive KVA.

What is the main byproduct within the generator when producing electricity? Heat is the main byproduct that is produced by the generator when producing electricity.

What determines the size of the generator? The sizing of the generator is based on the MVA

What is a short circuit ratio? A short circuit ratio is a ratio of the field ampere turns required to produce the rated voltage at no load and at rated frequency to the field ampere turns required to produce the rated armature current at a sustained short circuit.

What are electrical eddy currents? Electrical eddy currents are currents that circulate within a material as a result of electromagnetic forces induced by a variation of magnetic flux. Current flow is restricted by discontinuities in the material.

What is the weakest part of the generator? The weakest part of the generator is the insulation.

What factors affect the insulation within the generator?
> 1) Thermal aging (temperature/ time effects), 2) Overheating, 3) Excessive moisture, 4) Chemical attack, 5) Over voltage, 6) Vibration, 7) Over speed conditions, 8) Any presence of foreign material, and 9) Ionization effects.

What is a field ground? The field ground is the generator field windings that are isolated from a ground, electrically. The existence of one ground fault in the rotor windings will usually not harm the rotor. But should two or more winding ground faults develop may cause magnetic and thermal imbalances with localized heating and damage to the rotor forging. Field ground protection is usually tied together through the excitation system of the generator.

What is a kilowatt-hour? A kilowatt-hour example: A 100-watt light bulb uses 100 watts or 0.1 kilowatts of electricity. If this light were left on for an entire month the amount of electricity used can be figured by multiplying 0.1 kilowatts x 24 hours x 30 days x the cost of electricity (example 0.12 cents) which would equal $8.64 of electricity was used a month for the light bulb alone.

Blackout condition, without any outside running voltage, how would you know if the generator is close to synchronous speed? By looking at the rpm of the turbine/generator would give you an indication if the generator were close to synchronous speed. Two pole generators rotate at 3600 rpm to maintain 60 hertz while a four-pole generator rotates at 1800 rpm to maintain 60 hertz. This does not indicate the difference between system frequency and the generator frequency.

How is a three phase generator stator wound? In the photo below you can see how each phase is wired separately and in a fashion that puts each phase 120^0 out from the other phases.

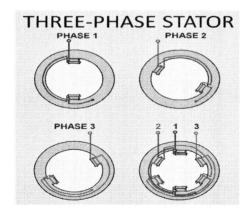

Instrumentation & Controls

This section will give you an idea of the different types of instruments used and how they work. With all the massive equipment and systems used to produce electricity it takes just as much equipment to monitor, evaluate and control it in order to produce electricity safely and efficiently. You will be exposed to some of the equipment found in a power plant, yet there is still a lot to learn that hasn't been mentioned.

What is absolute pressure? Absolute pressure is a pressure that is measured on a scale where zero is the total absence of any pressure.

What is an actuator? An actuator is a device that is used to drive a valve, damper, or some other piece of equipment through the use of an electric, pneumatic, or hydraulic motor (operator).

What is an airflow meter? An airflow meter is an instrument that indicates the amount of airflow through a system.

What is an ammeter? An ammeter is an instrument used for measuring the strength of amperes in an electric current.

What is a barometer? The barometer is an instrument used to measure atmospheric pressure. It was invented by Evangelista Torricelli, an Italian physicist in 1643. This instrument is used to determine height above sea level and predicting the changes in weather. There are two common types of barometers the aneroid barometer and the mercury barometer.

What is the difference between an aneroid and mercury barometer? The aneroid barometer is operated by the pressure of air on an elastic lid of an airtight metal box, which is under a vacuum. A pointer attached to the lid moves on a scale when a change in air pressure occurs. The mercury barometer is a glass tube filled with mercury and determines the air pressure by the height of mercury in the tube.

What is a bimetallic thermometer? A bimetallic thermometer is an instrument that measures temperature through the use of two different types of metals that expands and contracts at different temperatures to produce a readable temperature.

What is a block diagram? A block diagram is a drawing that uses symbols like circles or squares to represent major pieces of equipment in a drawing.

What is a bumpless transfer? Bumpless transfer is a term used to represent a smooth shift of control from auto to manual or vice versa with little or no fluctuation in the process variable.

What does a comparator do? A comparator is a device that picks up two different signals, compares them, and sends and output signal that is averaged between the two signals.

What is a compound gauge? A compound gauge is a gauge that can measure both pressure and vacuum.

How does a contact pyrometer work? A contact pyrometer is a device used to measure the surface temperature of an object through direct contact.

How does an optical pyrometer work? An optical pyrometer is a device that is used to measure radiant energy (temperature) without actual contact to the surface being measured.

What is a control circuit? A control circuit is a circuit that monitors, regulates, and can be used to protect equipment.

What is a bourdon tube? A bourdon tube is a pressure gauge consisting of an inlet tube, bourdon tube, lever & gear mechanism, a needle, and a casing. This type of gauge has been used for many years, only the construction and accuracy have been modified. The operation of a bourdon tube has not changed an oval tube in a semicircular shape closed at one end will straighten with internal pressure. The movement of the oval tube causes the needle to move to indicate actual pressure.

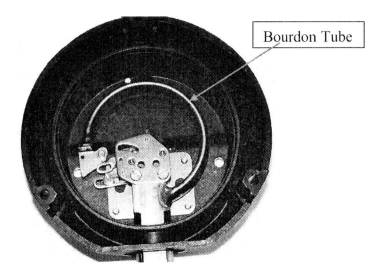

Bourdon Tube

What is a controller? A controller is a device that is to regulate a specific output of a process, through the use of a controlling element. Like a valve opening or closing with the use of an operator (motor, hydraulic, or pneumatic).

What is a control loop? A control loop is an arrangement of control elements that controls a process condition or variable. There are two general types of control loops, open loop control and closed loop control. A control loop will contain a sensing element, a measuring element, a controlling element and a final control element.

What does a control switch do? A control switch is a switch within a circuit that controls a final control element.

What is a conversion table? A conversion table is a table where different units of measure are compared to other units to indicate the equivalent value in different units of measure. Ex. One foot is equal to 12 inches. One gallon of water weighs 8.34 pounds. Two hundred twelve degrees Fahrenheit is one hundred degrees Celsius.

What is a converter? A converter is a device that changes a signal to another signal to be used by another device. Like changing an electrical signal to a pneumatic signal.

What is a corrective signal? A corrective signal is the final output signal from a controller within a control loop.

What is delta p? Delta p is a term used to represent differential pressure.

What is delta t? Delta t is a term used to represent differential temperature.

What do these signs represent(<, > =)? This sign < means less than, this sign > means greater than, and this sign = means equal to. These signs normally used in power plants represent flow or temperature requirements. Signs like ≤ or ≥ mean less than or equal to or greater than or equal to.

What are dependent variables? A dependent variable is a variable whose values are dependent on the values of an independent or chosen variable.

How does a derivative function work? A derivative function is a control function that detects how fast a process is changing, and produces a zero output signal when the input signal does not change.

What does deviation mean? Deviation is the difference between the set point of a process variable and the existing value of the process variable.

What is a diaphragm switch? A diaphragm switch is an electrical switch that is used to indicate level in a tank or vessel.

What is dielectric? Dielectric means not able to conduct electricity or is non-conductive.

What does a differential pressure cell do? A differential pressure cell is a device that measures the difference in pressure across an orifice or other type of flow restrictor. The output of the differential pressure cell is often used to measure the flow through a flow restrictor.

What is meant by direct proportion? Direct proportion is proportionality in which one variable increases by a specific multiple, another variable increases by the same multiple when the first variable decreases by a specific multiple, the second variable decreases by the same multiple.

What is an electrical diagram? An electrical diagram is a drawing that contains symbols that represent components of a process.

How does an electronic transmitter work? An electronic transmitter is an instrument that senses or measures a process and produces an output signal that is proportional to the value of the process variable.

What is electromagnetism? Electromagnetism is the relationship between electricity and magnetism based on facts that electric current produces a magnetic field and that a magnetic field can produce an electric current.

What is electromotive series? Electromotive or electromechanical series is the ranking of metals according to their tendency to lose electrons in a chemical reaction. Metals that lose electrons easily will react easily with other elements.

What is a loadstone? A loadstone is a black mineral that naturally has magnetic properties.

What is feedback? Feedback is a process in which a control system regulates itself by feeding back information to itself in order to control its own output.

What is a final control element? The final control element is a device that causes a change in the process variable being controlled.

What is a flow meter? A flow meter is an instrument that allows a precise amount of flow of a fluid to pass through while providing an indication of some type of measurement.

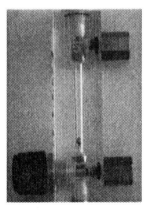

What is flow rate? Flow rate is the amount of a fluid whether gas or liquid that flows a specific distance in a specific amount of time.

What is a flow restrictor? A flow restrictor is a device that allows only a portion of a fluid to flow, restricting flow through the use of an orifice, or venture.

What does a flow transmitter do? A flow transmitter is an instrument used to determine the flow of a fluid through a pipe with a specific amount of time.

What is force? Force is pressure that can be exerted on a state of matter (liquid, solid, or gas) to cause movement or build pressure.

What is friction? Friction is the force of resistance between two objects when in contact.

What is friction head? Friction head is the difference between velocity and static head that results because of friction.

What can a gauge indicate? A gauge can indicate pressure whether positive pressure or negative pressure, temperature whether hot or cold, differential pressures, and differential temperatures.

When the needle in a gauge is bouncing around, what could this indicate? A bouncing needle usually would indicate that a pump might be cavitating (air bound). This is misleading because gauges on positive displacement pumps cause the discharge pressure gauge to bounce around because of the pulse the pump causes in the fluid.

Are all gauges the same? No! There are several different types of gauges depending on what is to be measured. Bourdon gauges are the most common, indicating pressures. Manometers are used to measure differential pressures.

What does a governor do? A governor is a device that is used to control the flow of a fluid to a specific device so that certain conditions can be maintained like speed of a turbine when loaded.

Why are some gauges filled with liquid? Some gauges are filled with liquid to keep the needle from bouncing back and forth causing the internal mechanism to wear out quickly and a gauge reading errors.

How does a gauge work? A gauge like a Bourdon gauge contains a closed end oval tube in a semicircular shape that straightens with internal pressure. The movement from the closed end of the oval tube is measured with the needle indication attached to the end of the tube. As the tube moves so does the needle, indicating the presents of pressure.

What is a ground relay? A ground relay is a protective device used to initiate an alarm should an abnormal flow of current to ground occur.

What is a ground resistor? A ground resistor is a device used to control the amount of current flowing to ground.

What is the purpose of a hand wheel on a valve? The hand wheel on a valve allows the operator to control the valve position manually in case of a malfunction with the valve actuator.

What is heat transfer coefficient? Heat transfer coefficient is the resistance of a material that makes up a heat transfer surface multiplied by the materials thickness.

What does a high or low selecting function do? A high selecting function is a control function where the highest of several input signals is selected to be the output signal. A low selecting function is a control function where the lowest of several input signals is selected to be the output signal.

What is a holding coil? A holding coil is a device that when energized will hold the coils contacts in the open or closed position.

How does a hydraulic actuator work? A hydraulic actuator is a device consisting of a case, fluid, a supply line to both sides of the actuator, and a piston that is located in between the two supply lines of the actuator. When fluid enters from the bottom of the actuator to push the piston up so the valve on the actuator would go open and if fluid entered on the top of the actuator the piston would be forced down to close the valve.

What is an independent variable? An independent variable is a variable that stands along and has been given specific values. It is a variable that can be manipulated.

What is an integral function? An integral function is a control function where the input signal is added to its output signal within a specific amount of time. This function is also known as a reset function.

What does inverse proportion mean? Inverse proportion is when one variable increases by so much and the other variable decreases by the same amount as the first variable or vise versa.

What does an inverter do? An inverter is a power converter that changes direct current (DC power) to alternating current (AC power).

What purpose does the inverter serve? The inverter can supply AC power to the plants control system from a DC power source in the event that AC power is lost to the control system.

What is meant by the left-hand rule? The left-hand rule is a method for determining the flow of the magnetic field around a conductor (wire) when current is flowing through it. This rule states that if the fingers of the left hand were wrapped around the conductor with the thumb pointing in the direction of the current flow, the fingers are said to be curled in the direction of flow of the magnetic flux lines.

What is a level controller? A level controller is an instrument that is used to maintain a specific level range within a vessel by opening, closing, or regulating a valve or providing a shutdown sequence to a control system or an electric motor or by adjusting the speed of a pump.

How do level transmitters work? There are several different types of level transmitters with each operating in a different manner.

1) **Floats** - a metal or plastic float that can provide two different functions, floating on top of the liquid or floating within the liquid. When the float moves either up or down because of a change is fluid level the transmitter generates an output signal that is proportional the fluid level in the vessel.

2) **Probes** - measure level by different electrical characteristics as a change in fluid flowing through it generating an electrical signal from the process.

3) **Paddle wheels** - a type of probe that has fins on its shaft and rotates when a fluid flows across the fins generating an electrical signal from the process.

4) **Nuclear detectors** - a low radiation source is mounted on a vessel on one side while a nuclear detector is mounted 1800 on the other side of the vessel, in line of the radiation source. The nuclear detector generates an electrical signal that becomes stronger as the radiation source becomes weaker (because of the fluid flowing in between the radiation source and the detector). The strength of the electrical signal generated can be used to determine the amount of a fluid flow or pluggage that may be developing within the system.

5) **Ultrasonic** - uses sound waves that are bounced off the liquid within a vessel to determine the level of the liquid (foaming effects the level indication).

6) **Hydraulic head** - measures the weight of a fluid to determine its level within a vessel.

7) **Differential pressure** - measures the differential static pressure at two different points with the difference being calculated by the transmitter then generating an electrical signal.

What does linear mean? Linear means that any type of device or motion where the effect is exactly proportional to the cause. Example: rotation and progression of a screw, current and voltage in a wire resistor at a constant temperature, and output versus input of a modulator or demodulator.

What are some different types of limit switches? There are several different types of limit switches, depending on the function they provide.

1) Mechanical limit switches that are moved enough to open or close a contact.

2) Magnetic limit switches that work off the opposite attract methods to open or close a contact.

3) Nuclear limit switches can determine if there is a fluid flow depending on the amount of radiation the detector sees, to open or close a contact.

What is the purpose for limit switches? A limit switch is an instrument used to determine the value of a specific process, like a valve or damper position being open, closed, or regulated (throttled). Limit switches can provide several functions in a control system like when the valve is fully open a start command on a pump is initiated and if the full open limit is not made or indicates not fully open, the electric motor to the pump will shut down if running or will not start if in operation.

How does a liquid filled thermometer work? A liquid filled thermometer operates on the principles that a liquid within a tube will expand and contract by a specific amount of change in temperature.

What is live zero? Live zero is the minimum signal a transmitter will produce to indicate the process variable being sensed or measured is at zero. This term is used to distinguish a minimum signal from a lost signal of a transmitter.

What is a load cell? A load cell is a load detecting and measuring element utilizing electrical or hydraulic effects, which are indicated and recorded remotely.

What does bias mean? Bias is a control signal that balances or controls two separate devices in a control loop. Ex. Controlling two variable speed fans at the same speed within the same circuit.

What is logic? Logic is the science of dealing with principles of thought and reasoning.

What is a logic circuit? A logic circuit is a circuit that is designed to perform complex functions by using elementary functions of mathematical logic.

What does a logic diagram indicate? A logic diagram is a drawing of an electrical control system that uses control logic to represent different circuits in a system.

What does a logic gate do? A logic gate is a control device that produces an output signal when a sequence of input signals is present.

What is mass flow rate? Mass flow rate is the amount of a fluid that flows through a given area within a certain amount of time.

What is a hot measuring junction? A hot measuring junction is a junction of two dissimilar metals in a thermocouple that is placed in a medium being measured to determine the temperature of the medium.

What is a cold reference junction? A cold reference junction is a junction of two dissimilar metals that completes an electrical circuit of a thermocouple and is usually located away from the hot measuring junction and/or process temperature.

What is a manometer? A manometer is an instrument for measuring pressures whether positive or negative with the use of a variety of fluids used in a tube to indicate pressure or differential pressure. Manometers are an accurate way of determining pressures; the manometers are usually used in low-pressure systems.

Where would find a manometer? Manometers can be found on the boiler to determine airflow through the boiler tubes, turbine lube oil reservoir, gland steam condenser, on the condenser, & in the control room. A manometer can be found anywhere it may be needed, whether calibrating another gauge or running tests on a specific system. Manometers can be filled with mercury, water, antifreeze, or some other type of liquid.

What is a mercoid switch? A mercoid switch is an enclosed capsule that contains liquid mercury in which it is used to complete an electrical circuit when the enclosed capsule is tilted because of a change in level. Basically used for level control systems to bring in alarms whether high or low.

What is a motor control switch? A motor control switch is a switch that allows the starting or stopping of a motor or similar device.

What is a multiplying function? A multiplying function is a control function where the input signals are multiplied together to produce an output signal.

What does offset mean? Offset means a deviation under normal conditions.

What is an ohm? An ohm is a unit of measure that measures electrical resistance that is equal to the resistance of a conductor which, with a potential difference of 1 volt across it, and passes a current of 1 ampere. The unit was named after a German physicist, Georg S. Ohm (1787-1854).

What does an ohmmeter do? An ohmmeter is an instrument used to measure resistance within an electrical circuit.

What does Ohm's law state? Ohm's law states that the voltage across a conductor equals the current flowing through it, times its resistance. Current, volts and ohms work together, if one of them is unknown, the other two can be used to find the third. Ex. Current = volts/ohms, Volts = current x ohms, and Ohms = volts/current.

What is an open circuit? An open circuit is a circuit that has no current flowing through it because the circuit is not complete for whatever reason.

How does an over current relay work? An over current relay is a protective device that breaks or opens a circuit when it detects an excessive flow of current.

What is parallax error? Parallax error is an error of how something is seen. Ex. If a gauge is read by looking at it in some other way other than straight on, an incorrect reading will be read.

What is a parallel circuit? A parallel circuit is a circuit that is designed with two or more parallel paths through which current can flow. Ex. With Christmas tree lights, when one light burns out the rest of the lights will stay lit.

What is a series circuit? A series circuit is a circuit that is designed with only one path for current to flow through. Ex. With Christmas tree lights, when one light burns out the rest of the lights will go out also.

How is percent defined? Percent is defined as a part of a whole number that is expressed in % (percent).

How does a photoelectric cell work? Photoelectric cells work on the principle of photoelectric effect, which is the emission of electrons from a surface when electromagnetic radiation, (light whether visible or ultraviolet) strikes it. There are three basic types of photoelectric cells, photo emissive cells, photovoltaic cells and photoconductive cells. When light strikes a photo emissive cell, the light causes electrons to be emitted and a current to flow. This type is used in television cameras and movie projectors for film soundtracks. Photodiodes are replacing the photo emissive cells because they are faster and respond to a large range of radiation. When light strikes a photovoltaic cell an electromotive force is generated between two different substances. This type is used in photography (exposure meters) and operating relays. The third type, the photoconductive cell contains a semiconductor which when exposed to light, its conductivity increases. This type is used to turn on streetlights and used for instrumentation purposes.

How does a pneumatic actuator work? A pneumatic actuator uses air and the use of a spring to allow the actuator to move in both directions. Air is introduced into a diaphragm on the actuator. The air pressure will either push the actuator down or up, depending on which side of the diaphragm the air is supplied to. A spring is usually located on the other side of the diaphragm, working against the pressure of the air on the other side. When the air pressure increases the valve will move one way, and when the air pressure decreases, the spring will move the valve back the other way.

What does pneumatic mean? Pneumatic means having something to do with a gas, such as air that is used to power or drive a piece of equipment.

How does a pneumatic transmitter work? A pneumatic transmitter is an instrument that is used to sense and measure a process variable and produce an air output signal that is linear to the value of the process variable.

What does hydraulic mean? Hydraulic means having to do with a liquid either at a steady state or in motion.

What does a positioner do? A positioner is an instrument that responds to a control signal sent from a controller to regulate or control the amount of air pressure that is going to an actuator.

What is the difference between negative and positive pressure? Negative pressure is any pressure that is lower than atmospheric pressure. Positive pressure is any pressure that is equal to or greater than atmospheric pressure.

What is pressure? Pressure is a force that can be exerted on an object and can be either positive or negative pressure. Temperature and area can influence this type of force. Pressure is equal to force divided by area.

What is pressure differential? Pressure differential is a calculated difference between two compared pressures.

What does process variable mean? Process variable is a process that is related to a condition such as temperature, level, flow, or pressure that is subject to change and affect a control process.

What is a process variable signal? A process variable signal is a signal that represents the actual value of a process variable.

What is proportional control? Proportional control is the simplest of all closed loop systems. In this type of control system the output of the controller is proportional to the deviation of the controlled variable from its desired value. This deviation is called the error signal.

What is a proportional function? A proportional function is a control function where the input signal is multiplied by a specified constant value to produce its output signal.

What does proportional mean? Proportional is a mathematical relationship between two variables where a change in one variable will cause a proportional change in the other variable. (Linear)

What does proportional plus rate control do? Proportional plus rate control makes adjustments for the rate of change in a process variable.

What does proportional plus reset control do? Proportional plus reset control is the proportional action that is repeated until the process variable returns to a set point.

What does proportional plus reset plus rate control do? Proportional plus reset plus rate control makes adjustments for the rate of change in a process variable and can be repeated until the process variable returns to a set point.

What does psi stand for? Psi is the acronym for pounds per square inch.

What does psia stand for? Psia is the acronym for pounds per square inch absolute.

What does psid stand for? Psid is the acronym for pounds per square inch differential.

What does psig stand for? Psig is the acronym for pounds per square inch gauge.

What is radius? Radius is the distance from the center of a circle to its circumference or the edge of the circle.

What is rate? Rate is a relation between two forms of units, like miles per gallon (mpg) or miles per hour (mph),etc.

What is a ratio? A ratio is the relation between two quantities expressed in the number of times one contains in the other. (A ratio of 10 to 1. (10:1))

What does a rectifier do? A rectifier is a device used to change alternating current into direct current.

What can a relay do? A relay is a device used to make contacts in one circuit that may be operated by a change in the same circuit or a different one circuit. It is a type of switch in which changes in the input are used to alter the output.

What is resistance? Resistance is a property of a material that makes it oppose the passage of an electric current.

What is a resistance temperature detector? An RTD is an instrument that is similar to a thermocouple. It measures temperature based on a principle that the resistance in a coil of wire changes when it is heated or cooled.

What is a schematic diagram? A schematic diagram is a drawing that identifies the components in an electrical system in which they are shown in their proper electrical sequence.

What does a sensing element, sense? A sensing element is an instrument that is used to translate an input signal from a process into mechanical motion or an electrical signal (current or voltage). It is basically the element within the control loop that senses the value of a process variable.

What is a set point? A set point is a value or desired range is to be maintained within a system.

What is a set point generator? A set point generator is an instrument that is used to generate a signal that corresponds to the desired value of a variable.

What does a set point signal do? A set point signal is a signal that is generated in a control loop that represents the desired value of a process variable.

What does short circuit mean? Short circuit is when an electrical circuit resistance drops to zero and the current reaches an extremely high value.

What does a signal generator do? A signal generator is part of a control loop that produces a continuous signal.

What is a single line diagram? A single line diagram is a diagram that uses a single line to represent the flow process through individual components in a system whether electrical or mechanical.

What is a solenoid? A solenoid can best be described as a coil of wire that contains a movable metal core. When the current passes through the coil of wire it creates a magnetic field causing the metal core inside of the coil to move in one direction or the other.

How does a solenoid actuator work? A solenoid actuator is an electric actuator that uses a magnetic field to move a rod or armature that is connected to a final control element.

What does span mean? Span is the difference between the minimum and maximum limits set of a process variable.

What is meant by set point? Set point is the desired value at which a process is to be maintained.

What does a square root extractor do? A square root extractor is a device uses the square root function and generates an output signal by taking the square root of the input signal.

What is a square root function? A square root function is a control function that takes the square root of its input signal to produce an output signal.

What is standard column? A standard column is a water column at a constant height and is used as a reference when measuring levels by static head.

How does a strip chart recorder work? A strip chart recorder is an indicating instrument that is used to graph process variables on paper through the use of pen and ink to give a continuous reading of the process variables.

What is subdivision? Subdivision is the markings on a gauge's face that correspond to a specific unit of measure.

How does a summing function work? A summing function is a control function where the input signals are added together to produce an output signal.

What is surge volume? Surge volume is a large volume of a fluid that must be available to meet changes in the demand of a system. Surge capacity is another word.

How does a tattletale gauge work? A tattletale gauge is a gauge that can indicate the highest or lowest temperature, pressure, or current reading of a measured variable since the last time the gauge was reset.

What is a thermocouple? A thermocouple is a device used to measure temperature. A basic thermocouple consists of two electrical conductors of dissimilar material that are joined at the ends to form a circuit. If the temperature at one end is higher than the other end, an electromotive force is generated which produces a voltage/current flow through the circuit. The total amount of flow produced depends on the temperature difference between the two junctions and the type of materials used for conductors.

How many different types of thermocouples are there? There are several different types of thermocouples; here are five thermocouples and their temperature range.

1) Copper - constantan, type T, -300 to 700^{0}F.
2) Iron - constantan, type J, 0 to 1400^{0}F.
3) Chromel - alumel, type K, 0 to 2300^{0}F.
4) Platinum - 10% rhodium, type S, 0 to 2700^{0}F.
5) Platinum - 13% rhodium, type R, 0 to 2700^{0}F.

How can you tell if a thermocouple has gone bad? When a thermocouple goes bad the temperature change will go to zero, become invalid, change rapidly, or not change at all.

What is a transducer? A transducer is a device that is used to convert a physical quantity into an electrical signal, either by proportion to quantity or according to a specified formula. Example; accelerometers, microphones, photocells, loudspeakers.

What is a transmitter? A transmitter is an instrument in a control loop that senses and measures a process variable and generates an output signal that represents the value of the process variable.

What is a trip circuit? A trip circuit is a protective device that will break or open a circuit that may be in danger, like motor overloads or bearing temperatures high, or a severe pressure differential.

What are variables? Variables are elements within a control circuit that are never constant, always changing and are monitored through the control circuit.

What is a thermowell? A thermowell is a pressure tight receptacle that protects a thermocouple from the contents that it is measuring.

Thermocouple screws into thermowell

Screws into pipe, shell etc…

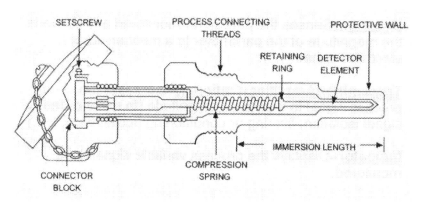

This thermowell has a cap, to be used if thermowell is empty or damaged.

What kind of chart is a truth table? A truth table is a systematic tabulation of all the possible input/output combinations produced by a binary circuit.

What is an AND gate? An AND gate is a logic gate designed to produce an output signal only when all input signals are present.

What is a NAND gate? A NAND gate is a logic gate designed to produce an output signal if one or more of the input signals are not present.

What is a NOR gate? A NOR gate is a logic gate designed to produce an output signal if none of its input signals are presents.

What is a NOT gate? A NOT gate is a logic gate designed to produce an output signal when it does not have any input signals.

What is an OR gate? An OR gate is a logic gate designed to produce an output signal when one or more of its input signals are present.

What is an XOR gate? An XOR gate is a logic gate that is designed with two input signals. If both input signals are the same (opened or closed) the output is zero, if both input signals are different then an output signal is generated.

What is vacuum? Vacuum is a pressure less than that of atmospheric pressure.

What is a vibrometer? A vibrometer is an instrument that is used to measure vibration.

What is velocity? Velocity is the rate of change in position in a direction, expressed as the distance traveled in a unit of time.

What is viscosity? Viscosity is a property of a fluid that makes it resistant to flow (thickness).

How can viscosity be affected? Viscosity can be affected in two ways, either by a fluids density or the temperature of the fluid.

What is a volumetric flow rate? Volumetric flow rate is a measurement of a volume of liquid that covers a certain distance within a specific amount of time.

What is a wet cell battery? A wet cell battery is a battery, which contains a liquid electrolyte, in contrast to the paste of a dry cell battery.

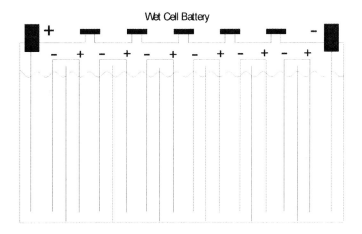

Wet Cell Battery

What is the difference between a primary cell and a secondary cell? A primary cell is a type of battery that will release energy by chemical reaction without being charged. A secondary cell is a type of battery that must be charged before it can release energy by chemical reaction.

What is the difference between ladder logic and loop logic? Ladder logic deals with logic that is digital, it is either on or off, open or closed. With loop logic, this logic deals with analog processes, like levels, ph, flows, temperatures, etc.

What is an amp-hour rating? Amp-hour rating is a unit of measure of an electric charge that is equivalent to a current of 1 ampere flowing for 1 hour. This is often used to show the quantity of electricity that can be delivered by a fully charged battery or accumulator.

What is an electric arc? An electric arc is an area of intense heat and light that is produced by the passage of electricity across a gap between two electrodes. This arc is able to span the gap between the two electrodes because the surrounding air becomes ionized and acts as a conductor. An electric arc occurs in welding to produce heat hot enough to fuse two metals together.

What is an electrode? An electrode is a conductor that supplies electric current, like in welding.

What is an electronic governor? An electronic governor is a governor that generally consists of three elements; a speed sensor, an electronic circuitry designed to develop an error signal, and a torque motor. Add more.

What is the difference between laminar flow and turbulent flow? Laminar flow is when all of a fluid is flowing in the same direction in which adjacent layers of the fluid do not mix, except on the molecular scale. With turbulent flow of a fluid, it is a type of fluid flow in which the particle motion at any point varies rapidly in magnitude and direction and the adjacent layers will mix.

How does a magnetic coupling work? A magnetic coupling is a variable speed coupling that uses electric current to generate a magnetic force that is used to transfer torque from the shaft of the driver to the shaft of the driven equipment. The stronger the current, the faster the driven equipment will rotate and vice versa.

How does a hydraulic coupling work? A hydraulic coupling is a variable speed coupling that uses fluid such as oil and centrifugal force to transfer the torque from the shaft of the driver to the shaft of the driven equipment.

How does a moisture separator work? There are many different designs of moisture separators yet they all seem to use centrifugal force to remove moisture within the vapor.

What does a multiplier do? A multiplier is a number that can be found on the face of certain instruments, which must be multiplied by the instrument indication in order to obtain a correct reading.

What is a manual trip? A manual trip is a device that will allow an operator to manually trip a device for whatever reason. It is a protective device.

What is differential pressure? Differential pressure is just that, it is the difference in pressures being compared for measurement.

What does a voltmeter do? A voltmeter is an instrument that is used for measuring voltages within a circuit (potential difference).

What is a pilot cell? A pilot cell is a single cell of a battery upon which readings are taken in order to give an indication of the state of the entire battery.

What instrument is used to measure the internal resistance of a fluid? The instrument is called a Saybolt viscometer and it measures the internal resistance of a fluid.

What are five factors that affect fluid flow? The five factors that affect fluid flow are head, viscosity, wall roughness, pipe length, and pipe size.

What are two types of flow rates? The two types of flow rates are volumetric and gravimetric (mass).

How many inches of mercury are equal to one psi? One psi is equal to 2" Hg, which is about 1.958 lbs of mercury.

Why is mercury used to measure condenser vacuum instead of water? One inch of mercury (Hg) is equal to 27.74 inches of water (H2O). Four inches of mercury is equal to 54.4 inches of water. As you can see to measure condenser vacuum using water would take an extremely large manometer.

What types of pressure can a bourdon tube gauge measure? Bourdon tube gauges can measure either pressure or vacuum and sometimes even both on the same gauge.

How does a linear variable differential transformer work? A linear variable differential transformer (LVDT) is an instrument that is used to transmit an electrical feedback signal to an electronic governor. The signal is proportional to the movement of a control valve.

What is a rotary variable differential transformer? A rotary variable differential transformer works similar to a LVDT, but rotational and is used for load reference signals.

What is a variable ratio transformer? A variable ratio transformer is a transformer whose voltage ratio can be altered by alternating the number of active turns in either the primary or secondary winding circuits.

What is a capacitor? A capacitor is an electrical device used for receiving and storing a charge of electricity. Capacitors store energy in their electric fields, causing the current to lead the voltage. Capacitance is measured in farads or microfarads. In a purely capacitive circuit, the current leads the voltage by 90^0.

How does a transmissometer work? A transmissometer is an instrument used to measure particulate matter in flue gas. It directs a beam of light through a stack or duct where a mirror on the other side reflects the beam back to the transmissometer. The amount of light returning is proportional to the particulate matter within the flue gas.

How does a pH meter work? A pH meter is a specialized milli-volt meter that is used to measure the potential difference between the reference electrode and the solution in which it is immersed in at a given temperature, and then gives a corresponding pH value.

How does a pitot tube work? A pitot is an instrument that measures total head (velocity head plus static head) through one port and static head alone in another port. It consists of a cylindrical probe that has two openings. The first opening is called the impact opening and it faces into the stream. The second opening is called the static opening and it faces perpendicular to the flow of the stream. When the pitot tube is placed within the flow of the stream and positioned in the upstream direction, the differential pressure measured between the two ports can be correlated to the flow if the static pressure and temperature of the flowing media are known.

What are the disadvantages of using nuclear density meters?
1) The signal is not linear with solids content unless a linearizer is used.
2) A nuclear license is required
3) Cannot distinguish between suspended and dissolved solids
4) The buildup of scale or solids within the pipe could generate errors.

What is the function of a signal multiplexer? A signal multiplexer is used in a Data Actuation System and provides an interface between the transmitted instrument signals and the computer. Provides signal conditioning, analog to digital signal conversion, and switching capability for multiple inputs.

What is the difference between an open control loop and a closed control loop? An open control loop usually has only two elements, a controlling element and a final control element, where as a closed control loop has a sensing element, a measuring element, a controlling element, and a final control element.

What is integral control? Integral control is based on the repetitive integration of the difference between the controlled variable and its set point over the time the deviation occurs.

What is derivative control? Derivative control is a control function of the rate of controlled variable change from its set point.

What is feed-forward/feed-back control? This is control logic that provides a fast response system that can compensate for changes within a calibrated curve. Ex. A change in load occurs, the feed-forward signal immediately compensates by positioning the fuel valve to meet the calibrated curve requirement. If the curve is correct, no error signal is generated allowing the feed-back loop to do nothing. But if an error signal were generated, the feedback loop would readjust the position of the fuel valve to eliminate any pressure error that might develop due to shifts in the calibration.

How does an electromagnetic filter work? An electromagnetic filter is a filter that uses magnetism to remove certain magnetic suspended solids from condensate/feedwater systems. The EMF consists of a pressure vessel, a magnetic coil that is shielded, sheers, retention plates and a power control unit. This type of filter provides many advantages like low-pressure drop through the filter, minimal required flush water, short backwash cycle and no chemical requirements.

What is a transistor? A transistor is a small device used in many pieces of electronic equipment that regulates the flow of electricity between two terminals, the emitter and collector, by means of variations in the electric current flow between a third terminal (base), and one of the other two. Transistors can lower the voltage while keeping the electric current strong in a circuit. The word transistor originated from two other words, transfer and resistor. Transistors can be used in two ways, to amplify a signal or as a switch.

What is the purpose of a resistor? A resistor is a device used in electronic equipment (computers, monitors, etc.) to control current or voltage in a circuit through the use of resistance.

What is the purpose of a diode? A diode is a two terminal device that is used to allow the flow of electricity in only one direction. It is similar to a check valve, but for electronic use. The "di" means two and the ode means electrode.

What does a spectrometer measure? A spectrometer is an instrument used to measure wavelengths of radiation, energies of particles, or other quantity that covers a range of values.

What do the colored bands on a resistor represent? A resistor contains colored bands on its shell to indicate how much resistance the resistor will provide. Each color represents a numeric value, there are 10 numbers: 0 is black, 1 is brown, 2 is red, 3 is orange, 4 yellow, 5 is green, 6 is blue, 7 is violet, 8 is gray, and 9 is white. To determine the resistance of a resistor follow this order; read the colored bands from left to right, the first colored band gives the first number according to the color code, the second colored band gives the second number, the third colored band represents the multiplier or divisor. Should the third band be a color between 0 and 9 in the color code, it states how many zeros to be added behind the first two numbers. If there is no fourth band, the resistor has a tolerance rating of + or - 20%. If the fourth band is silver, the resistor has a tolerance of + or - 10%. If the fourth band is gold, the resistor has a tolerance of + or - 5%. The third colored band can also be silver or gold. If it is, according to the color code, the first two numbers must be divided by 10 if the band color is gold or 100 if the band color is silver.

What is a galvanometer work? A galvanometer is an instrument used for measuring, detecting, and determining the direction of a small electric current.

How many different types of transistors are there? There are many different types of transistors that are used, here is a list of a few transistors: PNP point contact transistor, NPN point contact transistor, PNP junction transistor, NPN junction transistor, alloy transistor, grown-junction transistor, micro alloy transistor, Germanium mesa transistor and a silicon planar transistor.

What is an amplifier? An amplifier is an electronic circuit, which uses a device like a transistor or magnetic contrivance to increase the strength of a signal without altering its characteristics.

How many different types of amplifiers are there? There are several different types of amplifiers, depending on the purpose for it to be used. Audio amplifiers, voltage amplifiers, power amplifiers, operational amplifiers, differential amplifiers are just a few types that are normally used.

What is distortion? Distortion is any departure from the original shape because of applied stress or the release of residual stress in the material.

What is binary code? Binary code is a binary number system used in which each decimal digit is represented by four binary digits. This is what decimal digits are like in binary; 0=0000, 1=0001, 2=0010, 3=0011, 4=0100, 5=0101, 6=0110, 7=0111, 8=1000, 9=1001, 10=0001 0000.

What is a strain gauge? A strain gauge is an instrument that is used to monitor and/or measure strain and the effects of pressure.

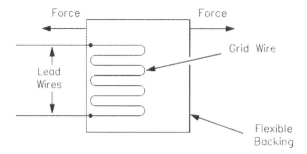

What is a RTD? RTD stands for Resistance Temperature Detector which functions similar to a thermocouple, but RTD's are more accurate in measuring temperatures.

How do RTDs work? Temperature increase, Resistance increase, Temperature decrease, Resistance decrease.

What is a variable area meter? A variable area meter is an instrument used to indicate the amount of fluid flow through a pipe. This instrument is a vertically mounted tube that tapers from the bottom up. A float device is positioned in this tube and moves upward as flow increases. A calibrated scale is positioned either on the tube or alongside the tube to indicate the amount of flow through the pipe.

What are the three most commonly used metals for RTDs? Copper, nickel, and platinum are the most commonly used metals for RTDs.

What are some common methods used to determine flow? There are several methods used to determine flow they are: variable area meters, pitot tubes, differential pressure, turbine meters, weigh scales, and open channel weirs.

What is an amplidyne? An amplidyne is a rotary magnetic amplifier of high gain, which is used in servo systems.

What is gain control? Gain control is a means for varying the degree of amplification of an amplifier, often a simple potentiometer.

What is a viscometer? A viscometer is an instrument that is used to measure a fluid's viscosity.

What is an error signal? An error signal is a signal representing the difference between the reference (designed value) and the actual value of a controlled variable.

What is under voltage no-close release? Under voltage no-close release, is a device that is used to act upon the trip coil of a circuit breaker in such a way as to prevent the circuit breaker from being closed if the voltage is below a certain value.

What is a thermal circuit breaker? A thermal circuit breaker is a device that is used to open a circuit by virtue of thermal expansion.

What are three types of pH flow sensors?

1) DIP - inserted into a vessel and can be removed for calibration or maintenance.
2) Flow through - is placed in a flowing stream of a fluid

3) Insertion - works similar to a flow through sensor, not needing sample lines. It is inserted through packing and is valved directly into the flowing fluid.

What is a saturated signal? A saturated signal is an electrical signal for instrumentation meaning that the signal is electrically pushed below zero or above 100%. This creates a dead band for the instrumentation to respond to causing a loss in response time. The way to fix this is to put the controller in manual and either increase or decrease the percentage on the controller for a minute or so and return to auto operation.

What are two categories of optical sensors? Intrinsic and extrinsic are the two categories of optical sensors. Intrinsic sensors, the optical fiber performs the measurement itself while the extrinsic sensor has a coating or a transducer at the tip of the fiber that performs the measurement.

What are three different types of variable frequency used? 1) Voltage - source inverter, 2) Current - source inverter, and 3) Pulse - width modulated design. In all 3 types, a solid-state rectifier converts the alternating current (AC) power to direct current (DC). Then an inverter transforms the direct current to an alternating current output at an adjustable voltage and frequency.

How does a radiation or nuclear detector work? A radiation source like Cesium 133 - 137 or Cobalt 60 is mounted on one side of a vessel and is directed through the vessel walls toward a radiation detector located on the other side of the vessel. As the fluid flows between the radiation source and the detector causes a change in the level of radiation that is detected and once a certain drop in radiation is detected at the detector an alarm would be annunciated indicating possibly pluggage developing within the vessel.

What is an electrometer? An electrometer is an electrical device used for measuring voltage by an electrostatic process without drawing any current from the source.

What is an extensometer? An extensometer is a device used for measuring expansion and contraction under tension and compression.

What is a hypsometer? A hypsometer is a device used for measuring the boiling point of water, which changes with air pressure and can be used to measure altitude.

What is a schlerometer? A schlerometer is a device that is used for measuring the hardness of a material.

What is weir? A weir is a structure that is placed in a channel of water similar to a dam except, a V-shaped notch is cut out in the center of the weir so that water may flow through it. The very bottom of the V shaped cut is the minimum amount of water flow while the very top of the V indicates the maximum amount of water flow. Since water travels at a fairly constant rate of speed, the flow of water through the weir is proportional to the depth of the water.

What is an inductor? An inductor is an electronic device used in an electric circuit because of its inductance. Inductance is measured in henries. An inductor stores energy in its magnetic field and causes the current to lag the voltage. In a purely inductive circuit, the current lags the voltage by 90^0.

What is a hydrometer? It is a graduated instrument that is used to find the specific gravity of a liquid. This instrument can be used to determine the concentration of an acid, strength of alcohol, or how much salt there is in a substance.

What is a dose meter? An instrument designed for measuring the dose of radiation received in a given time. The simplest form is a film badge.

What is the purpose neutron detectors? As its name implies, it detects neutrons. They are used in the management and control of a nuclear reactor, which is used to control nuclear reactions for the purpose of generating the right amount of heat needed to turn water into steam thus generating electricity.

What is are two types of neutron detectors? Neutron detectors can be counters or ionization chambers. Counters, which detect neutrons by sensing the individual ionizations they produce, are most useful when the neutron flux is low. Ion chambers are more useful at high neutron fluxes. They measure the electrical current that flows when neutrons ionize gas in a chamber. Both of these detectors are mounted on the exterior of the reactor vessel and only measures neutron leaking from the reactor.
Two other types of neutron detectors, a self-powered neutron detector and a miniature fission chamber, can be used to measure the neutrons inside the instrument tube of the fuel assemblies.

How does a neutron detector work? Many neutron detectors rely on a scintillation detector which responds to neutrons which enter the device. A scintillator is a material that absorbs energy and reemits it as light; when combined with a light sensor that absorbs the light and turns it into an electronic pulse, this reaction can be analyzed to reveal information about the original energy — in this case, the neutrons. It is also possible to use optical methods, and to use gas filled detectors. With a gas filled detector, the device does not detect the neutrons themselves, but rather the reactions they leave in their wake. In all cases, sophisticated software needs to be connected to the device to register and interpret results.

What are the four basic elements to process control?

1. Detector
2. Transmitter
3. Controller
4. Final Control Element

What is a frisker? A frisker is a device used to detect contamination. It has a meter and a handheld probe that is slowly passed over the area of interest while the meter is being watched.

What are excore detectors? The excore detectors are located outside of the reactor and detect fast neutrons that have escaped the core and are thermalized in the concrete in the area of the detectors. The concrete has a similar cross-section for fast neutrons similar to water. Concrete contains 7 to 8% water by weight. It also has a large proportion of oxygen, which acts as an additional moderator to slow down fast neutrons.

Neutron leakage is changed as reactor coolant system density is changed by temperature. As temperature decreases, neutron leakage lowers due to the higher water density. The higher density of water causes more fast neutrons to thermalized in the core, so fewer neutrons leaks all the way out.

What are in-core detectors? The in-core detectors are located within the reactor vessel and reactor core. This instrumentation gives indications about the reactor core axial and radial power distribution, provides information to calculate fuel burn-up within the core, monitors variations within the core, and helps to troubleshoot any abnormalities that may occur during startup and shutdown.

What is a hypsometer? An apparatus for measuring the boiling point of a liquid.

What is a scintillation counter? An instrument that detects and measures intensities of high energy radiation. Incoming particles strike a phosphor layer, and the flashes of light so produced are detected by a photomultiplier whose output current pulses are counted electronically.

Valves, Traps, Strainers & Dampers

As a power plant operator you will spend a lot of time with these 4 pieces of equipment. Most power plant operators will spend approximately half their time working with valves, traps, strainers or dampers. From experience, an operator needs to understand the basic purpose and function of each valve, traps, strainers and dampers. This equipment is easy to work with but sometimes you might have a problem that can only be solved if you know how these devices work and how they work together.

What is a valve? A valve is an instrument that is used to stop, start, or control (regulate) the flow of a fluid. There are many different kinds of materials that valves are very are made of like stainless steel, iron, and brass. These instruments are very precise; some valves can control the flow of a fluid with just a minute turn of a hand wheel.

How many different types of valves are there? There are many different types of valves used in today's industry. The reason for so many different valves is because each valve provides a different function. Some valves are made to isolate equipment, some control the flow of a fluid and some are used for safety reasons. Here are several different types of valves and the function they provide.

Gate valve – Used for isolation, either open or closed not throttled.

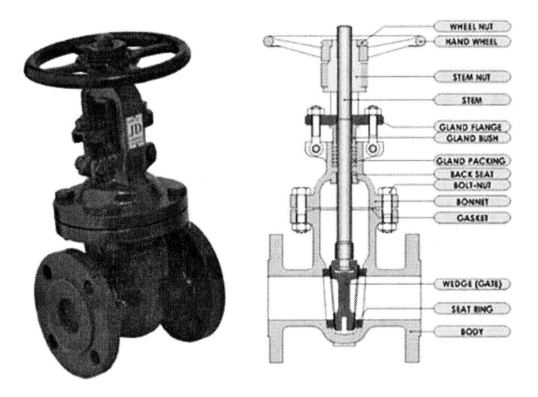

Globe valve – Used for controlling a fluid & isolation, it can be throttled.

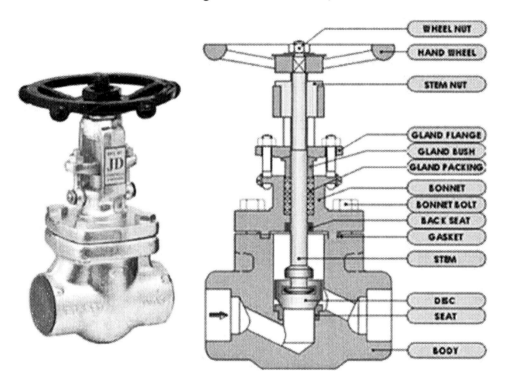

Butterfly valve – Used for control & isolation, it can be throttled. Normally used on low pressure systems.

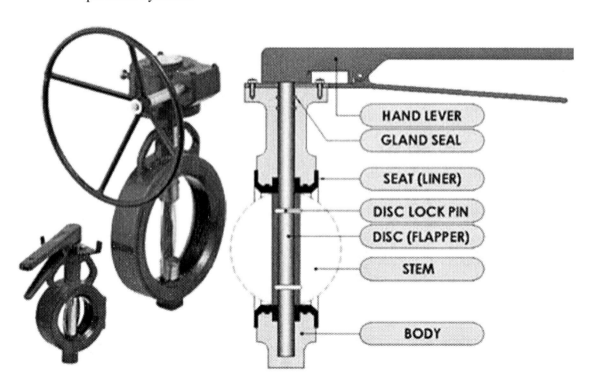

Check valve – Allows a fluid to flow in only one direction.

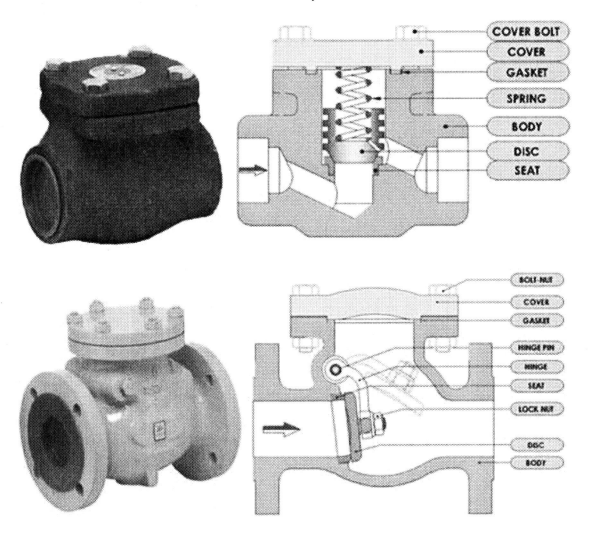

Safety valve – Normally closed, used on gas systems, opens quickly (pops open) on high pressure.

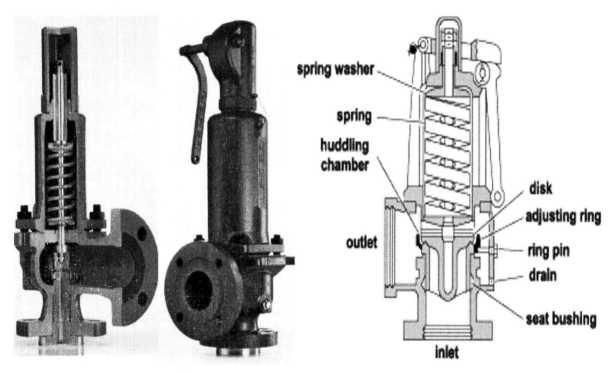

Relief valve – Normally closed, used on liquid systems, opens gradually on high pressure.

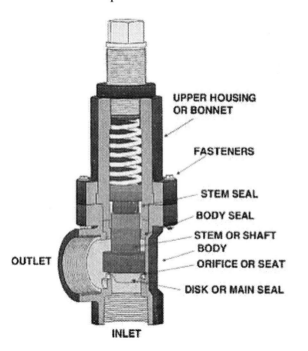

Safety/Relief valve – will be used on a system where liquids and gases may be present and may be relieved properly through this type of valve.

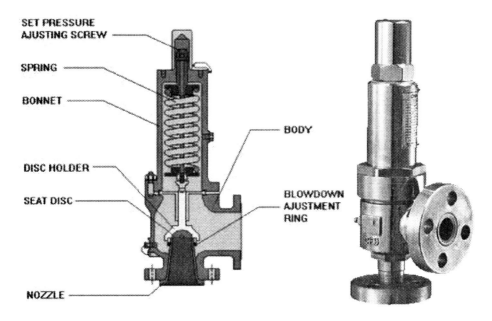

Diaphragm valve – Used in systems dealing with chemicals.

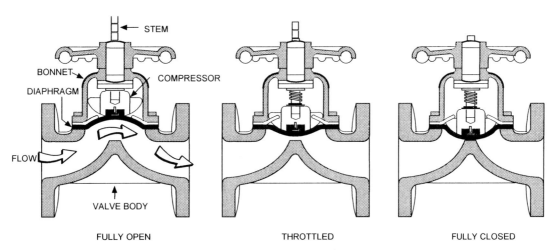

Plug valve – Usually used in systems with small piping, with low   pressure. Usually a ¼ turn will fully open/close valve.

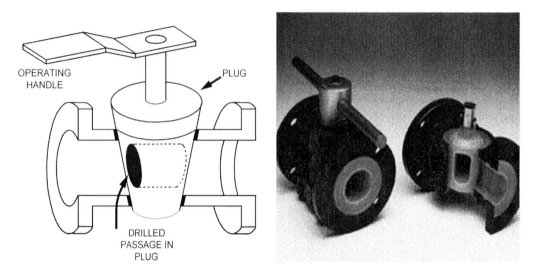

Ball valve – Same principal as the plug valve but shape of valve is spherical.

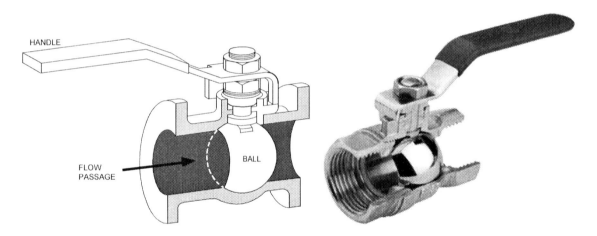

Can a gate valve be used to regulate flow? No. A gate valve is used for isolation purposes only. This valve is either opened or closed, it is not throttled. If this valve were opened only 25%, there would be an 80% flow rate through the valve. Unless there is no other means of regulating the flow, the gate valve could be used to try to control a flow and if then not for an extended period of time, because damage to the valve and valve seat will occur.

What is the purpose of a globe valve? A globe valve can be used for isolation like a gate valve but is mainly used to control (regulate) the flow of a fluid.

How well does a globe valve regulate flow? A glove valve is a very precise instrument. It is said, that for every 10% of the valve being opened there is an equal 10% of fluid flow increase.

When a gate valve chatters, what does this indicate? If a gate valve chatters it means that the valve seat and valve disc are damaged by erosion from the fluid flowing through the pipe. The chattering is the valve disc banging on the valve body/seat because of the excessive clearances between the valve disc and the body/seat caused by the erosion.

What is the purpose of a butterfly valve? A butterfly valve can be used just like a globe valve to control the flow.

How does a butterfly valve work? This type of valve has two half-moons shaped metal disc that is attached to the shaft of the valve that runs through the valve body. This valve moves only a total of 90^0 degrees.

photo courtesy of Milwaukee Valve Co.

What is the purpose of a check valve? A check is to allow the flow of a fluid to flow in only one direction. This type of valve needs very little maintenance, yet sometimes it might hang open so a light tap with a wrench should jar it loose so it will close.

How does a check valve work? A check valve is a valve disk that has a hinge on one side of it. When a fluid flows through the pipe and passes the check valve, the valve will float on top of the fluid, but when the flow stops moving and tries to flow backwards the check valve will be pushed down and held in position by the weight or pressure of the fluid to the valve seat where it will remain until the flow starts moving again.

How are check valves different from all other valves? Check valves are different from all other valves because they only allow fluid to flow in one direction.

What are three basic types of check valves? The three basic types of check valves are the swing type, the lift type, and the stop-check type.

What is a butterfly check valve? A butterfly check valve is a check valve that has two openings which look like half circles. On the backside of these half circles there are springs to keep them shut. These types of valves are for low-pressure applications. The valve body is so slim that you would think it was just another pipe flange. The pictures below should provide a better idea of what they look like.

Air Flow IN   Air Flow OUT

This is a butterfly check valve, where the two halves work independently and are spring loaded to ensure that both halves close when not in use.

What types of valves are used for isolation? Gate, Globe, Butterfly, ball, diaphragm valves.

What is the purpose of a safety valve? A safety valve is a specially designed valve used to relieve excessive pressure quickly in a gas type system to prevent a rupture or explosion in a system.

How does a safety valve work? A safety valve is a valve that will "pop" open should the system's pressure exceed the valve's set point. Once the pressure has dropped below that set point, the valve will close and reseat. This type of valve can become easily worn out if it opens to many times.

What determines the number of safety valves required for a main steam system? The number of safety valves is determined by the amount of steam removal it must be able to release in order to ensure that the system it is designed to protect is not damaged from over pressurization. These figures from the boiler and safety code stating that the number of safety valves is determined by the amount of maximum steam flow to the turbine. E.g. If the maximum steam flow to the turbine operates around 10 million lbs/hr the number of safeties required must be able to remove 1.5x that amount so another words, the number of safety valves required must be able to remove 15 million lbs/hr of steam from the main steam system.

Do all of the safety valves lift at the same time? No. The safety valves are designed to open up at specific pressures and close at specific pressures. They will open up one at a time until the pressure is controlled and slowly begins to come down. If all of the safety valves opened at the same time, you would have a very uncontrollable release of steam pressure which would cause the system to go from one extreme to another.

What is the purpose of a relief valve? A relief valve is a specially designed valve used to remove excessive pressure slowly from a liquid system to prevent a rupture or explosion in a system.

How does a relief valve work? A relief valve will slowly open should the system's pressure exceed the valve's set point and will close when the pressure falls back down to normal. This valve opens slowly because if it opened fast, a pressure shock (hammer) would occur.

What type of system would you find a safety valve? A gas type system like air, natural gas, hydrogen, carbon dioxide, nitrogen, etc.

What type of system would you find a relief valve? A liquid type system like water, oil, etc.

What does blowback mean? Blowback is a term used to describe the difference between the pressure needed to open the safety valve and the pressure at which the safety valve will close.

How can you manually test a safety valve? A safety valve is constructed with an arm on one side of the valve. It is this arm that allows you to test the safety valve. A rope can be tied around this arm to allow the operator to use a fulcrum to raise the arm up. As the arm raises the safety valve will pop open and the operator can allow the arm to lower and rest against the valve body.

What types of valves are used for control (regulation)? Globe, Butterfly, Ball valves.

What types of valves are used for low-pressure systems only? Gate, Globe, Diaphragm, Butterfly & Ball valves.

What types of valves are used for harsh chemicals? Diaphragm valves and some corrosion resistant valves made from stainless steels.

What is a hammer valve? A hammer valve is a valve whose handle is weighted. The weight in the valve handle is used to build inertia when opening or closing the valve. The valve handle swings in a circular motion and hits the motion stops on the valve stem.

How does a diaphragm valve work? A diaphragm valve is a valve that is constructed of a membrane made of rubber or other synthetic material that is used to provide a seal in order to stop the flow of a fluid when closed.

What are two types of diaphragm valves? A straight through diaphragm valve and a weir diaphragm valve are the two types of diaphragm valves used.

What is the difference between these two types of diaphragm valves? The biggest difference between these two types of diaphragm valves is the pressure drop across the valve. The weir type diaphragm valve has a high-pressure drop across the valve because of a reduced flow passage through the valve body.

If you try to open a diaphragm valve and it will not open, what could be the problem? Depending on the fluid involved with the valve. If the fluid has suspended solids and the membrane of the valve has a small hole in it, the fluid will fill up the membrane cavity not allowing the valve membrane to be drawn open.

What is a tandem blow-off valve? A tandem blow off valve consist of two valves, one near the boiler called the blowing valve and the second valve, the sealing valve, the one farthest from the boiler. The blowing valve should be opened last and closed first while the sealing valve is opened first and closed last. Use these valves either fully closed or fully open, never throttled these types of valves.

What is the purpose of a plug valve? A plug valve is used in small piping systems usually of low pressure that operate similar to that of a butterfly valve. They both move only a quarter of a turn to go from fully open to fully close.

What are two common types of plug valves? Two common types of plug valves are the tapered plug cock valve and the ball valve.

What does back seating a valve mean? There are several types of globe valves that have a seating area that is machined on the valve stem above the top of the valve disc. By back seating a valve means to keep fluid pressure off of the valve packing because of a packing leak by using the back seat on the valve stem to rest against the top part of the valve bonnet. It is possible, but highly not recommended; that the packing can be replaced will the valve is back seated. This procedure is not recommended because of the unknown condition of the back seat surface. Certain valves cannot be back seated so know which ones they are.

By what other means can a valve be opened or closed, other than by hand? Valves can be operated through a number of actuators other than by hand including electrically operated, pneumatic diaphragm operated, hydraulically operated and solenoid operated but not limited to.

What is packing? Packing is a material that is used in valves to prevent valve stem leak through. Graphite and other types of similar material are used.

What does packing do in a valve? Packing in a valve prevents the fluid inside the valve body from leaking out along the stem of the valve.

What does fail safe mean for a valve actuator? Fail-safe means that a valve actuator can be designed to default or presume a certain position when a loss of air or electricity to the valve actuator occurs. Fail safe open, means that if a valve actuator loses air or electricity the valve will become wide open. Some valves can be designed so that the actuator can leave the valve in the last position it is in should a loss of air or electricity occurs.

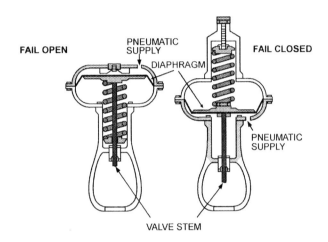

How can you tell if a valve is leaking through? There are a couple of indications that a valve in leaking through. If the temperature of a pipe is the same on both sides of the valve (This is easier if the fluid in the pipe in normally hot, like steam.). Sometimes you might be able to hear a flow through a valve or the valve may just chatter.

If you have two valves on a pipe in series, what valve would be the primary valve? The primary valve is the source valve; it is the closest to the source. The secondary valve is used for throttling the flow of a fluid.

What is the reason for using the secondary valve for throttling and the primary valve for isolation? The primary valve is used for isolation in case the secondary valve begins to leak through or should the packing blow out of the valve.

What would happen if a valve were over tightened? When a valve is over tightened, the valve disc is crushed into the valve seat damaging both parts of the valve.

When a valve has been fully opened, why is it good practice to turn the valve handle back a half of turn? Due to thermal expansion or contraction, the valve could bind up in the valve body making it very hard to open or close the valve.

Where would vents and drain valves be located on plant piping? Vent valves would be located on the highest point in a piping system, on top of the pipe itself. The drain valves are located on the lowest part in a piping system, usually under the bottom of a pipe.

What is a damper? A damper is a set of louvers used to control the flow of air or other gaseous fluid, and can be used both a fan or after the fan or in ductwork to balance flows through out a system.

How many different types of dampers are there? There are three basic types of dampers: Louver, round, and guillotine. Louvers are used for isolation or modulation. Round dampers are used for isolation or modulation while guillotine dampers are used for isolation only. Dampers are classified by their shape and/or configuration.

How does a damper work? Louver dampers can be operated in a parallel blade arrangement or an opposed blade arrangement. Round dampers can operate like a butterfly valve or can be cut into pie shaped vanes connected through linkage to control flow. Guillotine dampers are usually solid flat piece of metal that is used for isolation purposes through the use of a chain and/or gear operated linkage.

What is a strainer? A strainer is a device that is used to remove suspended solids from a fluid. A strainer is a basket of woven wire that can be made to filter out big impurities or very small impurities depending on the application of the fluid.

On what systems would you find strainers? Strainers could be found on just about any system that would suffer from damage because of suspended solids like a fire protection system, steam system, oil control system, oil purification system, etc.

How many different types of strainers are there? There are three very common strainers used throughout a plant, they are: a duplex strainer, rotary strainer, or Y-strainer.

What is the difference between a duplex strainer, a rotary strainer, and a Y- strainer? A duplex strainer is a two-basket strainer that allows one strainer to remain in operation while the other element is shut down. A rotary type strainer is a strainer that rotates with several baskets, allowing continuous operation. And a Y-strainer is a strainer that is located in-line of a pipe to filter out small amounts or solid impurities.

How are these strainers cleaned? The duplex strainer has two baskets, while one is in service the other can be cleaned out by several methods like back flushing. The rotary type strainer can be cleaned whenever the strainer differential builds up, and can be cleaned manually or by an automatic flushing system. The Y- strainer usually has a blowout line on the suction side of the strainer, so to clean the strainer the simple opening of a valve would allow the fluid in the pipe to push the solid material in the strainer to be forced out.

What is a trap? A trap is a device used to prevent the flow of one fluid like steam, while allowing another fluid to flow through like water. This setup would keep water from buildup inside a steam pipe.

How can you tell if a trap is not working properly? There are a few good indications that a trap is not working properly, water hammers develop and both lines at the trap are cool. On some types of traps there may be a test connection that can be cracked open to see if there is any water within the steam line.

How many different types of traps are there? There are several different types of traps, here are three basic types of traps: thermostatic, mechanical, and thermodynamic traps. (thermostatic bellows trap, float trap, inverted bucket trap, tilting disc trap)

On what type of system would you find a trap? Traps can be found on many systems that want to remove unwanted water from steam to prevent any water hammers or other damage. Traps can be found on such equipment as steam inlet piping to the turbine, auxiliary steam system, and extraction piping from turbine.

What is a steam trap station? A steam trap station is a setup where a trap, an inlet strainer to the trap, trap isolation valves and a bypass valve is located in a section of piping. A test connection and/or a strainer blowout valve may be present to check trap operation.

What are three basic types of steam traps? The three basic types of steam traps are the thermostatic, mechanical, and thermodynamic traps.

How does a mechanical trap work? A mechanical trap operates with the use of a float or other device that works off of the difference in densities between steam and water.

How does a thermodynamic trap work? A thermodynamic trap operates on the principles of a temperature/pressure relationship. These types of traps are sometimes called impulse traps.

How does a thermostatic trap work? A thermostatic trap operates when a difference in temperature occurs between certain fluids, like steam, water, and/or air.

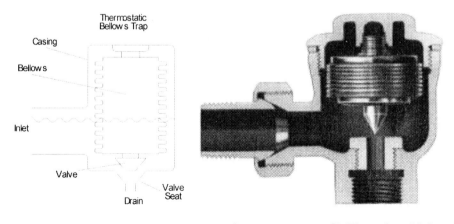

photo courtesy of Milwaukee Valve Co.

431

How does an orifice work? An orifice is used in ducts or piping that will cause a restriction of flow that will create a pressure drop that is proportional to the square of the velocity.

How does a float trap work? A float trap uses a float to maintain a level in the trap, not allowing any steam to flow through. As the level in the trap rises, the float moves upward allowing water to flow through the opening of the trap. And as the water level decreases, the float moves downward blocking the opening in which the water level is maintained and steam is not allowed to escape.

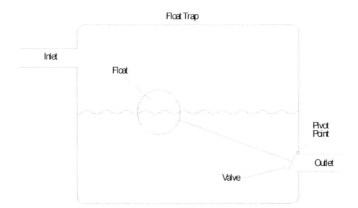

How does an inverted bucket trap work? An inverted bucket trap is just that, an inverted bucket with a small hole in its top and is connected to a rod with a valve at the end. The water at the bottom of the bucket acts as the seal. The water enters the trap from the bottom and flows out the bottom of the bucket and then flows in between the outside of the bucket and the traps inner wall until it finally flows out of the top of the trap valve. Should steam reach the trap, it will be collected in the top of the bucket, which will cause the bucket to float. As the bucket rises to the top, the valve attached to it will begin to close. As the steam condenses back into water, the bucket will lose its buoyancy and sink back to the bottom of the trap to repeat the process all over.

What is an impulse trap? An impulse trap is a thermodynamic trap that operates on the principles of a temperature/pressure relationship.

What does it mean when a trap is air bound? A trap is said to be air bound when air enters the trap and will not allow water or steam to enter.

How can a float trap avoid an air bound condition? To avoid an air bound condition in a float trap, the trap may be configured with a thermostatic bellows arrangement to remove any air that is trapped in the trap.

How does a tilting disc trap work? A tilting disc trap is a type of impulse trap that uses a disc to control flow through the trap. As water enters the trap it pushes the disc upward and allows the water to flow through the outlet. When water flows around the disc, it forms a low-pressure area around the disc. Should steam enter the trap, it will cause the water in trap to flash into steam, causing a sudden pressure increase because of the expanding steam. This expansion will push the disc down and stop the flow through the trap. The disc will open again as soon as the steam has condensed back into water; thus dropping the pressure around the disc.

Tilting Disc Trap

What are three types of orifice plates? 1) Concentric - the hole in the plate is located in the center. 2) Eccentric - the hole in the plate is located on the outside edge of the plate. 3) Segmental - a section of the plate is actually removed.

What is a needle valve? A needle valve is a specially designed slender pointed rod that operates within a circular valve seat and is used for fine quality control of a fluid. The end of this valve is tapered so that fluid control is more linear.

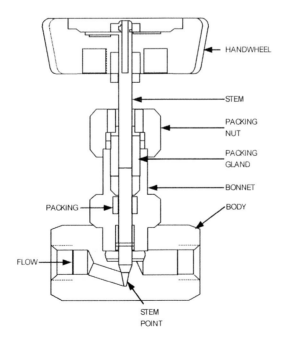

What is a double beat valve? A double beat valve is a cylindrical valve that is hollow and is used for controlling high-pressure fluids. The seats design for this valve is where the two ends that are exposed to the pressure are of slightly different in area, so that valve is almost balanced which allows the valve to be easily operated.

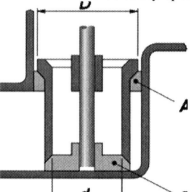

What is the difference between a rising stem valve and a non-rising stem valve? A rising stem valve is threaded through the hand wheel, as the hand wheel is rotated the stem will rise through the hand wheel. With a non-rising stem valve the hand wheel is attached to the valve stem, the threaded part of the stem is contained within the valve body where the valve disc is attached. As the hand wheel is rotated, the valve disc will move up or down the threaded part of the valve stem. A yoke bushing is what keeps the hand wheel from moving away from or towards the valve body.

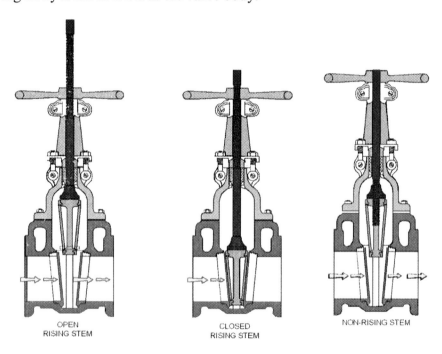

OPEN RISING STEM

CLOSED RISING STEM

NON-RISING STEM

Pumps

One of the most common pieces of equipment you will work with is a pump. It might be a small chemical pump designed to move an ounce of fluid per hour or it might be a massive several thousand horsepower pump designed to move 1.5 million pounds of water per hour. You will work with pumps of low pressure and pumps with high pressure as well as pumps with high or low fluid volume. Pumps are one piece of equipment you really want to become familiar with. Once you have the operation and basic functions of these pumps you will be able to troubleshoot and see problems forming before they become critical. Again it will take time and practice but it will pay off.

What is the purpose of a pump curve? A pump curve is a graph that shows various pump characteristics in relation to a pump's flow rate.

What could possibly happen if the pump suction pressure should drop below the minimum net positive suction pressure? If the pump suction pressure should drop below the NPSH, boiling of the fluid in the pump causing cavitation.

What does net positive suction head mean? Net positive suction head is the actual pressure at the suction of the pump. This is the lowest pressure at which the pump should operate safely without damage to the pump.

How can the net positive suction head of a pump be increased? The net positive suction head can be increased by increasing the pressure of the fluid at the suction of the pump. (Raising a tank level higher)

Why does a boiler feed pump require a higher minimum net positive suction? Since a boiler feed pump moves such an enormous amount of water, it must ensure that the fluid being pumped remains in a liquid state.

What parts of a pump can reduce flow when they wear out? The most detrimental part of a pump is the impeller. When the vanes on the impeller wear out, they no longer effectively meet the rated flow required for that pump.

What types of pumps are there? There are basically two types of pumps, centrifugal and positive displacement pumps.

What are positive displacement pumps? Positive displacement pumps are pumps that use a piston or other solid object with reciprocating motion to displace a fluid.

What are centrifugal pumps? Centrifugal pumps are pumps that use centrifugal force to displace fluids.

What is a jet pump? A jet pump is a hydraulic pump that accelerates a fluid through the use of a ventura to create a low-pressure area for the fluid to be moved. The most important characteristic of a jet pump is that flow is produced in a fluid without moving parts.

What is the difference between a single stage pump and a multistage pump? A single stage pump is a pump that increases the fluid pressure only once, yet with a multistage pump the fluid pressure is increased several times, each time increasing the pressure in each stage.

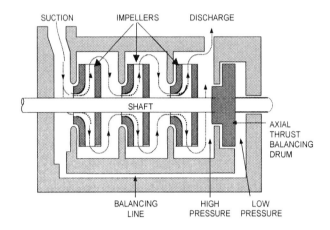

What are two major pump classifications that are based on operating principles? Positive displacement and centrifugal pumps are the two major classifications for pumps.

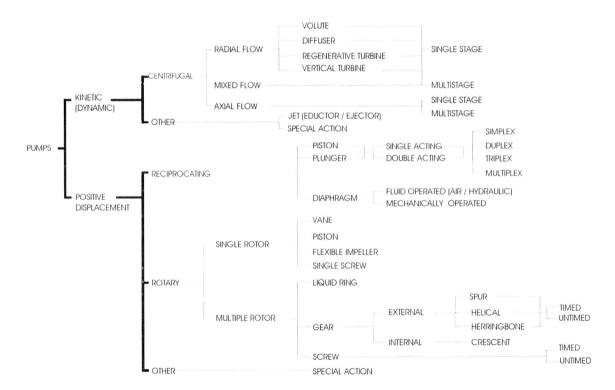

What is meant by pump capacity? The capacity of a pump measures the volume of a fluid discharged per unit of time. (cubic feet per second cfs or gallons per minute gpm)

What is suction lift? Suction lift occurs when the fluid level is below the center of the pump and the pump must use a suction action in order to pull the liquid level up to the pump.

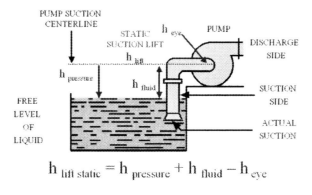

$$h_{lift\ static} = h_{pressure} + h_{fluid} - h_{eye}$$

courtesy of General Physics

What is the maximum theoretical suction lift of a pump? The maximum for certain pumps can be up to 33 feet while most pumps are limited to no more than 22 ft of suction lift.

What is static suction lift? Static suction lift is the vertical distance in feet from a pump's center to the level of the liquid to be pumped. The suction or intake of the pump may be several feet below the level of the liquid but the additional distance is not calculated when determining the static suction lift. The static suction lift is considered a disadvantage because of the use in additional energy to raise the fluid to the pump level.

What is total dynamic suction lift? The total dynamic suction lift is the static suction lift and other related factors to dynamic characteristics. Factors include velocity head and frictional losses in the suction pipe and connections. These affect the total inlet suction pressure and make the total dynamic suction lift a larger absolute value, yet it is still a disadvantage.

What does velocity head mean? Velocity head is the energy that is associated with the flow or motion of a fluid within a pipe.

What is suction head? Suction head is the pressure in the suction of the pump.

What is total dynamic suction head? Total dynamic suction head is the vertical distance in feet from the center of the pump to the level of the liquid minus the velocity head and all frictional losses within the suction piping.

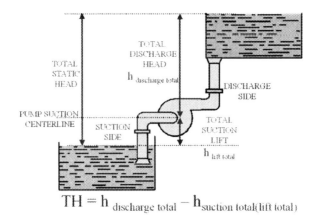

$$TH = h_{discharge\ total} - h_{suction\ total(lift\ total)}$$

courtesy of General Physics

What is static suction head? Static suction head is the vertical distance in feet from the center of the pump to the level of the liquid to be pumped. The liquid level is now above the center of the pump, which becomes an advantage because there is no energy needed to move the liquid toward the center of the pump.

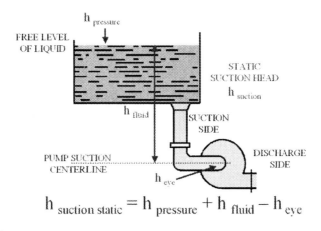

$$h_{suction\ static} = h_{pressure} + h_{fluid} - h_{eye}$$

courtesy of General Physics

How can the suction lift or suction head affect the discharge conditions of a pump? The pump discharge conditions are not affected and remain the same with either suction lift or suction head.

With centrifugal pumps, what unit is used to express dynamic head? Centrifugal pumps express dynamic head in feet.

With positive displacement pumps, what unit is used to express dynamic head? Positive displacement pumps express dynamic head in pounds per square inch (psi).

What does flashing mean? Flashing is an undesirable condition where water rapidly changes into steam.

What does discharge head mean? Discharge head is the total number in feet a pump will pump a liquid in a vertical direction.

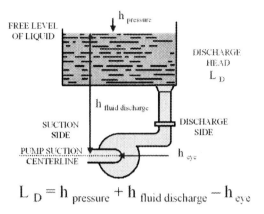

$$L_D = h_{pressure} + h_{fluid\ discharge} - h_{eye}$$

courtesy of General Physics

What is pump slip? Not all the energy goes into the pump so the difference in pump speed and driver speed is called the pump slip.

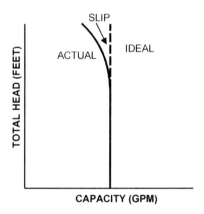

What are the centrifugal pump laws? The volumetric flow rate or capacity is directly proportional to the pump speed. The discharge head (pressure) is directly proportional to the square of the pump speed. The power required by the pump motor is directly proportional to the cube of the pump speed.

What is cavitation? Cavitation is an unwanted condition that can occur in centrifugal pumps. This condition occurs when the fluid in a pump begins to boil at the eye of the pump and where shock waves are created by the collapse of bubbles in the fluid as it reaches the outer end of the pump's impellers. This condition will lead to impeller pump damage should it continue for long periods of time.

What does N.P.S.H. stand for? N.P.S.H. is an acronym meaning Net Positive Suction Head.

How is a pump's capacity affected by cavitation? When a pump begins to cavitate, the capacity of the pump falls dramatically.

What are some ways to identify cavitation of a pump? Cavitation can be identified in a pump because when a pump is cavitating it will sound like marbles or pebbles are in the impeller of the pump. Another indication is a fluctuating discharge pressure on the pump. If the temperature of the fluid being pumped begins to increase, the NPSH pressure to prevent boiling and cavitation also increase.

How does cavitation damage a pump? Damage to a pump by cavitation occurs when the shock waves from the collapse of the bubbles in the boiling fluid hit the impeller. These shock waves leave small impressions on the pump impeller and the thermal stress that the pump is put through will diminish pump life.

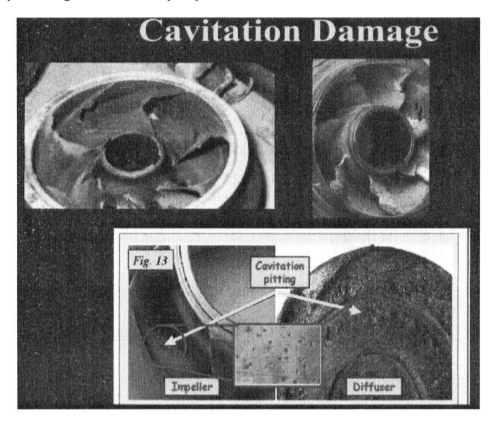

What is the volute effect? The volute effect occurs as water enters the volute of a pump, it begins to spread out in order to fill the larger area. The spreading out causes the fluid to decrease in speed in which the speed is converted into pressure.

What is the purpose of the impeller? The purpose of the impeller is to use centrifugal force to build pressure in a fluid being pumped. The impeller is a circular shaped device containing a series of curved vanes that extend outward from center.

What is a volute? A volute is the part of a pump that starts within the pump and widens until it reaches the discharge piping. As the water enters the piping from the pump's impeller, the water fills the larger area causing the water to slow down and the speed of the water is converted into pressure.

What is a reciprocating pump? A reciprocating pump is a pump that uses a back and forth motion to move a fluid.

What is a rotary pump? A rotary pump is a pump that displaces fluid using gears or screws that rotate.

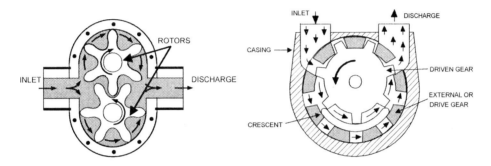

What is the purpose of an inter cooler and an after cooler? To cool the air before it is released into the air system, and also helping to remove moisture from the air.

What is a balancing drum? A balancing drum or piston is used in larger pumps to reduce the axial thrust of the pump by creating an opposing force to counteract the axial thrust. The balancing drum is located in the high pressure section of the pump and allows high pressure fluid to leak off to a low pressure area off the pump to create a pulling effect in the opposite direction of the axial thrust, thus minimizing the wear on the thrust bearing of the pump.

What is the purpose for priming a pump? The purpose for priming a pump is to remove any air that may be trapped inside of the pump. If the air were to remain in the pump, the pump would not operate properly.

What could possibly happen if the pumps are not primed? If a pump is not primed correctly, air could remain in the pump causing cavitation in the pump. Another way for a pump to become air borne is if the fluid it is pumping changes temperature / pressure configurations could cause the fluid to become a gas instead of a liquid.

What determines if a feedwater/condensate booster pump is used in a plant cycle? A condensate booster pump is designed into feedwater construction when it is determined that normal condensate pump discharge pressure is not high enough to flow through multiple heat exchangers and still maintain adequate net suction pressure to the feed pump. As the condensate flow passes through the piping and through the heat exchangers, the overall pressure will continue to decrease from the moment it leaves the discharge of the condensate pump.

What is an axial flow compressor? An axial flow compressor is a compressor that compresses a fluid along the shaft of the compressor.

What is a centrifugal compressor? A centrifugal compressor is a compressor that uses centrifugal force to compress a fluid in the compressor.

What is a positive displacement-reciprocating pump? A positive displacement - reciprocating pump is a pump that uses reciprocating motion that positively displaces fluid.

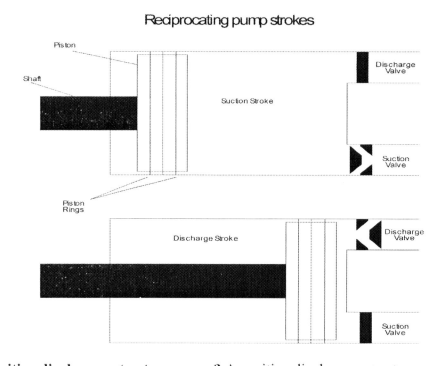

Reciprocating pump strokes

What is a positive displacement rotary pump? A positive displacement rotary pump is a pump that positively displaces fluid by using a rotating motion.

What are clearance pockets? Clearance pockets are small air chambers that are built into the walls of a compressor cylinder. They will reduce the flow of air from the compressor to the air receiver when the air pressure in the system is too high.

What is a compressor? A compressor is a device that compresses a fluid into a small volume to increase the pressure of a fluid.

How is a double acting compressor different from a single acting compressor? A double acting compressor is a compressor that performs two functions as it operates. On one side of the piston is the compression stroke while on the other side air is being drawn in to the cylinder of the compressor. With a single acting compressor, the fluid is only compressed when the piston moves in the only one direction and air is drawn into the cylinder when the piston moves in the other direction.

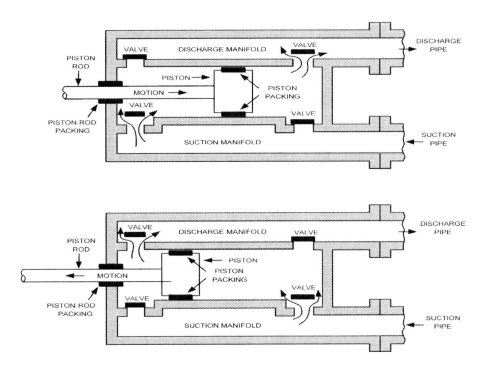

What are water jackets? Water jackets are water-cooled chambers located between the cylinders and the outer casing of a pump. These water jackets help remove the heats away from the cylinders by flowing cooling water through them.

What is a stroke? A stroke considered being one complete movement along the length of a cylinder.

What is a dynamic compressor? Dynamic compressors are compressors that move gas by a process of acceleration. Two types are centrifugal and axial flow compressors.

What are inlet valve unloaders? An inlet valve unloader is a device used to air entering a compressor to be forced back out of the inlet when the demand for air is low. This device is usually found in reciprocating compressors.

What does a centrifuge do? A centrifuge is a machine that is used for separating two substances with different densities. As the machine spins around, the heavier (denser) substance is pushed out towards the wall while the lighter substance is carried over. This machine can separate water from oil or cream from milk.

How does a positive displacement pump work? A positive displacement pump moves back and forth in motion to bring air into the compressor and push the air out of the compressor and into the air system. As the compressor moves in one direction, it pulls air into the compressor and when the compressor moves into the other direction it pushes the air out of the compressor.

How does a centrifugal pump work? A centrifugal pump uses the principles of centrifugal force to move a fluid, by spinning, rotating, or moves in a circle. The fluid enters the suction of the pump, then directly into the vanes of the impeller. As the impeller rotates in a circle, the fluid is pushed out to the outer end of the impeller until it reaches the volute. When the fluid reaches the volute, it flows into a larger area causing the speed in the fluid to slow down and converting its speed into pressure.

What is the function of a pump? The function of a pump is to build up pressure in a fluid causing it to move.

What does a diffuser do? A diffuser acts similar to that of a volute in a pump, in that it provides a space for fluid to spread out and to smooth the turbulence within the fluid once leaving the impeller.

What is a coupling? A coupling is a device that connects the driver (motor) to the equipment to be driven (pump). Couplings can be fixed to provide constant speed or variable speed to control the speed of the driven equipment. There are several different types of fixed couplings like jaw couplings, grid couplings, or gear couplings and there are several different variable speed couplings like hydraulic, pneumatic, or electromagnetic.

What is a flexible coupling? A flexible coupling is a coupling that is used to join two shafts together where it is impossible for them to be perfectly aligned. This type of coupling might be a steel spring, a rubber disk, or flexible bushings.

What is the purpose of a warm-up line on a pump? The purpose of a warm-up line on a pump is to keep a pump that is not in service at operating temperature so that thermal stress is minimized.

What does it mean to deadhead a pump? To deadhead a pump means to keep a fluid from flowing through the pump by blocking the discharge of the pump.

What is the purpose of a recirculation line on a pump? The purpose of the recirculation line on a pump is to ensure that the pump will always have a flow through it. This will keep the pump from being deadheaded, which could cause severe damage to the pump.

How could you tell if a check valve is not fully seated on a pump? Should a check valve not fully seat on a pump several indications are possible like the sound of a fluid flowing through the pump, the pump itself begins to spin backwards, or there may be a temperature indications on the pump or associated piping.

Where is the highest pressure of a pump? The highest pressure in a pump is located at the end of the impeller of the pump or the end of the last impeller of the pump.

How can a flow be controlled by a pump? A pump can control a flow by controlling the speed of the pump. The motor can be variable speed and/or the pump can be controlled by variable speed controllers like hydraulic couplings or electromagnetic couplings.

What are two general types of drives for pumps? Two general types of drives for pumps are constant speed or variable speed motors.

Positive displacement pumps can be divided into what two categories? Reciprocating pumps and rotary pumps are the two categories for positive displacement pumps.

How does a positive displacement rotary pump work? A positive displacement rotary pump uses the rotating motion of gears or a screw to build pressure in the fluid causing it to move.

What could possibly happen if a positive displacement pump was deadheaded? If a positive displacement pump should become deadheaded, the pressure in the pump could cause a rupture in the pump piping or the pump itself.

What are two major things to check on an electric motor? When checking on an electric motor, the ventilation and the bearings are the two most important.

What are two basic types of couplings? Fixed couplings and variable speed couplings are the two basic types of couplings.

How many different kinds of fixed couplings are there? There are three basic types of fixed couplings; they are the jaw coupling, the grid coupling, and the gear coupling.

How many different kinds of variable speed couplings are there? There are only two common types of variable speed couplings used; they are the hydraulic coupling and the electromagnetic coupling.

Other than couplings, what other types of equipment can be used to connect a motor with a pump? Other than the use of couplings, belts, chains, and gearbox (gear reducers) can be used.

What are sheaves? Sheaves are grooved pulleys that can be found on equipment that use V-belts, ropes, and serpentine or round belts.

How can changing the sheaves on a pump, affect the pumps capacity? By changing the sheaves on the pump or the driver will cause the capacity of the pump to change. The changing of the sheaves will either slow the pump speed down or speed the pump up. In doing so the volume of the fluid it is to move will either decrease or increase.

How does a gear reducer work? A gear reducer is a device that allows the driver to operate at one speed while the runner operates at another speed, by using several different sizes of gears used in combination.

What is the difference between fixed couplings and variable speed couplings? With fixed couplings, the runner and the driver are held a constant speed where as a variable speed coupling, the driver remains at a constant speed while the runner's speed is varied.

How does a hydraulic coupling work? A hydraulic coupling uses a fluid like oil to control speed. If a pump needs to increase speed, then more oil is pumped into the coupling and if the pump needs to decrease speed, then less oil is pumped into the coupling.

What is slip? Slip is the amount of difference in speed between the runner and the driver. The bigger the difference, the more slip; the smaller the difference, the less slip.

What is a scoop tube? A scoop tube is a device used on hydraulic couplings to control the speed between the runner and the driver, by scooping a little fluid out of the coupling or a lot of fluid out of the coupling.

How does an electro-magnetic coupling work? An electro-magnetic coupling works by increasing or decreasing the electrical field around the coupling. If a pump needs to be increased, then more electricity is used to make the magnetic field stronger and decreasing the amount of slip. If the pump needs to be decreased, then less electricity is used to weaken the magnetic field and increasing the amount of slip.

What are three devices used to move air? Three devices used to move air are compressors that compress gas into a small volume, a fan used to move large amounts of air at low pressure, and a vacuum pump used to remove air from a vessel.

What are two major uses of compressed air? Two major uses of compressed air in a power plant are house service air (general service air) and instrument air.

Why must instrument air be clean, cool and dry? Instrument air is air used for most of the instrumentation in a plant. The air is dry so that water or condensation does not develop in the instruments, which could cause problems in pneumatic controls.

What does reciprocating mean? Reciprocating means to move back and forth in motion.

How does a moisture separator work? A moisture separator works on the principles of water weighing more than air. Air flows through the separator usually in a spiral motion. This motion causes the water in the air to be thrown to the sides and is allowed to run down to the bottom of the separator while the air leaves through the top of the separator.

What types of heat exchangers are inter coolers and after coolers? The inter coolers and after coolers of a compressor are of a shell and tube type heat exchanger. With one fluid flowing through the tubes and the other fluid flows around the tubes.

Why does the air temperature increase when the air is compressed? The air temperature increases when it is compressed because the molecules of air are hitting each other more when compressed, this in turn generates heat.

How does a single stage compressor work? A single stage compressor, air is sucked in through the inlet to the compressor and compressed and then released the discharge. The air in this type of compressor is only compressed once.

How does a multi-stage compressor work? A multi stage compressor sucks air through the inlet to the compressor and compresses it, when the air is released, and it enters into another stage of the compressor where it is compressed again. When the air is released it goes to another stage, each time the air goes through a stage, the pressure of the air is a little higher than the last. All of the compression stages happen in one compressor, just in a different part of the compressor.

How is heat removed from an air-cooled compressor? Air cooled compressors have heat sinks or fins on the outside of the compressor casing to allow heat to flow away from the compressor.

What are two common types of compressed air dryers? Two common types of compressed air dryers are the refrigeration dryer and the desiccant dryer.

How does a refrigeration dryer work? A refrigeration dryer uses a cooling system to cool the air, causing it to condense and then the air is sent into a separator to ensure that all the moisture is out of the air.

How does a desiccant dryer work? A desiccant dryer uses a material that absorbs moisture. Air is allowed to flow through the desiccant; as it flows through the desiccant, moisture in the air is absorbed, leaving the air moisture free.

What is meant by surge capacity? Surge capacity is a volume of fluid that is stored to meet a greater than average demand without causing any problems to equipment.

How does a compressor on/off cycle work? When the air system needs no more air, a sensor will send a signal to the compressor and tell it to stop compressing. When the air systems pressure decreases to a predetermined value, the compressor will start compressing air again.

What are two pressure regulating devices used in a dynamic compressor? There are two basic methods for controlling airflow through a fan, inlet guide vanes and an inlet throttle valve.

How do inlet throttle valves work? Inlet throttle valves are usually placed in a system where the air is not from the atmosphere. This valve throttles open and closed to control the amount of air entering the compressor, thus controlling the compressors discharge.

How do inlet guide vanes work? Inlet guide vanes control the flow of air into the fan. If a little air is needed then the vanes are opened a little. If more air is needed, then the vanes are opened further.

What is the purpose of a fan? A fan is a device used to generate energy (pressure) in air, by using centrifugal force.

What are two basic types of fans? Axial flow fans and centrifugal fans are two basic types of fans.

What is an axial flow fan? An axial flow fan is a fan the moves air along the shaft of the compressor, like a steam turbine but only in reverse. The air would flow from the bigger blading towards the smaller blading.

What is a centrifugal fan? A centrifugal fan is a fan that uses centrifugal force to move air and creating pressure.

What are some basic methods for controlling airflow through a fan? There are three basic methods for airflow control through a fan, variable speed couplings, dampers, or inlet guide vanes.

What might affect friction head? Friction head can be affected by the viscosity and velocity of a fluid as well as the length, roughness, and diameter of the pipe the fluid is to flow through. Viscosity and velocity has the greatest effect on friction head.

What three factors makeup total head? The three factors that makeup total head are, static head, velocity head, and friction head.

What formula can be used to find velocity head? The formula used to find velocity head is, $h_v = v^2/2g$.

What are two common types of vacuum pumps? Two common types of vacuum pumps are the roots blower vacuum pump and the liquid ring vacuum pump.

How does a roots blower vacuum pump work? In this type of vacuum pump, there are two impellers that are shaped in a figure eight. These impellers are geared together with a small clearance between them, which allows them to operate synchronously in opposite directions. Seal water is applied to the clearance to seal the chamber. The impellers trap the air and discharge it through the outlet of the pump.

How does a liquid ring vacuum pump work? Within this pump, the rotor, a large inlet port and a small outlet port is positioned off center in the casing. This positioning allows more space above the rotor than below it. The casing is partially filled with water. When the rotor blades spins it pushes the water outward because of centrifugal force. This forms a liquid ring around the rotor assembly. The air enters the pump through the inlet port where it is trapped by the liquid ring in between the blades. As the blades move toward the discharge port, the air becomes compressed and finally is pushed out through the discharge port.

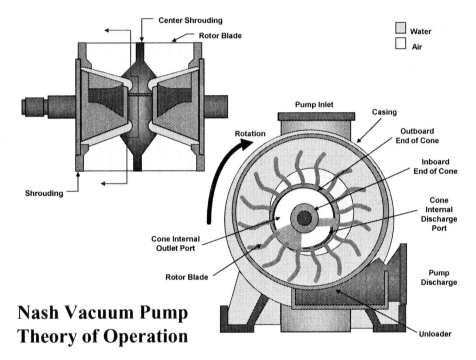

Nash Vacuum Pump Theory of Operation

Image courtesy of nash_elmo Industries

How are air compressors and vacuum pumps related? Both air compressors and vacuum pumps provide the same function, they both move air. The designs of the air compressor could be used as a vacuum pump, and the designs of a vacuum pump could be used as an air compressor.

What factors affect pressure? There are several factors that affect pressure; 1) type of fluid 2) temperature of the fluid 3) length of pipe 4) viscosity of the fluid 5) type of pump used 6) suction pressure.

What is static head? Static head is the amount of pressure a fluid in a vessel has that is measured from the top of the vessel to the bottom of the vessel where the piping is connected. The term "static" means <u>not moving</u>.

What is total head? Total head is made up of three factors: static head, velocity head, and friction head. With these three factors, total head can be found. $H = h_s + h_v - h_f$

What is friction head? Friction head means the resistance to flow.

What formula can be used to find total head? The formula to find total head is $H = h_s + h_v - h_f$. H is the total head, h_s is the static head, h_v is the velocity head, and h_f is the friction head.

What formula can be used to find friction head? Friction head is equal to static head + velocity head - total head. Friction head is proportional to velocity squared.

What formula can be used to find static head? Static head can be figured like this, 2.3 feet of water is equal to 1psi, pressure can be found by dividing the total head by 2.3 to obtain the pressure.

Can a fan cavitate? No. Compressed air will not damage the fan impeller by cavitating. Only a liquid can cause cavitation, air or any other gases will not.

What are wearing rings? Wearing rings are replaceable metal rings that are mounted between the impeller and the casing of the pump, forming a seal.

What are kinetic pumps? Kinetic pumps are pumps that convert the kinetic energy of a moving fluid into pressure. (centrifugal and jet pumps)

What is an eductor? An eductor is the type of pump that utilizes one fluid to move another fluid. (condenser vacuum pumps, air ejector, etc...)

What is a flexible impeller pump? A flexible impeller pump is a rotary type positive displacement pump that has an irregular shaped impeller that increases the impeller efficiency and allows the pump the ability to pump both thin and thick fluids.

What are six types of positive displacement compressors? 1) Reciprocating 2) Rotary screw 3) Sliding vane 4) Liquid ring 5) Rotary blower and 6) Diaphragm.

Are there any laws that apply to centrifugal pumps? The flow rate or pump capacity is directly proportional to the speed of the pump, the discharge head of the pump is directly proportional to the square of the pump speed, and the power consumed by the pump motor is directly proportional to the cube of the pump speed.

What is the difference between a rotary compressor and a rotary blower? The rotary blower is designed for high volume rather than high pressure like the rotary compressor.

Why are multi-stage pumps used instead of single stage pump in certain systems? With a multi-stage pump, it can generate an extremely large flow rate and at a pressure high enough to maintain that flow. With a single stage pump, one pressure and one flow rate.

What is the point of maximum efficiency of a pump? The point of maximum efficiency of a pump is the point at which the pump provides the maximum amount of flow for the least amount of power used.

What is the difference between pumps in series and pumps in parallel? Pumps that are in series use the water amount of fluid to increase pressure within the fluid. A multi-stage pump does just this, takes a specific amount of fluid and pushes it through multiple stages of the pump with each stage of the pump adding more and more pressure to the fluid. Pumps in series the pressure of the fluid increases but not the volume.

As with pumps in parallel, each pump has a specific capacity it will push out. Pumps in parallel are designed to have the same capacities in order to ensure that the pumps work together and not against each other. Pumps in parallel do not increase pressure, but they greatly increase flow through a system.

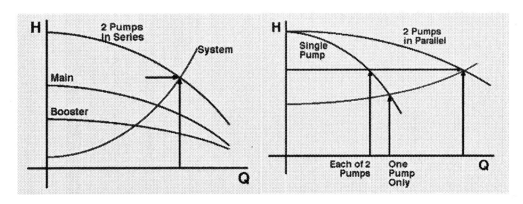

Series pump curve   Parallel pump curve

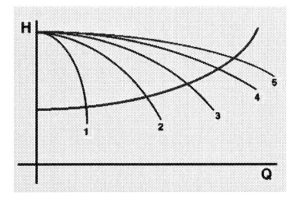

Multiple parallel pump curves

What are some effects of a lower than minimum flow in a boiler feed pump that has a minimum flow requirement? This condition would lead to higher turbulence and increased friction, which would generate excessive heat, which could cause water to flash into steam and possibly damage the pump.

Bearings, Lubrication & Seals

In this section you will see how bearings support and align the loads that are on them. You will see how many different types of bearings there are and how they function. The design of the bearings will give you an indication of what the purpose for that bearing is going to do.

You will be exposed to several different types of lubrication systems used to keep bearings normally operating. You will see how these lubrication systems function and the differences between each of them.

Seals are mentioned here to give you an idea of what a seal is and what its purpose is for. You will understand how bearings, seals and the lubrication system work together in order to keep the equipment running safely and efficiently.

What are bearings? A bearing is a mechanical device that is used to minimize friction between two surfaces while providing support and guidance for the moving part.

How many types of bearings are there? There are many different sizes and types of bearings, yet they all provide the same function, minimize friction and align moving parts. These bearings can range in weight to a single gram to several tons depending on the application.

What is meant by lubrication? Lubrication is the process of providing a substance such as oil, grease, or graphite between moving parts in order to reduce friction and heat buildup between them, thus extending the life of the parts.

How many different types of lubrication systems are there? There are four basic oil lubrication systems that are used. 1) Oil ring lubrication 2) Constant level oiler 3) Forced lubrication system and 4) Drip feed lubrication. There are also grease fittings and grease cups for lubrication as well.

Why is lubrication important? Lubrication provides a means of reducing friction and heat between two or more moving parts. Without proper lubrication moving parts would rapidly wear out and/or overheat because of friction. Without lubrication the motor of a pump would have to work harder to overcome the friction causing the life of the motor to rapidly decrease.

What are seals? A seal is a device that is used to keep a fluid in or a fluid out or to keep different fluids from coming into contact with each other.

How does a seal work? A seal on a turbine shaft is used to keep steam inside the turbine and/or keep air outside of the turbine by sealing the shaft where it penetrates the turbine. A seal on a generator is used to keep the hydrogen from inside the generator from escaping out of the generator and keeping air from mixing directly with the hydrogen. A seal on a pump is used to keep the fluid inside the pump. A sealed light switch keeps the arc from the switch from any hazardous gases in the air that might ignite around the arc.

What are the five principles of a seal? The five principles of a seal are to:

1. Retain Lubricant
2. Exclude contaminants
3. Separate fluids and/or gases
4. Withstand pressure differences.
5. Minimum wear / Maximum Performance

What factors determine the type of seal to use? There are many factors that determine the type of seal to be used which are:

1. Type of lubrication
2. Seal mounting (vertically/horizontally)
3. Thermal conditions
4. Mechanical and/or chemical factors
5. Sealing surface dealing with alignment and velocity
6. Exposure to the elements

How many types of seals are there? There are hundreds of different types of seals yet most of them provide the same function, here is a list of a few: Steam seals, water seals, labyrinth seals, carbon seals, packing seals, grease seals, oil seals, vacuum seals, pressure seals, mechanical seals, gasket seals, bushings, ceramic seals.

How do bearings work? Bearings work by maintaining the proper alignment between the rotating and stationary parts without allowing the two to come in contact. They also allow lubrication to move throughout the bearing between stationary and rotating parts. Bearings also control the direction of movement by limiting certain movements.

What is a bearing race? A bearing race is the inner and outer steel support rings for the bearing elements.

What is babbitt? Babbitt is an alloy of copper, tin, or other similar alloy used in bearings to reduce friction. This alloy is white in color and is soft and is slick when rubbed. It is composed of copper, tin, antimony, and lead. The melting point of babbitt is low. The addition of lead to babbitt greatly increases its lifespan. Babbitt is named after Isaac Babbitt (1799-1862) an American Inventor.

How does babbitt protect a bearing? The babbitt is softer than the bearing is, therefore the babbitt is allowed to wear out or become damaged instead of the bearing.

What is axial and radial movement? Axial movement is movement along a shaft. Radial movement is movement that seeks to move outward from the shaft.

On what equipment would you find a thrust bearing? Thrust bearings can be found on any equipment that generates a significant amount of thrust along a shaft.

What are bearing pedestals? Bearing pedestals are the supports that are used for the turbine bearings and turbine casing.

What are two basic types of bearings? There are two general types of bearings, a sliding surface bearing and a rolling contact bearing.

What is the difference between a sliding surface bearing and a rolling contact bearing? A rolling contact bearing is a bearing in which one surface rolls over another surface with lubricant in between the two surfaces. And a sliding surface bearing is a bearing in which two surfaces slide over each other with lubricant between the two surfaces.

What are four functions that all bearings perform? All bearings are designed to carry loads, distribute loads evenly, reduce friction, and to position moving parts.

What would happen if a bearing were not properly lubricated? Without proper lubrication to a bearing, the friction on the bearing would begin to increase and continue to increase until the bearing would burn-up and seize, or disintegrate. Bearing temperature will increase dramatically and vibration may develop.

What is the purpose of the bearing housing? The purpose of the bearing housing is to contain or position the bearing.

What is the purpose of a squealer ring? A squealer ring is used on a thrust bearing and is designed to make contact and produce a loud squealing noise before the wear on the thrust bearing becomes unsafe. A squealer ring can be found on many of the older units.

What type of bearing is a thrust bearing? A thrust bearing is a bearing that limits the movement of a shaft axially. This type of bearing does not support the shaft in any other way than limiting its axial movement.

 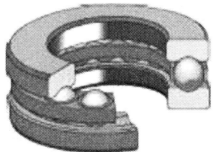

What are two types of thrust bearings? Two types of thrust bearings are the Tapered Land type thrust bearing and the Kingsbury thrust bearing. These are two of the most common thrust bearings used.

What is a needle roller bearing? Needle roller bearings are very small cylindrical shafts that are long and very narrow. The design and arrangement allow for high speed operation while maintaining lubrication retention. Needle bearings are a type of bearing that are contained in a flanged, antifriction cup. This design allows the individual rollers to bear directly on the shaft.

What is friction? Friction is the resistance to motion that exists between two bodies in contact with each other.

How does a rolling contact thrust bearing work? A rolling contact thrust bearing only prevents axial movement. A rotating ring is attached to the shaft where they rotate together. Rolling elements fit over the shaft to meet the rotating ring while a stationary ring attached to the housing also fits over the shaft is placed against the rolling elements. The rolling elements are placed in between both the rotating ring and the stationary ring. Should the shaft try to move, the rotating ring pushes against the rolling elements, in turn, which push against the stationary ring. Lubrication is provided in between all of the moving parts in order to reduce friction and wear.

How does a rolling contact bearing work? A rolling contact bearing consists of three basic parts, the outer ring, the inner ring, and the rolling elements. The outer ring is mounted in the housing, which is attached to a structure that is stationary. The inner ring is mounted on a shaft in which the shaft and inner ring rotate together. The rolling elements are positioned between the two rings, which maintain an equal distance between the rings, thus preventing radial movement.

How does a sliding surface bearing work? A sliding surface bearing consists of two basic parts, the housing and the bearing surface. This type of bearing prevents radial movement but allows axial movement.

How does a sliding surface thrust bearing work? A sliding surface thrust bearing works in the same manner as the sliding surface bearing except this bearing limits the axial movement with the use of a collar secured to the shaft. The bearing surface surround the shaft without contact yet is positioned so that they come in contact with the collar. Lubrication is provided in the bearing housing, seal located where the shaft penetrates the bearing housing and keeps the lubricant from leaking out. Should the shaft try to move axially, the collar will push against the bearing surface to limit the axial movement. The lubrication prevents the metal-to-metal contact between the collar and the bearing surface.

How does a tapered roller bearing work? A tapered roller bearing prevents both axial and radial movement. Rolling elements keep an equal distance between the two rings in order to prevent radial movement. Should the shaft try to move axially in <u>one direction</u>, the rolling elements will meet the outer ring to prevent movement. This type of bearing keeps the shaft from moving axially, but only in one direction.

What is film lubrication? Film lubrication is the formation of a film or wedge of oil that separates the moving parts from the bearing.

What is an oil wedge? An oil wedge is when a shaft rotates and oil is dragged under the shaft. This causes the shaft to operate on a film of oil, separating the shaft from the bearings surface.

What instrument is used to measure bearing temperature? A pyrometer is an instrument used to measure bearing temperature. This instrument can be electronic or mechanical.

What is a common way to determine if a bearing is running hot? A common way to check a hot bearing is to touch it carefully. If a pyrometer is handy, use it instead, usually less painful. Also, if a bearing is running hot, it might give off an unusual smell or make a strange noise that is not normally heard.

What instrument is used to measure bearing vibration? A vibrometer is the instrument used to measure bearing vibration.

What are two common types of lubricants used in a power plant? The two most common types of lubricants used in a power plant are grease and oil.

What are two common methods used for greasing a bearing? Two common methods for greasing a bearing are a grease gun or a grease cup.

How does an oil ring lubrication system work? In an oil ring lubrication system, the lower half of the housing contains an oil reservoir in which there is an oil ring positioned around the shaft. Half of the ring is always suspended in the oil reservoir. As the shaft begins to rotate, the ring will also begin to rotate. Oil will stick to the oil ring as it rotates around the shaft depositing the oil on to the shaft. The oil will then spread throughout the bearing where lubrication is achieved. A sight glass or dipstick can be used to indicate the oil level within the oil reservoir. Should the oil reservoir need oil, the oil reservoir can be removed and can be filled with oil.

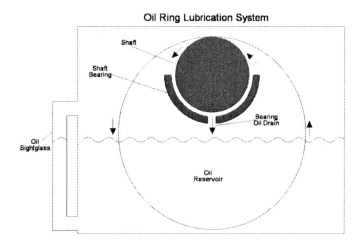

How does a constant level oiler work? A constant level oiler consists of an oil cup that is connected to the bearing housing by a tube or pipefitting. The oil level in the cup and the oil level in the housing remains the same because of the fittings connecting them together. An oil reservoir containing an opening at its bottom is placed within the oil cup. Gravity is what helps the constant level oiler to work. Should the oil level in the housing drop, the oil will begin to flow from the oil cup to the bearing housing and will continue to flow until the levels in the oil cup and the bearing housing are the same. The oil reservoir is usually transparent so that the oil level can be checked easily.

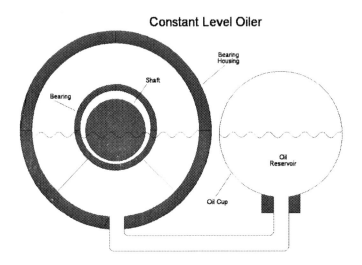

Constant Level Oiler

How much does a barrel of oil weigh? A barrel of crude oil weighs approximately 306 lbs.

What will happen if too much grease is added to a bearing? If too much grease is added to a bearing the bearing could overheat.

How does a grease cup work? A grease cup is a lubricating device consisting of an externally threaded cylindrical cup. By screwing down the internally threaded cap forces grease into the bearing. The grease cup is fixed on the bearing and can be refilled when empty by simply unscrewing the cap and adding more grease to the cup.

How can oil be damaged? Oil can be damaged by excessive heat that it is not designed for, impurities within the oil, and improper conditioning of the oil.

Why is it important never to return oil to its original container once it has been removed? Oil that has been removed from its original container may have been contaminated somehow and pouring it back into the original container would contaminate the rest of the good oil in the container.

What happens when oil gets to hot? The hotter the oil becomes the faster it will breakdown. If the oil gets too hot, metal-to-metal contact may occur at the bearings and the bearing can become overheated and damaged.

How does a forced lubrication system work? A forced lubrication system uses an oil pump to continuously circulate oil to the bearings. Oil is collected from the bearings in a reservoir where it is cooled and filtered before returning to the oil pump and back to the bearings.

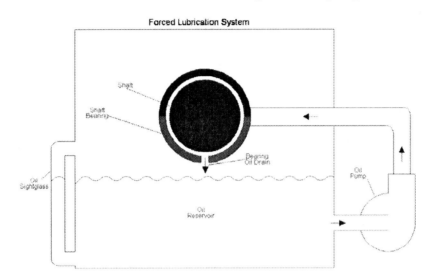

How does a drip feed lubricator work? A drip feed lubricator lubricates a bearing by allowing oil to slowly drip into the bearing. A reservoir is located above the bearing. At the bottom of the reservoir is a small hole that allows the oil to drip slowly and constantly into the bearing, providing lubrication. The drip rate can be controlled from the reservoir through the use of an adjuster knob located on the top of the reservoir. Should maintenance be required or the reservoir needs to be filled, a shutoff lever is used to stop the flow of oil to the bearing.

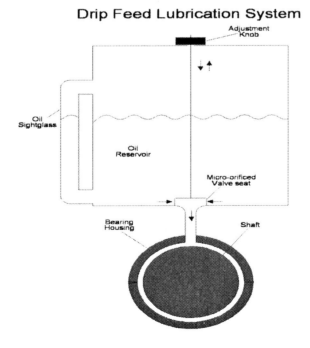

How many gallons are in a barrel of oil? A barrel of crude oil will contain 42 gallons of oil.

What is axial thrust? Axial thrust is the force that is created along the shaft of a pump or motor that tries to push the shaft in one direction, normally in the direction of a low pressure.

What is a labyrinth seal? A labyrinth seal is a seal made up of grooves to provide a maze or tortuous path to restrict the passage of a fluid like air or steam. Labyrinth seals do no prevent leakage; they only reduce or control leakage.

How are bearings named? Bearings are named for the type of movement that they are designed to prevent. Example, thrust bearing is used to prevent thrust along a shaft.

What is a carbon seal? A carbon seal is a seal that is used in sealing a shaft in which springs hold the carbon rings against the shaft.

What are some indications of bearing damage? Some indications of bearing damage are an increase in temperature of either the surface of the bearing or the temperature of the oil leaving the bearing. Vibration is a major indication that something is severely wrong with a bearing. Bearing on the left is from an Induced Draft fan, on the right a generator bearing.

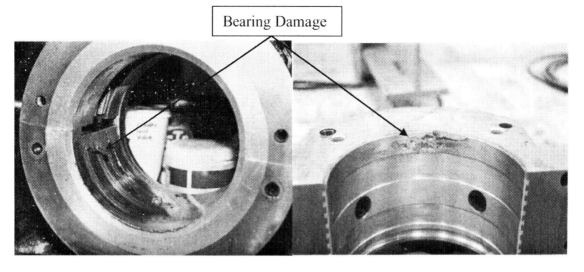

Bearing Damage

What is a linear ball bearing? A linear ball bearing is a bearing that consists of several hardened steel balls that roll in a spherical groove in between the surface of two concentric rings so that the load is carried by the steel balls.

What is an oil-less bearing? An oil-less bearing is a self-lubricating bearing, like plastic bearings, they are lubricated naturally.

What is hydrodynamic lubrication? Hydrodynamic lubrication is a lubrication process where a thick film of oil is placed between two moving surfaces keeping them at a substantial distance in which the load is supported by the hydrodynamic oil film pressure.

What are four clues in detecting a damaged bearing? 1) vibration, 2) high bearing metal temperature, 3) high bearing oil drain temperature, and 4) pieces of babbitt floating in the oil.

What is viscosity index? Viscosity index is the measure of an oil's ability to become less dense (thinner) when heated and to become denser (thicker) when cooled. The higher the oil's viscosity index, the lower its tendency to change its viscosity when heated or cooled.

Miscellaneous

In this section, I wanted to add different things that still had something to do with power plants and energy, different ways to produce energy, to protect the environment and the history of where we came from.

You will see how scientist worked and experimented in order to find success in their achievements. From the simple experiments to the splitting of the atom you will see how we advanced in technology. There is also a list to show you how much energy different appliances will use. You will find out how to conserve energy by changing the lighting in the home.

There is so much information available on power generation it is hard to place everything in one book. But the information in this book is to stimulate your mind and help you broaden your knowledge as well as enhance your way of thinking.

How is the containment building constructed? The containment building is what houses and protects the reactor vessel and its subsystems. It is designed to withstand internal (LOCA, HELB, etc.) and external pressures (missiles, airplane, etc.) if needed. Its foundation is buried very deep into the ground. It is formed from concrete and vast amounts of rebar (steel) for reinforcement. Depending on when the unit was constructed, the containment building can be anywhere from 3' to 6' thick with concrete & rebar. It is constructed to keep the radiation from getting out into the environment should an accident occur and is considered the last barrier protecting the environment from the radiation within the building.

In 1988, Sandia National Laboratories conducted a test of slamming a jet fighter into a large concrete block at 481 miles per hour (775 km/h). The airplane left only a 2.5-inch deep gouge in the concrete. Although the block was not constructed like a containment building missile shield, it was not anchored, etc., the results were considered indicative. A subsequent study by EPRI, the Electric Power Research Institute, concluded that commercial airliners did not pose a danger.   Wikipedia

The Turkey Point Nuclear Generating Station was hit directly by Hurricane Andrew in 1992. Turkey Point has two fossil fuel units and two nuclear units. Over $90 million of damage was done, largely to a water tank and to a smokestack of one of the fossil-fueled units on-site, but the containment buildings were undamaged.   Wikipedia

The following picture is to give you an idea of what the side view of a containment building wall would look like if you could see inside the wall.

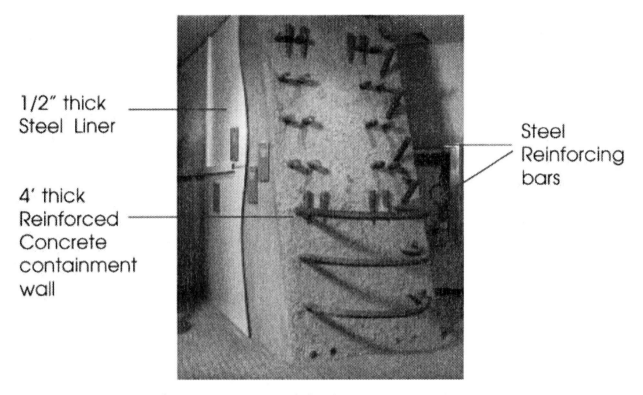

1/2" thick Steel Liner

4' thick Reinforced Concrete containment wall

Steel Reinforcing bars

Construction of Containment Wall

What are the four basic elements of fire protection?

1. Prevent fires from starting.
2. Detect
3. Suppress
4. Limit damage

What is core damage frequency? Core damage frequency (CDF) is a probabilistic risk assessment (PRA) statistic that estimates the frequency of accidents that cause damage to the nuclear reactor core. Core damage is the worst thing that can happen to a nuclear reactor, as due to damage of this structure, the operator might lose control of the nuclear reaction, possibly leading to a meltdown.

An assessment of permanent or temporary changes in a nuclear power plant is performed to evaluate if such changes are within risk criteria. Risk measures, such as core damage frequency and large early release frequency (LERF), determine the risk criteria for such changes.

This risk analysis allows decision making of any changes within a nuclear power plant in accordance with legislation, safety margins and performance strategies.

In a 2003 study commissioned by the European Commission, the average core damage frequency of nuclear reactors was estimated at once in 20,000 reactor years, or 5×10^{-5}. In a more recent 2008 study performed by the Electric Power Research Institute, the estimated core damage frequency for the United States nuclear industry is estimated at once in 50,000 reactor years, or 2×10^{-5}.

What is capacity factor? This is the ratio of its actual output of electricity for a period of time to its output if it had been operated at its full capacity. The capacity factor is affected by the time required for maintenance and repair and for the removal and replacement of fuel assemblies. The average capacity factor for U.S. reactors has increased from 50% in the early 1970s to over 90% today. This increase in production from existing reactors has kept electricity affordable. www.chemcases.com/nuclear/nc-10.html

What is large early release frequency? Large Early Release Frequency (LERF) is the frequency of those accidents leading to significant, unmitigated release from containment in a time frame prior to effective evacuation of the close-in population such that there is a potential for early health effects.

What is the difference between operating instructions (OI), operating procedure (OP), abnormal operating procedure (AOP) and emergency operating procedure (EOP)? Operating instructions (OI) are system specific directions for equipment manipulation. Operating procedures (OP) are overall plant integrated plan to accomplish power and mode changes. Abnormal operating procedures are pre-planned responses to anticipate operational events. AOPs give directions when to trip the unit. Emergency operating procedures (EOP) are procedures designed to optimize site responses after a post trip.

What is Maintenance Rule? The maintenance rule was issued, in part, in response to the number of plant transients and trips caused by equipment performance problems. Therefore, the number of trips and transients due to ineffective maintenance serves as a measure of the efficacy of the maintenance rule. In addition, equipment performance parameters provide an indication of maintenance effectiveness. However, the extent to which the maintenance rule contributes to any trends in these areas cannot be definitively separated from other factors that influence licensee performance. The MR started its first detailed Rule issued in 1988 and in 1991 the final detailed Rule was issued. The maintenance rule, 10 CFR 50.65, "Requirements for monitoring the effectiveness of maintenance at nuclear power plants," was issued on July 10, 1991, and became effective on July 10, 1996.

What does a blackout look like from space? This is a bogus picture from the blackout in the Northeastern part of the United States in 2003. Even though this picture is bogus it does provide an interesting picture to what might happen in the future. The animated map of the United States shows what states were truly affected by the blackout.

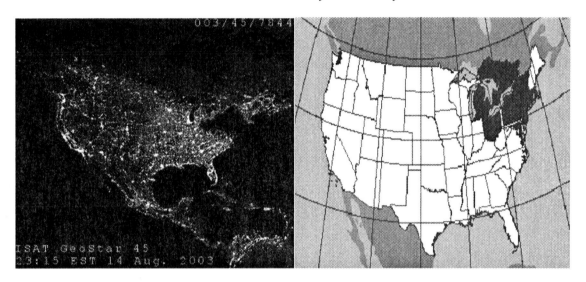

What is the Hall Effect? The production of a potential difference in a conductor or semiconductor carrying an electric current while it is in a strong transverse magnetic field.

Where did the word "nuclear" come from? The word nuclear, adjective of nucleus, drives from the Latin word nucleus (earlier nuculeus), diminutive of nux 'nut', thus 'kernel'. German Kern 'nucleus', is 'core'.

What are heavy metals? Heavy metals are chemical elements with a specific gravity that is at least 5 times the specific gravity of water. The specific gravity of water is 1 at 39^0F. There are 23 heavy metals: antimony, arsenic, bismuth, cadmium, cerium, chromium, cobalt, copper, gallium, gold, iron, lead, manganese, mercury, nickel, platinum, silver, tellurium, thallium, tin, uranium, vanadium and zinc.

How are heavy metals dangerous? In minute amounts, some of these elements are needed for good health; however in large doses they can quickly become toxic to living organisms. Heavy metal toxicity can result in damaged or reduced mental and central nervous system function, lower energy levels, damage blood composition, lungs, kidneys, liver and many other vital organs. Long term exposure may result in slowly progressing physical, muscular and neurological degenerative processes that mimic Alzheimer's disease, muscular dystrophy and multiple sclerosis. Allergies are not uncommon and repeated long term contact with some metals or their compounds may even cause cancer. When heavy metals enter the body, the body is unable to remove them through the body's waste system. Through continuous consumption of these heavy metals will cause the body to increase levels in heavy metals, eventually becoming more and more toxic in the human body thus leading to medical issues later on. The most commonly encountered toxic heavy metals are: arsenic, lead, mercury, cadmium and iron.

What is meant by potential difference? The difference in electric potential between two points in a circuit carrying current, measured in volts. It is equivalent to the work done (or received) in moving a unit positive charge between two points in an electric field, and is alternatively known as voltage.

What is dosimetry? Radiation dosimetry is the calculation of the absorbed dose in matter and tissue resulting from the exposure to indirectly and directly ionizing radiation. It is a scientific subspecialty in the fields of health physics and medical physics that is focused on the calculation of internal and external doses from ionizing radiation.

Dose is reported in gray (Gy) for the matter or sieverts (Sv) for biological tissue, where 1 Gy or 1 Sv is equal to 1 joule per kilogram. Non-SI units are still prevalent as well, where dose is often reported in rads and dose equivalent in rems. By definition, 1 Gy = 100 rad and 1 Sv = 100 rem.

What part do HEPA and charcoal filters play in the nuclear industry? HEPA and charcoal filters are used to make sure that radioactive particles are trapped and contained before any exhausted gases are discharged into the atmosphere. The HEPA filter traps actual particles while the charcoal filters catch any radioactive atoms still in the gases.

What is a design basis accident? The Design Basis Accident (DBA) for a nuclear power plant is the most severe possible single accident that the designers of the plant and the regulatory authorities could imagine. It is, also, by definition, the accident the safety systems of the reactor are designed to respond to successfully, even if it occurs when the reactor is in its most vulnerable state. Courtesy of NRC

When was the Chart of the Nuclides created? One of the earliest charts was published in 1935 by Giorgio Fea. Emilio Segre published a more detailed chart in 1946. The first edition of the present Chart of the Nuclides was compiled by G. Friedlander and M. Perlman and published by General Electric Company in 1946.

How many nuclides are known? At present, there are about 3100 known nuclides and 580 known isomers.

How can you find nuclear power plants by location or name? The Nuclear Regulatory Commission (NRC) has a web site that will give you this information. It will not only show the different regions of the United States regards to nuclear power plant but the web site also list all the nuclear plants in alphabetical order. Their web site is: www.nrc.gov/info-finder/reactor/

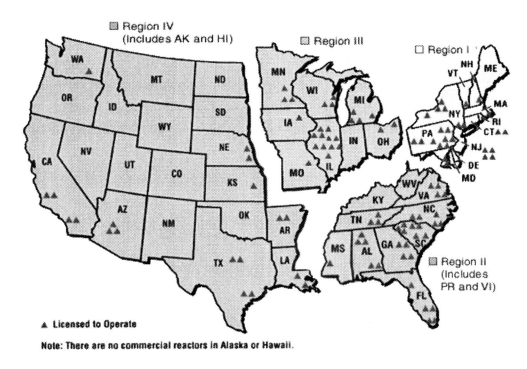

Courtesy of the NRC

What is unique about Palo Verde Nuclear Plant? The Palo Verde Nuclear Plant near Phoenix, AZ is the only generating station in the world that is not located adjacent to a large body of water. It uses treated sewage from several nearby municipalities to meet its cooling water needs, recycling 20 billion U.S. gallons of wastewater each year.

What is a radioisotope thermoelectric generator? In 1956, the U.S. Atomic Energy Commission began developing direct conversion devices to convert the energy from decaying isotopes into electricity. The first isotopic power generator was built in 1959. It generated 2.5 watts of power and used Po-210 as fuel, it also weighed 4 lbs. A radioisotope thermoelectric generator (RTG) is an electrical generator which obtains its power from radioactive decay. In such a device, the heat released by the decay of a suitable radioactive material is converted into electricity by the Seebeck effect using an array of thermocouples. RTGs can be considered as a type of battery and have been used as power sources in satellites, space probes and unmanned remote facilities. RTGs are usually the most desirable power source for unmanned or unmaintained situations needing a few hundred watts or less of power for durations too long for fuel cells, batteries and generators to provide economically and in places where solar cells are not viable.

What is SSTAR? SSTAR is an acronym meaning, "small, sealed, transportable, autonomous reactor". It is intended as a fast breeder reactor that is passively safe, has a self-contained fuel source of U-238 and an operating service life of 30 years, which could provide a constant power source between 10 and 100 megawatts. The 100 megawatt version of SSTAR would be approximately 15 x 3 meters in size and weighing at 500 tons. In order to reach the desired 30 year life span, the design calls for a moveable neutron reflector to be placed over a column of fuel. The reflector's slow downward travel over the column would cause the fuel to be burned from the top of the column to the bottom. Because the unit will be sealed, it is expected that a breeder reaction will be used to further extend the life of the fuel.

More importantly, this design is tamper resistant, which would prevent the leasing country from opening the reactor to use the generated plutonium for nuclear weapons. The tamper-resistant features will include radio monitoring and remote deactivation. The leasing country will therefore have to accept the capability for remote foreign intervention in the facility. As of today, there are no prototypes of SSTAR, yet but one is expected to be ready around 2015.

What is NEPA? The National Environmental Policy Act of 1970. NEPA requires that the NRC regulate and assess the impact of nuclear power on the total environment in terms of alternatives and the need for increased power. In an interpretation of NEPA, the U.S. Court of Appeals for the District of Columbia in 1971 made a far-reaching ruling (commonly referred to as the Calvert Cliffs decision). This ruling resulted in the NRC revising and strengthening its regulations to make a more rigorous implementation of NEPA. Here are some aspects of this ruling:

1. Environmental aspects must be considered at each stage in the licensing process. The environmental reports are circulated to government agencies and other interested persons.

2. At each licensing stage a cost-benefit analysis is made of environmental costs vs. economic and technical benefits. Consideration must be given to alternatives that will minimize environmental impacts.

3. Even if federal or state agencies certify that their environmental standards are satisfied, the NRC independently evaluates the total environmental impact. The NRC can, if it chooses, require controls more strict than those of a local agency.

4. Each nuclear power plant that was under construction but not granted an operating license was required to submit an environmental impact statement showing why its construction permit should not be suspended until a complete environmental and impact review was made. This in effect opened the possibility that plants already under construction might have to be redesigned or even abandoned.

What is health physics? A scientific field that focuses on protection of humans and the environment from radiation. Health physics uses physics, biology, chemistry, statistics, and electronic instrumentation to help protect individuals from any damaging effects of radiation.

What is plutonium? A heavy, man-made, radioactive metallic element. The most important isotope is Pu-239, which has a half-life of 24,000 years. Pu-239 can be used in reactor fuel and is the primary isotope in weapons. One kilogram is equivalent to about 22 million kilo-watt hours of heat energy. The complete detonation of a kilogram of plutonium produces an explosion equal to about 20,000 tons of chemical explosive. All isotopes of plutonium are readily absorbed by the bones and can be lethal depending on the dose and exposure time.

What does off-gas mean? Off-Gas is the un-dissolved ozonated air collected from the reaction tanks or de-ozonizing filters that can cover various manufactured parts, such as some computer parts. It is also a term used to describe the process of removing such ozonated air. It is potentially dangerous and is advised that people air out such contaminated items.

What gases are monitored in a confined entry? This would depend on the type of vessel to be entered, the type of substance it handles, any possible off-gasing, any type of cleaning material possibly being used to clean the confined entry but the number of variables can become numerous depending on type of vessel, etc. For example, carbon dioxide, nitrogen, oxygen, hydrogen sulfide may be just a few to be monitored.

What are the human responses to various O2 levels?

20 to 21%	Normal oxygen in atmosphere
19.5%	Becoming oxygen deficient
16%	Impaired judgment and breathing
14%	Faulty judgment and fatigue
10 to 12%	Nausea, headache
8 to 10%	Rapid fatigue, loss of consciousness
6%	Labored/difficulty breathing, death within minutes

An oxygen level above 21% does not affect human health in a negative manner but it can create other hazards in the environment.

What is meant by lower flammable limit (LFL)? The lower flammable limit (LFL) (lower explosive limit) describes the leanest mixture that is still flammable, i.e. the mixture with the smallest fraction of combustible gas.

What is meant by lower explosive limit (LEL)? Lower Explosive Limit (LEL) The explosive limit of a gas or a vapor is the limiting concentration(in air) that is needed for the gas to ignite and explode.

What is meant by permissible exposure limit (PEL)? Permissible Exposure Limit (PEL or OSHA PEL) is a legal limit in the United States for exposure of an employee to a chemical substance or physical agent.

What is a FSAR? Final Safety Analysis Report, a document required by the NRC to license a nuclear plant for construction or operation.

What is contained in the FSAR? The FSAR consists of 14 chapters covering some of the major topics like: site, building, system design basis, descriptions, accident analyses, etc.

What is a significant event report (SER)? A report concerning one or two isolated events involving operations, radiation exposure, etc. They usually have no recommended action except to review for applicability and consider actions as necessary.

What is a significant operating experience report (SOER)? A report concerning numerous related or reoccurring problems. They usually have recommended actions to take to prevent their happening again.

What is a licensee event report (LER)? A method of reporting certain events to the NRC. The LER program is required by 10CFR50.73. The LER is a 30 day written report to the NRC for certain events including: any plant shutdown required by tech specs, any violation of tech specs, any airborne radioactivity release exceeding 2 times MPC for one hour, any liquid release exceeding 2 times MPC for one hour.

What is tribology? Tribology is the science and technology of interacting surfaces in relative motion. It includes the study and application of the principles of friction, lubrication and wear. The word "tribology" derives from the Greek τρίβω ("tribo") meaning "(I) rub" (root τριβ-), and λόγος ("logos") meaning 'principle or logic'.

What is the difference between conventional flow and electron current flow? When Benjamin Franklin made his conjecture regarding the direction of charge flow (from the smooth wax to the rough wool), he set a precedent for electrical notation that exists to this day, despite the fact that we know electrons are the constituent units of charge, and that they are displaced from the wool to the wax -- not from the wax to the wool -- when those two substances are rubbed together. This is why electrons are said to have a *negative* charge: because Franklin assumed electric charge moved in the opposite direction that it actually does, and so objects he called "negative" (representing a deficiency of charge) actually have a surplus of electrons.

By the time the true direction of electron flow was discovered, the nomenclature of "positive" and "negative" had already been so well established in the scientific community that no effort was made to change it, although calling electrons "positive" would make more sense in referring to "excess" charge. You see, the terms "positive" and "negative" are human inventions, and as such have no absolute meaning beyond our own conventions of language and scientific description. Franklin could have just as easily referred to a surplus of charge as "black" and a deficiency as "white," in which case scientists would speak of electrons having a "white" charge (assuming the same incorrect conjecture of charge position between wax and wool).

However, because we tend to associate the word "positive" with "surplus" and "negative" with "deficiency," the standard label for electron charge does seem backward. Because of this, many engineers decided to retain the old concept of electricity with "positive" referring to a surplus of charge, and label charge flow (current) accordingly. This became known as *conventional flow* notation:

Conventional flow notation

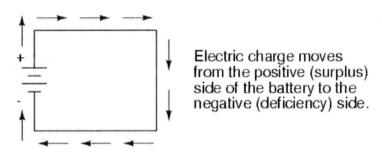

Electric charge moves from the positive (surplus) side of the battery to the negative (deficiency) side.

Others chose to designate charge flow according to the actual motion of electrons in a circuit. This form of symbology became known as *electron flow* notation:

Electron flow notation

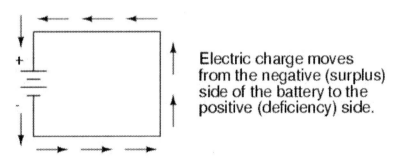

Electric charge moves from the negative (surplus) side of the battery to the positive (deficiency) side.

In conventional flow notation, we show the motion of charge according to the (technically incorrect) labels of + and -. This way the labels make sense, but the direction of charge flow is incorrect. In electron flow notation, we follow the actual motion of electrons in the circuit, but the + and - labels seem backward. Does it matter, really, how we designate charge flow in a circuit? Not really, so long as we're consistent in the use of our symbols. You may follow an imagined direction of current (conventional flow) or the actual (electron flow) with equal success insofar as circuit analysis is concerned. Concepts of voltage, current, resistance, continuity, and even mathematical treatments such as Ohm's Law (chapter 2) and Kirchhoff's Laws (chapter 6) remain just as valid with either style of notation.

You will find conventional flow notation followed by most electrical engineers, and illustrated in most engineering textbooks. Electron flow is most often seen in introductory textbooks (this one included) and in the writings of professional scientists, especially solid-state physicists who are concerned with the actual motion of electrons in substances. These preferences are cultural, in the sense that certain groups of people have found it advantageous to envision electric current motion in certain ways. Being that most analyses of electric circuits do not depend on a technically accurate depiction of charge flow, the choice between conventional flow notation and electron flow notation is arbitrary . . . almost.

Many electrical devices tolerate real currents of either direction with no difference in operation. Incandescent lamps (the type utilizing a thin metal filament that glows white-hot with sufficient current), for example, produce light with equal efficiency regardless of current direction. They even function well on alternating current (AC), where the direction changes rapidly over time. Conductors and switches operate irrespective of current direction, as well. The technical term for this irrelevance of charge flow is *nonpolarization*. We could say then, that incandescent lamps, switches, and wires are *nonpolarized* components. Conversely, any device that functions differently on currents of different direction would be called a *polarized* device.

There are many such polarized devices used in electric circuits. Most of them are made of so-called *semiconductor* substances, and as such aren't examined in detail until the third volume of this book series. Like switches, lamps, and batteries, each of these devices is represented in a schematic diagram by a unique symbol. As one might guess, polarized device symbols typically contain an arrow within them, somewhere, to designate a preferred or exclusive direction of current. This is where the competing notations of conventional and electron flow really matter. Because engineers from long ago have settled on conventional flow as their "culture's" standard notation, and because engineers are the same people who invent electrical devices and the symbols representing them, the arrows used in these devices' symbols *all point in the direction of conventional flow, not electron flow*. That is to say, all of these devices' symbols have arrow marks that point *against* the actual flow of electrons through them.

Who determines operator's license requalification's?

Each license is subject to a number of conditions (10 CFR 55.53) whether or not they are stated in the license. For example:

- Licensed operators and senior operators are required to observe all applicable rules, regulations, and orders of the Commission (10 CFR 55.53(d)).

- Licensed operators and senior operators are required to maintain their proficiency (10 CFR 55.53(e) and (f)) and to complete their facility licensee's requalification training and examination program (10 CFR 55.53(h)) as described in 10 CFR 55.59.

- Licensed operators and senior operators must have a medical examination by a physician every 2 years (10 CFR 55.21). If, during the term of the license, an operator or senior operator develops a permanent physical or mental condition (10 CFR 55.25) that may adversely affect the performance of their duties, the facility licensee must notify the Commission within 30 days of learning of the diagnosis. When appropriate, the NRC may issue a conditional license in accordance with the requirements in 10 CFR 55.33(b).

- Licensed operators and senior operators are prohibited from using, possessing, or selling illegal drugs (10 CFR 55.53(j)) and from performing licensed duties while under the influence of alcohol or any prescription, over-the-counter, or illegal substance that could adversely affect their performance. They are also required to participate in their facility licensee's drug and alcohol testing programs (10 CFR 55.53(k)) established pursuant to 10 CFR 26.

Operator License Renewal Process

If an operator or senior operator applies for renewal at least 30 days before the expiration date of the existing license, the license does not expire until the NRC determines the final disposition of the renewal application (10 CFR 55.55). The renewal process (10 CFR 55.57) requires the applicant to complete NRC Form 398 and submit it to the applicable NRC regional office with the following information: written evidence of the applicant's experience under the existing license, a certification from the facility licensee that the applicant is a safe and competent performer who has satisfactorily completed the requalification program for the facility (10 CFR 55.59), and certification on NRC Form 396 that the applicant's medical condition and general health are satisfactory (10 CFR 55.23). The NRC regional office will renew the license if, on the basis of the application and certifications, it determines that the applicant continues to meet the regulatory requirements (10 CFR 55.57(b)).

Does the nuclear industry share information with each other? YES, the nuclear industry does share information about their individual site events. When the nuclear industry first started, sharing information was never thought of…until Three Mile Island. TMI opened the eyes of the nuclear industry and determined in the benefit of mankind that information must be shared between the nuclear sites in order not to repeat the lessons learned from other sites.

How are nuclear reactors defended against adversaries? Commercial nuclear power plants are heavily fortified with well-trained and armed guards. They also have layered physical security measures, such as access controls, water barriers, intrusion detection and strategically placed guard towers. Together, these make up the plants' response to the Design Basis Threat – usually called the DBT. The DBT is developed from real-world intelligence information and describes the adversary force – coming from both ground and water – the plants must defend against. DBT specifics are not public in order to protect sensitive information that could aid terrorists. The NRC regularly reviews the DBT and adds new requirements when necessary.

Category I Fuel-Cycle Facilities

There are two NRC-licensed Category I Fuel-Cycle Facilities in the U.S. that make reactor fuel for nuclear plants. Since these plants handle nuclear material that could be targeted by adversaries, they also must defend against a DBT similar to that for nuclear power plants.

How are nuclear plants prepared to respond to emergencies? No matter how small the risk, the NRC requires all nuclear power plants to have and periodically test emergency plans that are coordinated with federal, state and local responders. The goal of preparedness is to reduce the risk to the public during an emergency.

In an emergency, the NRC and the licensee would activate their Incident Response Programs. Licensee specialists would evaluate the situation and identify ways to end the emergency, while the NRC would monitor the event closely, keeping government offices informed. If a radiation release occurred, the plant would make protective action recommendations to state and local officials, such as evacuating areas around the plant.

- Emergency Planning Zones (EPZs)

Each nuclear power plant has two EPZs. Each EPZ considers the specific conditions and geography at the site, and the community. The first is the Plume Exposure Pathway EPZ, which has a radius of about 10 miles from the reactor. People living there may be asked to evacuate or "shelter in place" during an emergency, to avoid or reduce their radiation dose. The second is the Ingestion Exposure Pathway EPZ. This has a radius of about 50 miles from the reactor. Protective action plans for this area aim to avoid or reduce the radiation dose from consuming contaminated food and water.

How are other sources of radioactive material protected?

Securing Materials

Radioactive materials are used in many beneficial ways, including medical, academic and industrial uses. Cancer treatment is just one way that radioactive materials benefit the public. Despite these benefits, some materials can potentially harm people and the environment if misused. For these reasons, their security, including use and handling, is strictly regulated in the United States by the NRC.

- "Dirty bombs"

A "dirty bomb," also called a "radiological dispersal device" (RDD), combines explosives, such as dynamite, with radioactive material. A dirty bomb is NOT a nuclear weapon. Most dirty bombs would not be highly destructive and would not release enough radiation to kill people or cause severe illness. Instead, a dirty bomb is a "Weapon of Mass *Disruption*" that could cause panic and fear, and require costly cleanup. Some materials licensed by the NRC could possibly be used in a dirty bomb, which is why they are strictly regulated.

How does NRC Rulemaking Work? Immediately after the 9/11 terrorist attacks, the NRC advised nuclear facilities to go to the highest level of security. After that, the NRC issued a series of mandates – called Orders – to further strengthen security. The NRC is taking a multifaceted approach to security enhancements in the post-9/11 threat environment. The NRC has raised the security of existing nuclear power plants while also requiring new security features in the design of new reactors that may be built in coming years.

Most recently, three new rulemakings provide additional security enhancements.

- One rule, issued by the NRC in March 2007 after extensive public comment, modifies and enhances the Design Basis Threat.

- A second rule, which was issued for public comment in 2006, proposes enhancements to the physical security at nuclear power plants. Among other things, the proposed rule addresses access controls, event reporting, security personnel training, safety and security activity coordination, contingency planning, cyber and radiological sabotage protection.

- A third rule, still in the early stages, will propose additional aircraft impact assessments for new power reactor designs.

- Rules – or regulations – and their enforcement are how the NRC protects people and the environment. Nuclear power plants must adhere to the rules or risk serious repercussions – up to closing a plant down. A new rule may be proposed by the NRC's five-member Commission, because of a petition from the public or as suggested by the NRC staff based on research or actual events. Once developed, a proposed rule is published in the *Federal Register* for a public comment period, usually 75 to 90 days. Once the comment period has closed, the NRC staff analyzes the comments, makes any needed changes, and forwards the final rule to the NRC Commissioners for approval. If approved, the final rule is published in the *Federal Register* and usually becomes effective in 30 days. Courtesty of the NRC

What is Yucca Mountain? A mountain in the SW Nevada desert about 100 mi (161 km) northwest of Las Vegas. It is the proposed site of a Dept. of Energy (DOE) repository for up to 77,000 metric tons of nuclear waste (including commercial and defense spent fuel and high-level radioactive material) presently held nationwide at commercial reactors and DOE sites. The project arose from the 1982 Nuclear Waste Policy Act requiring the DOE to construct a permanent underground nuclear-waste storage facility. Proponents of the use of Yucca Mt. as a repository claim that the area some 1,000 ft (300 m) beneath the mountain is the most viable site available, arid and remote with a deep water table, and that gathering the radioactive material in one location would allow for safer and more efficient and cost-effective protection. Opponents, including the state of Nevada, cite the potential for seepage into area groundwater, the danger of transporting waste to the facility, and the likelihood of the degradation of the storage containers and the occurrence of earthquakes and climate change over thousands of years. In 2002 President George W. Bush officially designated Yucca Mt. as the site for the nuclear waste repository, but regulatory hurdles and certain legal challenges must be surmounted before the facility can be constructed and opened.   Courtesy of Columbia Encyclopedia

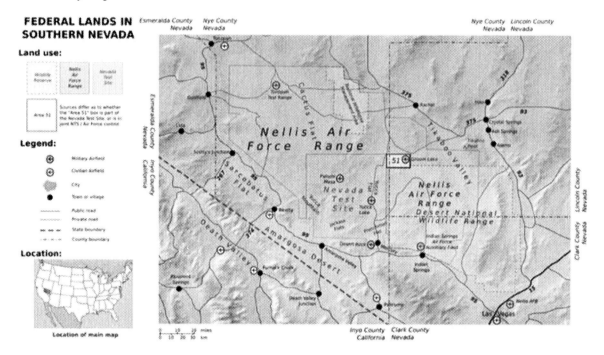

Courtesy of Wikipedia

What is electricity? Electricity can be described as the flow of electrons through a conductor…in simple terms.

What is the difference between a conductor and an insulator? A conductor allows electrons to flow very easy through the conductor. An insulator is just the opposite, it opposes the flow of electrons through itself. A superconductor is a conductor that has very low resistance through the material.

What is the difference between the generation system, transmission system and distribution system? The generation system is the system that actually generates the electricity to be put out on the electrical grid. The transmission system takes the electricity from the generation system and transfers the electricity in mass quantities throughout the local electrical grid. The distribution system receives electricity from the transmission system and distributes the electricity to the local area so that it can be used by local residents and businesses.

How do different types of power generation compare with nuclear?

Comparison to Nuclear Power Generation

Type	Nuclear	Coal	Natural Gas	Hydro	Petroleum	Renewables	Cogeneration
Capital Cost	High	Medium	Low	High	Low	Medium - High	Medium
Variable Cost	Low	Low	High	Low	High	Low	Medium
Operational Flexibility	Low	Low	Medium - High	High	Medium - High	Low	Low
Time for construction	Long	Long	Short	Long	Short	Short	Short
Environmental impact	Low - Medium	High	Low	Low - High	Medium	Low	Low
Fuel Availability	Plentiful	Plentiful	Some Concerns	Limited	Some Concerns	Plentiful	Some Concerns
Availability to sites	Limited	Limited	Flexible	Limited	Limited by air quality permits	Limited by resource availability	Flexible
Longevity	30 to 60 years	30 + years	30 + years	50 + years	30+ years	Depends on Type	30 + years
Dependability	High	High	Medium - High	High	Medium - High	Depends on Type	Medium - High
Outage Time	Low	Medium - High	Medium - High	Low	Medium - High	Depends on Type	Medium - High
Reliability	High	Medium - High	Medium - High	Medium	Medium - High	Depends on Type	Medium - High
Pollution	If released radiation	CO2, SO2, NOx, Ash, Mercury, etc.	CO, CO2, NOx,	Thermal	CO, CO2, NOx, SO2	Depends on Type	Depends on fuel type

What is the ASP Analysis Program? The NRC established the Accident Sequence Precursor (ASP) analysis program in 1979 in response to the Risk Assessment Review Group report (see NUREG/CR-0400, dated September 1978). The primary objective of the ASP Program is to systematically evaluate U.S. nuclear power plant operating experience to identify, document, and rank the operating events that were most likely to lead to inadequate core cooling and severe core damage (precursors), if additional failures had occurred. To identify potential precursors, NRC staff reviews plant events from licensee event reports (LERs), inspection reports, and special requests from NRC staff. The staff then analyzes any identified potential precursors by calculating a probability of an event leading to a core damage state.

What does the U.S. electrical grid look like?

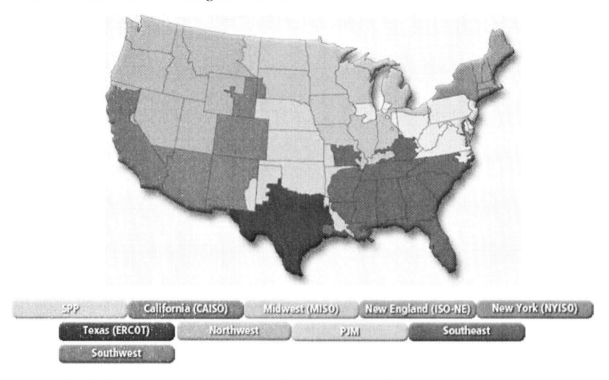

The U.S. is broken up into many different sections of the nation's electrical grid. You've heard the saying "never put all your eggs in one basket"…same rule applies here. If we only had one electrical grid to cover the entire U.S., whenever a problem should occur, it could jeopardize the entire country. With this setup, we greatly reduce the chance of causing a nationwide electrical outage (blackout).

What is an atomic clock and why is it important? They are the most accurate time and frequency standards known, and are used as primary standards for international time distribution services, to control the frequency of television broadcasts, and in global navigation satellite systems such as GPS.

What happens to a magnet if it gets too hot? A magnet will lose its strength because the heat causes the atoms to become excited and they are unable to line up and create a magnetic field. This point at which the magnet losses its strength is called the Curie Point.

What is the difference between diamagnetism, paramagnetism and ferromagnetism? Diamagnetism refers to the materials that are not affected by a magnetic field. Paramagnetism refers to the materials like aluminum or platinum which become magnetized in a magnetic field but their magnetism disappears when the field is removed. Ferromagnetism refers to materials that can retain their magnetic properties when the magnetic field is removed.

What do the poles of a magnet look like (electrostatic fields)? The South Pole is "negative" and the North Pole is "positive".

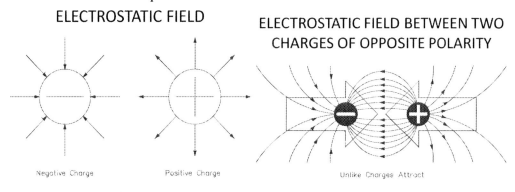

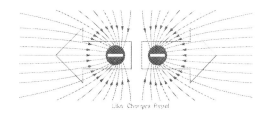

What is field density? For an electric or magnetic field, the number of lines of force passing at right-angles through unit area of it.

What is a Geiger counter? An instrument for detecting and measuring charged particles and thus radioactivity.

What does ergonomics mean? Ergonomics is a science that seeks to modify working conditions so that people can better adjust to their working environment.

What are the six simple machines? The six simple machines are a pulley, a lever, a wheel/axle, an inclined plane, a wedge, and a screw. Through these simple devices, man has found ways of moving massive amounts of material with little effort. These are some of the oldest tools known to man and strangely today so of the most commonly used.

What are three classes of levers? A first class lever is a pair of pliers where its pivot point is between the effort and the load. A second-class lever is like a nutcracker, where the load is between the pivot point and the effort. And a third class lever is like tweezers where the effort is in the middle while the load and the pivot point are on the outside.

How is steel pipe rated? Steel piping is rated by the standard temperature and pressure conditions for which specific steel piping is designed.

What is meant by the electrical grid?

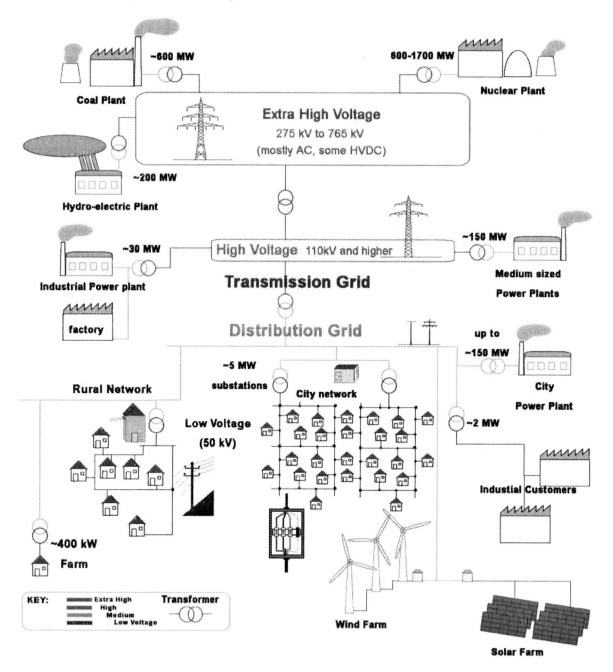

What determines what type of pipe connector to be used? The pipe connector to be used is determined by the application of the connector. There are three types of connectors, threaded, flanged, and welded. A threaded connector is usually used in systems that use small piping with a low pressure. A flanged connector is usually used in a system that has moderate pressure. And a welded flange connector is used in high temperature and high-pressure systems.

What determines the thickness of the pipe to be used? The thickness of a pipe to be used is determined by the amount of pressure the fluid in the pipe will be under. Low-pressure fluids can use thinner piping than a high-pressure fluid.

How much can a pipe expand when heated? A steel pipe that is 100 ft long carrying superheated steam can expand as much as 12 inches in length when heated from room temperature to over $1,000^0$F.

What determines the kind of pipe to use? The type of pipe to be used in a plant is determined by the type of fluid the pipe will carry. A circulating water pipe that carries cold or warm water at low pressures can be made of cast iron, because it is inexpensive and very dependable. And a system that carries air that is used for controls and instrumentation is usually made of copper, brass, or stainless steel tubing. Piping that carries steam is usually made of an alloy material because of the high temperatures and pressures it can handle.

Why should low temperature piping be insulated? Low temperature should be insulated to prevent condensation from building up on the outside of the pipe.

What is the purpose of piping supports? The purpose of piping supports is to allow for thermal expansion and contraction of the piping as well as to allow any other movement the piping might experience. There are several types of supports for piping, some rigid and some that work like a shock absorber with the use of a spring.

Why are loops and bends included in some piping? Loops and bends are added in piping construction to allow room for thermal expansion and contraction along with hydrostatic shock.

What does the recycling symbols on plastic containers mean?

(These symbols are usually found on the bottom of a container.)

1	**PETE**	Polyethylene Terphthalate	soft drink bottles
2	**HDPE**	High-Density Polyethylene	milk and water jugs
3	**V**	Vinyl / Polyvinyl Chloride (PVC)	Shampoo bottles
4	**LDPE**	Low-Density Polyethylene	plastic bags
5	**PP**	Polypropylene	squeeze bottles
6	**PS**	Polystyrene	fast-food packaging
7	**Other**	All other resins and multi-layered material.	

How many ways can electricity be created? The first thing to remember about energy is that it cannot be created nor destroyed by any means, just converted from one form to another. Electricity can be made in a number of ways, from power plants, batteries, static electricity, solar power, wind power, wave power, nuclear power. There is electricity all around us, sometimes without you even realizing it.

What are some natural types of energy? There are four general types of energy that occur naturally, geothermal, solar, and ocean wave energy.

Geothermal energy is the heat that is contained within the Earth's crust and is extracted from the Earth in the form of steam or hot water. This type of energy can be used to provide homes or businesses as well as for the production of electrical power.

Solar energy is radiant energy that is transmitted from the sun to the Earth. This energy from the sun can be harnessed to produce electricity that is inexpensive, yet it is very costly to collect and store this type of energy.

Ocean wave energy is energy that is contained within the waves of the ocean and is used to produce electricity.

Hydro energy is energy that is used through the falling of water flowing through a turbine.

Wind energy is energy that is harvested from the wind using wind type generators or turbines.

What is spontaneous combustion? Spontaneous combustion is a self-igniting combustion. It is a process where a substance will burst into flames as a result of heat being produced by oxidation. Due to oxidation, the heat that is produced cannot be dissipated fast enough into the air, so the temperature of the material will become so hot that it will reach it's ignition point and burst into flames.

What is a polarized outlet? Polarity is when a property has two opposite poles like a magnet or a battery. When interchanging the positive and negative conditions of the terminals to which it is applied would be called reverse polarity. On certain outlets you can see that one slit is bigger than the other slit. This is called a polarized plug because you can only plug in the appliance one way.

What is a GFCI outlet? A GFCI (Ground Fault Circuit Interrupter) outlet is a specially designed outlet that is made for use where the electrical outlet could possibly get wet, greatly increasing the odds of getting shocked. This outlet is constructed to monitor the flow into and out of the outlet. Should a sudden rush of current try to flow through the outlet the sensing element would pick it up and trip the electrical outlet before it has time to electrocute someone. These outlets cost a few dollars more but they provide a great safety feature and by the NEC is required by electrical code standards.

What are "noble metals"? Noble metals are metals that are highly resistant to oxidation and corrosion, like gold and platinum.

What are "heavy metals"? Heavy metals are metals that have a specific gravity of 5.0 or greater. Metals like lead, copper, mercury, cadmium, platinum, and gold are heavy metals and are toxic to organisms. Heavy metals are poisonous and can accumulate and persist in living organisms causing heavy metal poisoning. Example, if a heavy metal is dumped into an ocean and eaten by fish, the fish will in turn be eaten by humans causing a transfer of heavy metals from the fish to the human body.

What is dynamo-electric? Dynamo-electric means having to do with conversion of mechanical energy into electrical energy, or vice versa.

What is Hero's steam engine? A Greek, Hero of Alexandria was a brilliant inventor. He created a toy sphere made of metal that hung between two brackets so it could spin freely. The metal sphere was filled with water and then heated over a fire. As the water boiled into steam, the sphere began to rotate on its axis because of the steam escaping from two outlets on the sphere.

What are PCB's? PCB's or Polychlorinated Biphenyls belong to a class or compounds containing a variable number or chlorines. The toxicity will vary depending on the actual position of the chlorine atoms within the structure. PCB's have been around for 40 years before they were linked to extensive global contamination and human illness occurred. Because of the low flammability of PCB's, they became extensively used for insulating and cooling of electrical equipment such as transformers and capacitors. PCB's physical appearance can be in the form of an oily liquid to a white crystalline solid. Even though resistant to deterioration, the PCB's can be slowly decomposed of by soil bacteria. Yet the amount of time required is based upon the chemical makeup of the PCB. PCB's are highly toxic to biological life forms, the higher the chorine content the greater its toxicity. Exposure can lead to eye irritation, liver damage, jaundice, vomiting, abdominal pain, discoloration of the skin and nails, changes in the immune system, cancers and respiratory distress. (Harte, Holdren, Schneider, Shirley)

What are CFC's? CFC's is an acronym meaning Chloroflurocarbons. These compounds of chlorine and fluorine were once used as aerosol propellants and refrigerants and in foam packaging. It is now known that Chloroflurocarbons deplete the ozone layer and act as a greenhouse gas.

Will hot water freeze faster than cold water? Of course a cup of hot water will not freeze faster than a cup of cold water, but if the hot cup of water is allowed to cool down to the same temperature as the cold cup of water and both were put in the freezer, the water that was heated earlier would freeze faster than the cup of cold water. Reason, when water is heated, the oxygen within the water is driven out of the water in the form of bubbles. It is the oxygen that acts as an insulator to the water.

What is a laser? The term laser is an acronym for Light Amplification by Stimulated Emission of Radiation. This device will produce a narrow high intensity light that is so powerful it can vaporize the best heat resistant materials. The light from a laser contains only one color and is said to be coherent because of the concentration of photons. Lasers can be used for many different purposes like a knife or to fuse metals together, remove diseased tissue from the body, send communication signals, bar code scanners, printing, target designators, etc.

What is TRI? TRI stands for Toxic Release Inventory. It is a government mandated publicly available compilation of information on manufacturing facilities in the U.S. that release over 300 different toxic chemicals and 20 categories of chemical compounds that are released into the environment. The EPA keeps up reports on these facilities and makes them available to the public.

What is eutrophication? Eutrophication is a process where the supply of nutrients in a body of water such as a lake or pond is increased. As plant growth is accelerated in the water due to increased nutrients, plant overcrowding will occur, as the plants die the decaying vegetation will slowly deplete the oxygen in the water thus causing fish to die.

Does cold water weigh more than hot water? Cold water weighs more than hot water or is denser than hot water. If a container was divided into two sections by a plastic separator and one side was filled with cold water that was dyed red and the other side was filled with hot water that was dyed blue. When the divider is pulled out of the container you will see that the cold water will sink to the bottom of the container and the hot water will rise to the top of the container.

What is sensible heat? Sensible heat is heat that is added or removed from a material that causes a change in temperature.

What is latent heat? Latent heat is the energy needed to produce a change of state in a substance; temperature remains constant during the change. Latent heat is released when the substance changes back to its former state. These various heat changes are measured in Calories or Joules. The specific latent heat of a substance is the difference in ENTHALPY of the two states. The word latent derives from the Latin word latens meaning "lying hidden".

What is latent heat of vaporization? Latent heat of vaporization is the amount of heat needed to convert one pound of a substance from a liquid state at its boiling point to a gas state without an increase in temperature.

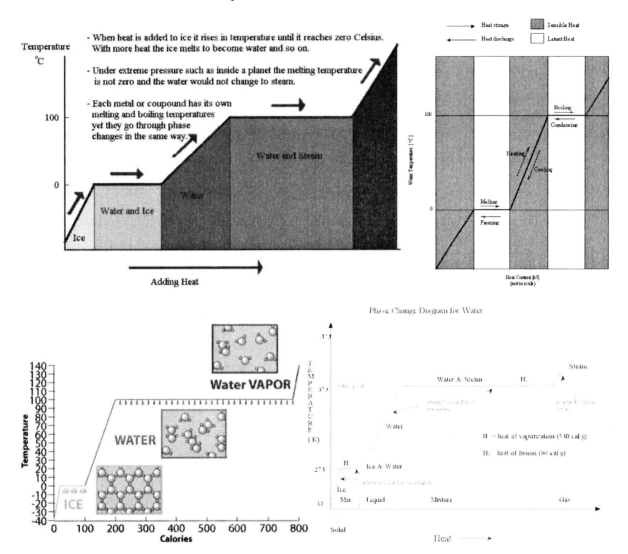

What is the most efficient component in a power plant? The most efficient component in a power plant is the generator or transformer approximately 98%.

What is the only other substance (other than water), that is less dense as a solid than it is at a liquid? Bismuth is the only other substance that is less dense as a solid than it is at a liquid.

What are ferromagnetic materials? Ferromagnetic materials are materials that can become highly magnetic in a weak magnetic field such as iron, steel, or nickel.

What are paramagnetic materials? Paramagnetic materials are materials that are only slightly attracted by a magnet and will not retain magnetism. Examples: aluminum, platinum, and oxygen.

What are diamagnetic materials? Diamagnetic materials are materials like bismuth and antimony that are actually repelled by magnets.

What are nonmagnetic materials? Nonmagnetic materials are materials that cannot be magnetized or attracted to a magnet. Examples: copper, brass, water, glass, wood, and paper.

How many different types of magnets are there? There are many different types of magnets. Some are permanent magnets while others are temporary magnets. Temporary magnets are those materials that do not stay magnetized all the time like a magnet attracted to a nail will cause the nail to become magnetic but when the magnet is removed from the nail, the nail is no longer magnetized. A permanent magnet is always magnetized, like the refrigerator magnets in your home.

What is the difference between the magnetic field of a permanent magnet and a temporary magnet? A temporary magnet becomes magnetized when moving charged particles, when electrons flow through a coil of wire. A permanent magnet remains magnetized because of the spinning electrons within the atoms of the magnet. Some of the strongest permanent magnets are ceramic magnets that are made under high temperatures and pressures containing powders of various metal oxides.

What kind of material is used for the core of a transformer? There is a material called Stalloy, which is an alloy of iron combined with 4% silicon it is commonly used in the core of a transformer.

What is a radiometer? A radiometer is an instrument that is used for measuring the intensity of radiant energy in any part of the electromagnetic spectrum. A radiometer looks similar to a clear light bulb that contains a set of vanes that is blackened on one side. The vanes are suspended on an axis and operate in a vacuum. When the vanes are hit with radiant energy the vanes will begin to spin, the stronger the energy the faster the vanes will spin.

What does ferrous mean? Ferrous means containing iron or is derived from ion.

What is pyrometry? Pyrometry is the process of measuring exceedingly high temperatures through the use of a pyrometer, which is an instrument that can function in heat far hotter than that tolerated by ordinary thermometers. Most pyrometers measure temperature by detecting the rise in electrical resistance in a metal, by the increase in the intensity of light, or by similar electrical or radiation techniques.

What is a potentiometer? A device used to obtain an accurate measurement of the electromotive force or voltage of an electrical cell. A potentiometer can be used to calibrate electrical instruments and/or used as a variable resistor. A volume control knob on a radio is a type of potentiometer.

What is graphite? Graphite is a soft, black mineral having a metallic luster and is a form of carbon. Graphite is used in pencils, lubricants, and in many other devices.

How does a heat pipe work? A heat pipe is a device that can move heat while keeping the temperature relatively constant. A heat pipe is made up of a sealed metal tube with a porous lining that holds boiling liquid. To move heat in a heat pipe, one end is immersed in the heat source, so that the liquid's temperature is brought to it's boiling point, this forces the vapor to move to the cooler end of the pipe. As the vapor reaches the cooler end of the pipe it condenses which in turn releases heat. Heat pipes are very efficient because they can transfer large amounts of heat over long distances and keep the temperature constant without using an external power supply.

What is a hygrometer? A hygrometer is a device used to measure relative humidity.

What is Kerosene? Kerosene is a mixture of hydrocarbons that can be derived from oil, coal, and tar yet most kerosene is produced from distilled petroleum. Kerosene is colorless and is mainly used as a fuel for jet engines, heating, lighting, a solvent, and thinner for paint.

What is a magnetometer? A magnetometer is an instrument for measuring the intensity (strength) and the direction of magnetic forces.

What is magnetism? Magnetism is a force that occurs naturally in certain substances and can be transferred to or induced in others. Properties of magnetism are its complementary forces of attraction and repulsion and its capacity to align itself on a north-south axis.

What is magneto-hydrodynamics? Magneto-hydrodynamics is a method of generating electricity by passing a high velocity stream of plasma (gas at a very high temperature) across a magnetic field. The principle is the same as that of the electric generator, except that a plasma stream is used rather than a coil of wire that acts as the conductor. Magneto-hydrodynamics is still in the experimental stages.

How does a linear accelerator work? A linear accelerator is a device that accelerates charged particles in a straight line under a vacuum by alternating the negative and positive impulses from an electric field.

What is oil? Oil is any of several kinds of greasy fluids that are lighter than water, that will burn easily, and will not mix or dissolve in water yet will dissolve in alcohol. Mineral oils such as gasoline and kerosene are used for heating and lubricants. Fixed vegetable oils change when they absorb oxygen like linseed, tung, and olive oil. Volatile vegetable oils usually have a distinct odor and flavor such as peppermint, and turpentine.

What is permalloy? Permalloy is an alloy that can be temporarily magnetized by an electric current. Permalloy is an important component in transformers. It is composed of nickel and iron.

What is a perpetual motion machine? A perpetual motion machine is a hypothetical machine that once is set in motion it could go on forever by creating its own energy, until an external force stops it or the machine is just worn out.

What is a petrochemical? A petrochemical is a chemical that is made or derived from petroleum or natural gas.

What is sublimation? Sublimation is the act of changing a solid into a gas, or a gas into a solid, without going through a liquid state, like dry ice.

What is a Faraday cage? The concept of a Faraday cage is logically attributed to Michael Faraday, an 19th Century pioneer in the field of electromagnetic energy. Faraday studied the work of earlier scientists such as Benjamin Franklin and theorized that electromagnetic waves naturally flowed around the surface of conductive materials, not through them. For example, if a metal box containing a mouse were placed directly in the path of an electrical current, the electricity would flow over the box but not into the compartment with the mouse. The mouse would not be electrocuted. Such a box would be considered a Faraday cage.

The important concept to remember is that a Faraday cage acts as a shield against the effects of electromagnetic energy. When a car is struck by lightning, the metal frame becomes a Faraday cage and draws the electricity away from the passengers inside. A microwave oven door has a screen which prevents electromagnetic energy from escaping into the room. Electronic parts which generate radio frequencies are often protected by Faraday cages called RF shields. Even a concrete building reinforced with lead or rebar can be considered a Faraday cage.

How does a synchrocyclotron work? A synchrocyclotron is a cylindrical shaped particle accelerator that is designed to accelerate charged particles by changing the frequency of the electric field so that it synchronizes with the motion of the particles to compensate for the increasing mass acquired by the particles as they approach the speed of light.

How does a synchrotron work? A synchrotron is a particle accelerator where doughnut shaped rings made of magnets that surround a vacuum and produces a magnetic field that increases in strength as the accelerated particles increase in velocity. This will keep the accelerated particles in a circular orbit where their radii remains constant. Since this type of particle accelerator requires no magnet in the center, the rings of the accelerator can be built several miles in diameter.

What are synthetic fuels? Synthetic fuel is any combustible matter that can take the place of crude oil and / or natural gas. Sources for synthetic fuels include animal and plant matter as well as coal, oil, shale, and bituminous sands.

How can a 42-gallon barrel of crude oil produce 44 gallons of petroleum product? A 42-gallon barrel of crude oil can produce 44 gallons of petroleum product because of the reduction in the density of the crude oil during the refining process. Here is a breakdown in gallons of all the products produced from a barrel of crude oil, 1) gasoline 19.36, 2) distillate fuel oil 9.20, 3) jet fuel 3.86, 4) residual fuel oil 2.44, 5) still gas 1.93, 6) petroleum coke 1.81, 7) liquefied refining gas 1.72, 8) asphalt & road oil 1.34, 9) petrochemical 1.26, 10) naphtha type jet fuel 0.34, 11) lubricants 0.46, 12) special naphthas 0.17, 13) miscellaneous products 0.13, 14) kerosene 0.13, 15) finished aviation fuel 0.08 and 16) waxes 0.04.

What is torsion balance? Torsion balance is a device used for measuring the force of a twisting motion. These measurements are registered on a calibrated gauge.

What is a transducer? A transducer is a device used for converting an input of energy of one form into an object of another. Example: Light bulbs, microphones, and loud speakers.

What is fiber optics? Fiber optics is a branch of physics that deals with the transmission of light pulses along a bundle of glass or plastic fibers that have refractive properties that allow them to transmit light around curves and into inaccessible locations. Fiber optics is used in communication systems, computer systems, medical equipment, and is used in many other applications.

What is an alloy? An alloy is a metal that is created by mixing and fusing more than one metal together or a metal and a non-metal together. This depends on what the metal will be used for, like hardness, lightness, and / or strength. Example, brass is made from copper and zinc and bronze is made from copper and tin.

What is asbestos? Asbestos fibers are small and odorless, often invisible except through a microscope, and indestructible in most uses. They can be transported on clothes and other materials, and they have aerodynamic features that allow them to be suspended and re-suspended easily in the air and to travel long distances. Once released, asbestos fibers are difficult to detect and contain, and they readily enter the surrounding air. People are thus exposed not only at the time and place of release, but long after the release and far from its source. There is constant renewal of risk because asbestos fibers reenter the atmosphere repeatedly over time.(Harte, Holdren, Schneider, Shirley) Asbestos is a name for various fibrous minerals that occur as flexible fibers that are not combustible or heat conductive, which makes them useful insulating materials. As great as asbestos is as an insulator, it is dangerous if inhaled. Asbestos particles cause lung cancer and asbestosis. Once these particles enter the lungs they become permanently attached which after prolong exposure to asbestos particles will reduce the lung capacity making it harder for an individual to breathe.

What instrument is used to measure wind speed? An anemometer is an instrument used to measure wind speed. There are several types of anemometer, the most common being a rotating type mechanism that has cups mounted on a vertical shaft. An acoustic anemometer depends on the speed of sound in the wind.

What is a particle accelerator? A particle accelerator is a tool of science used to accelerate electrically charged subatomic particles to high velocities. These particles can be focused to interact with other subatomic particles or to breakup atomic nuclei, so more can be learned about the fundamentals of natural matter. These particle accelerators use electromagnetic fields to accelerate the subatomic particles either in a straight line or in a circular motion. Accelerators are rated by the kinetic energy they produce, which is measured in electron volts (eV).

What is capillarity? Capillarity is the tendency of the surface of a liquid to rise or fall within a narrow tube as in a capillary tube. In the case of adhesion, these forces are weaker than the attraction of the molecules that are on the wall of the tube; a concave meniscus is found (which means the liquid level is higher where the liquid actually touches the tube). Surface tension pulls the rest of the surface upwards and the level rises until the weight of the column balances the surface tension. In the case of cohesion, with a liquid like mercury, where the molecules of the liquid are more strongly attracted to each other than to the walls of the tube and forming a bubble at the top of the mercury.

What is cohesion? Cohesion is the attractive force holding the atoms or molecules of a single substance together. Cohesion is a word derived from the Latin word cohaesus meaning pressed together. Example, a drop of water does not lay flat on a surface because of cohesion.

What is condensation? Condensation is the changing of a substance from a gaseous state to a liquid state. This change usually occurs when warm air meets a cold surface or mixes with cold air. Condensation is the atmosphere is result of dew, fog, and clouds. Condensation can be seen every day, like the drops of water that form on the outside of a soda can.

What is a galvanometer? A galvanometer is an instrument that is used for detecting, measuring, and determining the direction of a small electrical current.

What is dispersion? Dispersion is an optical phenomenon where a beam of white light is separated into its component colors when it passes through a triangular glass prism. All the colors of white light have a different index of refraction; meaning the colors bend at an angle when passing from one medium to another.

What is an electric arc? An electric arc is an area of intense light and heat that is produced by the passage of electricity across a small gap between two conductors. Electricity is able to jump across the small gap because it ionizes the surrounding air, which then becomes a conductor. Electric arcs are used in welding to fuse metals together and in spotlights or floodlights.

What is the difference between concave and convex? Concave is like a hallow lens that curves inward, convex is like a hallow lens that curves outward.

How does a photoelectric cell work? A photoelectric cell is a device that either produces a current or allows current to flow when light shines on it. Devices like this are used for turning on lights, opening doors, burglar alarms, and many other uses.

What is anelectric? Anelectric means something that cannot be electrified by friction.

What is stainless steel? Stainless steel is a corrosion resistant steel alloy containing more than 10% chromium.

What is chromium? Chromium is found in nature is on the periodic table of elements. Chromium metal is extremely resistant to chemical attack, which accounts for its use in stainless steel and chrome plating. (Harte, Holdren, Schneider, Shirley)

What is flash point? Flash point is the temperature at which the vapor given off from a volatile liquid will ignite spontaneously in air, in the presence of a flame.

How does atmospheric pressure affect the boiling point of a liquid? The boiling point is the temperature at which the pressure of a liquid is equal to the pressure of the atmosphere acting upon the surface of the liquid. The boiling point of water at sea level is 212° F. If the pressure were increased on the surface of a liquid, the boiling point would also increase. If the pressure were decreased on the surface of a liquid, the boiling point would also decrease.

Does natural gas have a natural smell? Natural gas does not have a natural smell. The odor of natural gas is an artificial smell that is injected into the natural gas to give it a unique smell. This unique smell does not smell like any other substance, this makes it easy to find a gas leak.

What are three criteria that the odorant of natural gas must have? The odorant in natural gas must 1) not be absorbed by the soil (underground leaks would be hard to detect) 2) must be harmless to life and 3) must be non-corrosive.

What is natural gas made up of? Natural gas is mostly made up of methane (CH_4) 94%, followed by Propane (C_3H_8) 3.5%, Ethane (C_2H_6) .8%, Carbon Dioxide (CO_2) .7%, trace material at .6%, and Nitrogen (N) .4%.

What is in a gallon of gasoline? A gallon of gas represents roughly 100 tons of plant matter, the amount that exists in 40 acres of wheat. Burning that gallon puts 20 pounds of carbon dioxide into the air. The annual consumption of gasoline in the U.S. is about 131 billion gallons of gas, which is equivalent to 25 quadrillion pounds of prehistoric biomass and releases some 2.6 trillion pounds of carbon dioxide. This gets worse when you consider the other fuels, coal, natural gas, and oil that we consume. Since 1751, roughly the start of the Industrial Revolution, humans have burned the amount of fossil fuel that would have come from all the plants on Earth for 13,300 years. (Discover Magazine)

How does a lightning rod work? A lightning rod or lightning arrester is a device that protects electrical equipment from lightning by leading the lightning strike to the ground and not the equipment. The lightning rod was invented by Benjamin Franklin in 1753.

What does passivate mean? Passivate means to treat a metal by forming a protective coating on its surface and to reduce its chemical activity.

What determines where a power plant is built? There are several factors that will determine where a plant is built, like the type of coolant to be used, if pollution will be a problem, the excessive amount of noise when running, the type of fuel available, response from the community, impact on the environment, etc.

What are some advantages and disadvantages of a nuclear power plant? A nuclear power plant can provide inexpensive power for years to come yet it is expensive to build and maintain and the danger of a serious mishap is always present. Another disadvantage is what to do with the nuclear waste the plant produces.

How does a split phase motor work? A split phase motor splits a single phase current into two separate phases. The magnetic field of the two coil pairs increase at different times and decrease at different times causing a rotating magnetic field. These types of motors are usually fractional horsepower motors that are used in refrigerators and washing machines.

What is voltaic? Voltaic means having to do with, or to produce an electric current by a chemical action.

How does a single phase motor work? A single-phase motor depends on the principle of magnetic induction. In this type of motor since a single phase is used to operate the motor, the design of the coils within the motor are special. The fact that current is strongest in one coil pair at a different time than it is in the other coils, the magnetic field seems to rotate around the inside of the stator. This rotating magnetic field induces a voltage within the rotor causing current to flow in the rotor creating another magnetic field. The interaction between the two magnetic fields is what causes the rotor to turn.

How does a three-phase motor work? A three-phase motor is a motor that is energized by three electromotive forces that are 120^0 out of phase from each other.

What is the difference between a synchronous motor and an induction motor? A synchronous motor is an AC motor where the rotor's field is constant. The rotor's field provided by a permanent magnet or by an electromagnet whose power is supplied from an external source. The rotor of a synchronous motor turns at a speed the same as the speed of alternation or rotation of the stator fields. And as with an induction motor, the rotor's field is produced by induction. The alternating or rotating magnetic field of the stator induces a voltage in the rotor conductors. Current flow in the rotor conductors, in turn, produces the rotor field. The rotor of an induction motor will turn at a speed slower than the speed of alternation or rotation of the stator's field.

What is the greatest advantage of using a synchronous motor? The greatest advantage for using a synchronous motor is the lack of slip present making the motor highly efficient.

What are several different ways for measuring energy? There are about as many ways for measuring energy, as there are ways in converting energy. Here is a list of some terms used for measuring energy; watts, volts, voltage, amps, vars, amperes, ohms, mhos, henry, amp-hour, kilowatt hour, psia, psig, psid, psi, Fahrenheit, Celsius, Kelvin, Rakine, velocity, farad, hertz, ft-lbs, lumen, maxwells, scfm, BTU's, calories, coulomb, impedance, lux, flux, in Hg, in H_2O, mph, mps, quad, mega, kilo, giga, Gilbert, decibel, Tesla, weber, joule, emf, curie, cfpm, eer, seer, BTU/Kwhr, becquerel, rads, rems, ssu, siemens, oersted and BTU/lb. As you can see, there are quite a few ways for measuring energy, but this list has only scratched the surface, it is only the beginning.

How is static electricity affected by weather? Static electricity can be affected by weather. During the summer it is rarely that you get shocked by static electricity yet when winter rolls around you seem to get shocked by static electricity several times a day. During the summer season, the weather is fairly humid. This humidity allows the static charge to be released before building up a painful charge. Humid air conducts static electricity better than dry air. But during the winter the dryer climate especially inside the house or building will cause the static electricity to build up until it painfully discharges through your fingertips to a metal object.

How do batteries work? (Electrons will travel from the negative terminal of the battery to the positive terminal.) A wet cell battery uses a liquid electrolyte containing positive and negative charged particles. A chemical reaction attracts the positive and negative particles to the two electrodes (rods made of different substances are placed within the electrolyte). When a conductor (wire) is attached between the two electrodes, electrons will flow from the negative terminal to the positive terminal causing current to flow through the conductor (wire). But with a dry cell battery a paste type electrolyte is used instead of a liquid and this paste is made up of chemicals that reacts to a central rod with the result that some of the atoms are attracted to one end of the rod, creating a positive and negative anodes at each end of the battery. When a conductor connects both electrodes together, a chemical reaction occurs in the paste electrolyte. Here the electrons travel from the negative electrode through the conductor (wire) to the positive electrode, creating a flow of current through the conductor.

How does a rechargeable battery remember? The term memory refers to the ability of (NiCd) nickel cadmium batteries to remember the point at which they were previously charged. Within the cells of a battery, the elements nickel and cadmium exist as crystals. These crystals dissolve in a hydroxide solution as the battery discharges and then reform into crystals when the battery is recharged. If the battery is not fully discharged, some of the crystals don't dissolve and will continue to grow, eventually reducing the surface area and therefore reducing the voltage and performance of the battery. If this process continues long enough, the sharp edges of the crystals can grow through the metal plate that separates the negative and positive ends of the battery, causing an electrical short. To prevent most crystals from becoming too large to cause damage, drain the NiCd batteries completely down every one to two months.

What is a nuclear battery? A nuclear battery or radioisotope thermal generator (RTG)is a battery that produces an electrical current from the energy of radioactive decay either directly by collecting beta particles or by using a simple principle in which a semi-conductor type material can establish current when one side is heated while the other side is being cooled. This effect is known as the thermoelectric effect. The benefit of using these batteries is simple... they last a long time. NASA has used these batteries in their space programs knowing that these types of batteries will last many years. Courtesy of AtomicInsights.com

What does TEFC mean for a motor? TEFC means "Totally Enclosed Fan Cooled", meaning this type of motor is neither watertight or air tight, allowing no free air circulation through the motors interior. The heat from the motor is removed by a finned motor frame in which a shaft driven fan, blows outside are around the motor frame. There are additions that can make these types of motors explosion proof.

What types of metals increase the strength of steel? This is a very small list of different elements that could be mixed with steel to change its characteristics.

> **Molybdenum** – increases strength, elastic limit, resistance to wear, impact qualities and hardness.
> **Chromium** – increase strength, hardness, toughness, and resistance.
> **Nickel** – increases strength and corrosion resistance.
> **Tungsten** – is similar to molybdenum.
> **Vanadium** – increase strength, toughness, and hardness.
> **Copper** – in small amounts improves corrosion resistance.

> (A steel beam, no matter how strong it is will bend under its own weight if enough heat is applied to it. Also, steel will corrode faster in the presence of oxygen as the temperature rises.)

How does an explosion proof switch work? An explosion proof switch is a switch that is similar to a light switch yet is sealed tightly to contain any sparks within itself to prevent any type of combustible gases to ignite.

What are four types of fire hazards?

1) **Pyrophoric** – a material that will ignite spontaneously in air at or below 130^0F, like Diborane.
2) **Flammable chemicals** – chemicals that will ignite at temperatures below 100^0F, like Toluene.
3) **Combustible chemicals** – chemicals that will ignite at 100^0F or higher, like Cyclohexanone.
4) **Oxidizers** – chemicals that cause or support fire in other materials, like hydrogen peroxide.

What is a flame arrester? A flame arrester is a primary component in a safety can. The flame arrester sits inside the spout or fill opening, preventing fire flashback to the can contents. The flame arrester is constructed of wire mesh or perforated metal designed with a large surface area to permit heat dissipation and a free flow of liquid during filling and pouring. As heated air passes through the openings of the flame arrester, the heat transfer to the arrester's metal surface by heat convection, transferring uniformly throughout the arrester by heat conduction. Thus, hot air entering the can is cooled. Plus, by absorbing and dissipating the heat, the flame arrester keeps the vapor temperature below its flash point.

What is an orifice? An orifice is a device used to create a differential pressure by restricting the flow of a liquid.

What are three things that determine a flow through an orifice?

1) Rate of fluid flow 2) Size of the orifice 3) Shape of the orifice

How harmful is antifreeze? Antifreeze is a solution, which lowers the freezing point and raises the boiling point of water. Most commonly contain ethylene or propylene glycol, or methanol with a few percent of corrosion inhibitors such as phosphates. These chemicals are very poisonous to humans and animals.

What is backlash? Backlash is the motion that is lost between two elements of a mechanism; the amount the first element has to move before communicating its motion to the second element.

What is fulcrum? A fulcrum is the place where a lever is rested or supported. Example, a see saw, the fulcrum is between the effort and the load.

What is the efficiency of a typical power plant? The efficiency of a typical power plant can range from 30% to 40% efficiency depending on several factors.

What are three ways to create pressure? Three ways to create pressure is mechanical, hydraulic, and heat.

What factors influence the rate of heat transfer by conduction? The greater the temperature difference the faster the heat flows. The type of material is a factor that determines the rate of heat transfer. The thickness of the material. And, finally the surface area exposed will affect the heat transfer rate.

How does temperature affect molecules? Molecules are affected by temperature in that, the higher the temperature of a fluid or other material the molecules move faster and faster causing the fluid or material to expand and the cooler the temperature of a fluid or material the slower the molecules move causing the fluid or material to contract. At -273.15°C an object is said to be at absolute zero, where an object will not emit radiation (contains no more heat) and all molecular motion stops.

What factors can affect the heat rate of a power plant? There are three major factors that can affect the heat rate of a power plant; the temperature of superheated steam entering the turbine, the amount of vacuum in the condenser, and the temperature of the feedwater. A decrease in superheated steam temperature causes an increase in heat rate and an increase in superheated steam temperature will cause a decrease in heat rate. If the condenser vacuum decreases, the turbine will operate less efficiently because less energy is removed as the steam passes through the turbine. And the closer the feedwater temperature is to its boiling point, the less energy is required to turn the water back into steam. The higher the feedwater temperature, the lower the heat rate and the greater the efficiency of the power plant.

What is heat transfer coefficient? Heat transfer coefficient is an expression of the resistance of the material that makes up a heat transfer surface multiplied by the thickness of the material.

What are hydrocarbons? Hydrocarbons are a large group of organic compounds that contain only hydrogen and carbon, obtained from coal tar and petroleum. Gasoline consists of a mixture of hydrocarbons.

What are emulsions? Emulsions are a mixture of liquids that will not dissolve in each other.

What are two common types of hand held fire extinguishers? Stored pressure extinguishers and pressure cartridge extinguisher to use a stored pressure extinguisher a pin must be pulled to release the handle. When using a pressure cartridge extinguisher the cartridge must be punctured to pressurize the cylinder.

When would foam be used to fight a fire? Foam would be used in a fire that contained chemicals such as gasoline, oil, diesel fuel, jet fuel, etc. The reason foam is used is because using water to put out a chemical fire causes the burning chemical to continue burning but is now able to flow with the water causing another fire hazard.

Why should you always back away from a fire after it has been extinguished? Backing away from an extinguished fire will keep you aware of the conditions of the burned material. If the burned material should flare up again you will be in a position to re-extinguish the fire and keep your backside from being burned.

What are the four different types of fires?
> A – a fire that involves wood, paper, rags, or similar material.
> B – a fire that involves chemicals.
> C – a fire that involves electricity.
> D – a fire that involves flammable metals such as magnesium, lithium, sodium, potassium, titanium, zirconium and phosphorous.
> K- a fire involving commercial kitchen equipment.

How come we don't use water to put out a metal fire (class D)? Metal fires burn extremely hot which causes water when sprayed onto a metal fire to break down into its basic components (H_2O), hydrogen and oxygen. This combination of elements now separate and added to the fire makes the fire more explosive. (Like adding gasoline to a fire.) Dirt or sand is usually used by the fire department or if possible they will let the fire burn itself out.

What guidelines should be followed when using a fire extinguisher? Being sure you have the right type of fire extinguisher. Make sure the fire extinguisher is fully charged. Pull pin from handle, squeeze handle to provide a short burst from extinguisher to clear nozzle of any obstructions. Approach the fire with the wind at your back, to begin to extinguish the fire use a sweeping motion back and forth aimed at the base of the fire. If a fire is too big to fight find someone to help you.

What is renewable energy? Renewable energy is energy that will never run out like solar energy from the sun, wind for the wind generators, water or tidal energy for the hydroelectric plants, and geothermal energy from the earth's crust.

What are pylons? Pylons are tall towers that hold up the electrical transmission lines to carry the electricity across the land.

What is a three-phase fault? A three-phase fault is an extremely rare type of fault that involves all three phases. This type of fault is the most severe as far as the levels of fault current are concerned.

When did electricity revolutionize the world? The electrical revolution began in the late 1700's and built momentum in the 1800's as more was learned through experimentation. Even as of today, the world's scientists are still discovering the answers to many unanswered questions. The word electricity originated from the Greek word elektron meaning "amber". Amber was one of the first substances to be triboelectified.

What are synch-check relays? A synch-check relay is a protective relay that will not allow a circuit breaker to be closed unless the frequency difference, voltage magnitude difference, and voltage angle across the open circuit breaker are within acceptable limits.

What is insulator flashover? Insulator flashover is when an electrical discharge occurs over the surface of an insulator.

What are arcing horns? Arcing horns are a modification made to a disconnect switch to increase the switches current interrupting capability. It is a metal projection that is placed at the upper and lower ends of a suspension type or other type of insulator, in order to deflect an arc away from the insulators surface.

What is the flashover test? The flashover test is a test that is applied to an electrical apparatus to determine the amount of voltage that occurs between any two parts or between a part and a ground.

What is a fault? A fault is an open, grounded, short, or break in an electrical circuit.

What determines the height of a transmission tower? The height of a transmission tower is determined by the amount of power is being carried through the power line. The higher the power rating, the higher the tower needs to be. The amount of power in the power line will also determine the space needed between each phase.

What are shield wires? A shield wire is the wire positioned at the top of the electrical transmission tower to protect the transmission line from lightning strikes by allowing the lightning to go to ground safely.

What is the difference between a phase to ground fault and a phase-to-phase fault? A phase to ground fault or line to ground fault is known as a high-impedance fault. High impedance faults can hardly be detected by relays since the fault current may be very low. Fault magnitudes can range from barely being noticeable to a magnitude of a three-phase fault. These faults occur between a phase and a ground. A phase-to-phase fault or line-to-line fault can occur between two phases from wind blowing the phases together or a kite has gotten caught between the phases. The power imbalance has a detrimental impact on generators. Fault currents are higher than a phase to ground fault. If a ground is involved with a phase-to-phase fault, the fault is said to be a double phase to ground fault.

What determines the design of a transmission tower? The design of a transmission tower is determined by many factors like, the voltage to be carried, and the size of the conductor, minimum clearances, and climate conditions.

What type of material is used to construct transmission lines? In the early days transmission line conductors where mostly made of copper. Today's transmission lines are made of aluminum because it is lighter and less expensive as compared to copper. Since the strength of aluminum is not great, the aluminum conductors are reinforced with steel cable (ACSR).

What is a synchronous machine? A synchronous machine is an AC machine that must rotate at the same speed as the system's rotating magnetic field. The generator is a synchronous machine.

What is asynchronous machine? An asynchronous machine is an induction machine. If the rotor of an induction machine rotates faster than synchronous speed, the machine is said to be an induction generator. But if the induction machine's rotor rotates slower than synchronous speed, the machine is said to be an induction motor.

What is a synchronous condenser? A synchronous condenser is a synchronous machine that operates like a synchronous motor. The MW to turn the machine's shaft is drawn from the power system. The full capability of the machines excitation system is then available for voltage control purposes. On a hydroelectric generator, it can operate in synchronous condenser mode. It is when the unit's water turbine is typically de-watered and the unit's rotor would turn as if it was motor.

What are two main reasons for using AC power? The first reason is that a generator naturally produces an alternating current output and secondly, using AC power minimizes the loss of power flowing through high power transmission lines do to heat through the use of transformers, with DC power this is not possible. Also, large AC motors and generators are easier to build and cheaper and easier to maintain than DC equipment.

How much power can a transmission line carry? Transmission lines can carry voltages over 1,000 KV. The higher transmission line voltage is, the lower the current is through the line and the lower the power losses through the line. Several utilities are now using high voltage Direct Current (DC) transmission lines.

What would cause a transmission line to sag? A sagging transmission line is caused when the conductors in the transmission line are overheated due to carrying excessive amounts of current than it was designed for. Normally transmission lines sag, but excessive sagging is dangerous.

How are the electrical systems of the United States joined together? Many years ago most electrical systems worked independently, so when a problem occurred (loss of power) only that system was affect and had to cope with the problem by itself. In the last several decades, electrical systems have joined together (interconnected) to increase reliability and reduce the loss of total generation. The NERC (North American Electric Reliability Council formed in 1968 is a voluntary group of people who either generate, market, or transmit electricity. This counsel provides guidelines to provide reliable operation to the electrical system in North America.

What are the four major interconnections in North America? The four major interconnections in North America combine the United States, Canada, and a small part of Mexico. The first interconnection is the eastern interconnection, which is the largest of the four and can provide 500,000 MW at peak load. The second interconnection is the Western interconnection that is the second largest interconnection and can provide over 100,000 MW at peak load. The third forms the majority of Texas and does not cross-state lines, which is called the ERCOT interconnection and can provide around 50,000 MW at peak load. And the smallest of the four interconnections is the Hydro-Quebec interconnection and can provide approximately 30,000 MW at peak load.

How can two interconnected systems receive or send power from one another? In North America, each interconnection maintains their own version of a 60Hz frequency, which means that each of the four interconnections are not synchronized with the other interconnections. For example, the Western interconnection frequency may be at 60.02Hz, while the eastern interconnection frequency is at 59.96Hz. Only DC transmission lines connect the major interconnections together because DC current operates independent of frequency (there is no frequency with DC current).

What do the IEEE numbers represent? The IEEE numbers represent a protective device numbering system that is used to identify specific protective devices. There are approximately 100 numbers used to represent protective devices. Here is a list of those numbers: There are 7 numbers missing, 5 and 95 through 100.

1. Master Element
2. Time-Delay Starting or Closing Relay
3. Checking or Interlocking Relay
4. Master Contactor
6. Stopping Device
7. Anode Circuit Breaker
8. Control Power Disconnecting Device
9. Reversing Device
10. Unit Sequence Switch
11. Reserved for Future Application
12. Over-speed Device
13. Synchronous-speed Device
14. Under-speed device
15. Speed or frequency latching device
16. Reserved for future application
17. Shunting or discharge switch
18. Accelerating or decelerating device
19. Starting to running transition contactor
20. Valve (solenoid)
21. Distance relay
22. Equalizer circuit breaker
23. Temperature control device
24. Reserved for future application
25. Synchronizing or synchronism check device
26. Apparatus thermal device
27. Under-voltage relay
28. Flame detector
29. Isolating Contactor
30. Annunciator relay
31. Separate excitation device
32. Directional power relay
33. Position switch (limit)
34. Master Sequence device
35. Brush operating or slip-ring short relay
36. Polarity or polarizing voltage device
37. Undercurrent or under-power relay
38. Bearing protective device
39. Mechanical condition monitor
40. Field relay
41. Field circuit breaker
42. Running circuit breaker
43. Manual transfer or selector device
44. Unit sequence starting relay
45. Atmosphere condition monitor
46. Reverse-phase or phase balancing current relay
47. Phase sequence voltage relay
48. Incomplete sequence relay
49. Machine or transformer thermal relay
50. Instantaneous over-current or rate of rise relay
51. A/C Time Over-current Relay
52. A/C Circuit Breaker
53. Exciter or DC Generator Relay
54. Reserved For Future Application
55. Power Factor Relay
56. Field Application Relay
57. Short-Circuiting or Ground device
58. Rectification Failure Relay
59. Over-voltage Relay
60. Voltage or Current Balance Relay
61. Reserved for Future Application
62. Time-delay stopping or opening relay
63. Liquid or gas pressure or vacuum relay
64. Ground protective relay
65. Governor
66. Notching of jogging device
67. A/C directional over-current relay
68. Blocking relay
69. Permissive control device
70. Rheostat
71. Liquid or gas level relay
72. D/C circuit breaker
73. Load resistor contactor
74. Alarm relay
75. Position changing mechanism
76. D/C over-current relay
77. Pulse transmitter
78. Phase angle measuring or out of step protective relay
79. A/C reclosing relay
80. Liquid or gas flow relay
81. Frequency relay
82. D/C closing relay
83. Automatic selective control or transfer circuiting device
84. Operating mechanism
85. Carrier or pilot-wire receiver relay
86. Locking out relay
87. Differential protective device
88. Auxiliary motor or motor generator
89. Line switch
90. Regulating device
91. Voltage directional relay
92. Voltage and power directional relay
93. Field changing contactor
94. Tripping or trip-free relay

What is a DC tie? A DC tie is a facility that connects two interconnections together by converting AC to DC and matching the AC frequency to the interconnection receiving the power from the other interconnection. This facility includes an AC to DC converter, a short bus or length of transmission line, and a second converter to convert DC back into AC.

What are the benefits of the interconnected system? One reason is the advantage to interconnection reduces the total generation capacity required; reduced power production costs and provides enhanced reliability. The second reason is the NERC was established to facilitate communication and coordinate operating policies among the four interconnected systems.

What is metallurgy? Metallurgy is the science and technology of metals.

How does a circuit breaker know when to trip? A circuit breaker used in a house contains a permanent metal strip that heats up and bends when current passes through it. Should the circuit become overloaded, the metal strip inside the breaker will bend enough to "trip" the circuit breaker preventing the flow of electricity to the circuit.

What are three ways that a circuit breaker can be closed? A circuit breaker can be closed by air pressure, springs, or a solenoid.

What is the phonetic alphabet? The phonetic alphabet is an alphabet that uses words to represent a letter in the alphabet to help clarify the correct letter being used. **A** - Alpha, **B**- Bravo, **C**- Charlie, **D** - Delta, **E** - Echo, **F** - Foxtrot, **G** - Golf, **H** - Hotel, **I** - India, **J** - Juliet, **K** - Kilo, **L** - Lima, **M** - Mike, **N** - November, **O** - Oscar, **P** - Papa, **Q** - Quebec, **R** - Romeo, **S** - Sierra, **T**- Tango, **U** - Uniform, **V** - Victor, **W** - Whiskey, **X** - X-ray, **Y** - Yankee, and **Z** - Zulu.

What are two advantages that circuit breakers have over fuses? A circuit breaker can be reset every time it is tripped, a fuse has to be replaced and a circuit breaker provides protection against arching, where as a fuse will generate an arch every time it is replaced.

What are four major types of circuit breakers available that are distinguished by their insulating medium used? Oil, Sulfur hexaflouride (SF_6) gas, air and a vacuum are the four insulating mediums used in circuit breakers.

How can an arc in a circuit breaker be beneficial? When the circuit breaker contacts begin to separate, an arc if formed between them. As the circuit breaker opens, it causes an abrupt change in current. The abrupt change in current leads to high voltage spikes. The arc is beneficial by slowing down the change in current, thereby reducing the magnitude of the resultant voltage spike.

What are the three goals of a circuit breaker? 1) establish an arc 2) dissipate arc energy so that the circuit breaker is not damaged and 3) extinguish the arc before the circuit breaker contacts reach their full open position.

How can a glass fuse help troubleshoot wiring problems? In many of the older house still use the glass type fuse as a circuit breaker. If the fuse should blow, depending on what the sight glass looks like would determine if the circuit was overloaded or a short circuit is present. If the metal strip inside the fuse is cleanly melted, then the circuit was overloaded. If the window of the fuse is discolored, then there is a short circuit present within the electrical system.

What is the difference between an endothermic and exothermic reaction? An endothermic reaction is where heat energy is absorbed in and exothermic reaction is where heat energy is given off (released).

What are the 10 signals? The <u>10 signals</u> is a numeric code used to represent a message in short hand. There are many different codes in existence as there are agencies that use this type of communication.

10-1 Receiving poorly	10-20 Unit location
10-2 Receiving well	10-21 Call _____ by telephone
10-3 Stop transmitting	10-22 Cancel last message
10-4 Received message (OK)	10-23 Arrived at scene
10-5 Relay message to_____.	10-24 Assignment completed
10-6 Busy	10-25 Meet _____.
10-7 Out of service (off air)	10-26 Estimated time of arrival is
10-8 In service (on air)	10-27 Request information on license
10-9 Repeat last message	10-28 Check registration
10-10 Negative ("no")	10-29 Check records
10-11 In service	10-30 Use caution
10-12 Stand by	10-31 Pick up
10-13 Report_____conditions	10-32 Units requested
10-14 Information	10-33 Emergency! ___needs help
10-15 Message delivered	10-34 Correct time
10-16 Reply to message	10-35 Request a Supervisor
10-17 Enroute	10-36 Send ambulance
10-18 Urgent	10-37 Vehicle accident
10-19 Contact_____.	10-38 Assist the person at

What is corona? Corona is an undesirable condition that occurs on electrical energized equipment when the surface potential (voltage) is so large that the dielectric strength of the surrounding air breaks down and becomes ionized. Corona is undesirable because of energy losses that it causes. The symptoms of corona are a visible ring of light and a hissing sound. The cure for corona is to use corona rings; they reduce the gradient of the electric field and reduce the chances of corona occurring.

How much energy does a lightning bolt have? A bolt of lightning can produce anywhere from 10 to 100 million volts of electricity per strike with an average of 30,000 amperes.

What is ozone? Ozone can be represented in two forms, one good and the other very bad. Ozone (bad form) is a major ingredient in smog and aids in the air pollution around the World.

The bad ozone lies in the lower section of our atmosphere, the one we breathe. The good ozone high up in the atmosphere helps us on Earth by absorbing the harmful ultraviolet radiation coming from the Sun. This protection reduces the amount of UV rays into the Earth, which can cause skin cancer. The bad ozone when exposed can cause chest pain, coughing, wheezing, lung and nasal congestion, labored breathing and eye and nose irritation. (Harte, Holdren, Schneider, Shirley)Ozone is made up of 3 atoms instead of normally two atoms; this makes ozone extremely toxic. An amount of less than one part per million is poisonous to humans.

Is ozone good or bad? Ozone is a beneficial component of the upper atmosphere (20-30 miles high) and is formed naturally in the atmosphere by a photochemical reaction. Yet, ozone is a major air pollutant in the lower atmosphere. Ozone is also used for oxidizing, bleaching, deodorizing, and disinfecting.

What are some things that can be done to reduce ozone pollution? Here are some things that can be done to help reduce ozone pollution. Carpool, ride a bus, or if possible ride a bike. Drive your most fuel efficient car available. Refuel only during the evenings and don't top of the tank. Try to limit the usage of volatile liquids (varnish, paints, degreasers). Use gas powered lawn equipment during the evenings. Avoid rapid acceleration. Keep your vehicles tuned-up. Avoid excessive idling (drive through, traffic). Walk or ride a bike for short trips when possible. Don't burn trash or brush on days where the ozone could be higher than normal.

What is the greenhouse effect? The greenhouse effect is the surface warming of the Earth that results when the atmosphere traps in the Sun's heat. Greenhouse gases like water vapor, carbon dioxide, methane and other gases absorb the radiation from the Earth and hold this heat in the atmosphere instead of letting it go into space. Without these gases the Earth would freeze and life could not exist. The problem with the greenhouse effect is that there is to many greenhouse gases. All of these gases are causing the Earth to slowly warm-up above normal average temperature of the Earth.

By recycling one aluminum can, how much energy can be saved? For every aluminum can recycled, a 1/2 gallon of gasoline could be saved.

What is inertial confinement fusion? Inertial confinement fusion is an experimental process of creating energy by focusing a laser beam onto a pellet containing deuterium and tritium producing controlled thermonuclear energy.

How is pressure affected by area? A force that is acting over a small area will exert a greater pressure than the same force acting over a large area. Pressure is defined as the force per unit area; it is calculated by dividing the amount of force by the area over which it is acting.

How many different types of energy are there? There are six basic types of energy, mechanical energy, chemical energy, heat energy, nuclear energy, electrical energy, and radiant energy.

Mechanical energy is potential or kinetic energy such as turbines or dynamo.

Chemical energy is energy stored in an atom or molecule and is released by a chemical reaction such as a battery or gasoline.

Heat energy is the energy that is produced by the movement of a substance's molecules; the faster the molecules move, the hotter the substance will become, such as water in a kettle.

Nuclear energy is the energy that is stored in the nucleus of an atom, and when released creates large amounts of energy such as a nuclear explosion or a nuclear power plant.

Electrical energy is the energy produced by electrons moving from atom to atom such as a motor, light bulb.

Radiant energy is energy that contains rays, waves, or particles, like electromagnetic radiation such as infrared radiation, light, ultraviolet radiation, X-rays, gamma rays, and cosmic rays.

How many types of force are there? Force can change four factors of an object; size, shape, direction, or speed. There are six types of force that will affect at least one of the four factors will affect an object, they are:

- **Pushing** - in which changes in direction and/or speed is affected.
- **Pulling** - in which changes in direction and/or speed is again effected.
- **Compressing** - in which the object's size and shape is affected.
- **Twisting** - in which an object is effected the same way as pushing or pulling but in a spiral motion.
- **Bending** - in which the shape and size of an object is affected.
- **Stretching** - in which the size and shape of the object is effected by a form of pulling.

What nation produces the most greenhouse gases? The most greenhouse gases are produced by a single nation the United States at approximately 20%.

What is a heat pump? A heat pump is a machine operating on a reversed heat engine cycle to produce a heating effect. Energy from a low temperature source; earth, lake, or river is absorbed by the working fluid, which is mechanically compressed, resulting in a temperature increase. Then the high temperature energy is transferred into a heat exchanger.

Whatever happened to the Edison Electric Light Company? The Edison Electric Light Company that was created in 1889 has become one of the largest electric companies in the World, through mergers it is now known as the General Electric Company.

What is an over-compounded motor? An over-compounded motor is a DC motor that has a series of windings that are specially designed to allow the speed of the motor to increase as the load increases.

What is tribo-electrification? Tribo-electrification is a phenomenon of the separation of electrical charges through surface friction. Example, rubbing a glass rod with silk, the silk becomes negatively charged while the glass rod becomes positively charged. Amber was the first known substance to be tribo-electrified.

What is an AC reluctance motor? An AC reluctance motor is simply a synchronous motor with the magnetized rotor replaced by an un-magnetized piece of steel so shaped that it has a number of preferred positions into which it will settle for any given primary field configuration. A preferred position is one in which the resistance of the magnetic circuit (which is called reluctance) is a minimum hence the name of this type of motor.

What is cathodic protection? Cathodic protection is the protection of a metal structure against electrolytic corrosion by making it the electron receiver (cathode) in an electrolytic cell by means of an impressed e.m.f. or by coupling the metal with a more electronegative metal.

What is flash point? Flash point is the temperature at which a material will give off vapor and will ignite when exposed to an ignition source.

What is a fuse? A fuse is an electrical protective device that is used to prevent excessive current from flowing through a conductor causing damage to equipment and possibly a fire because of wire overheating. A fuse, consist of a piece of fusible material which is in the circuit allowing current to flow through it. Should the current exceed the fuses maximum amp rating, the fusible material will melt and break the circuit causing the flow of current to stop.

How are fuses rated? Fuses are rated by the maximum current they will carry continuously and/or the minimum current at which it can be relied upon to break or open the circuit.

What is the difference between free vibration and forced vibration? Free vibration is when a force is applied to an object causing vibration; the vibration is free to continue until the object comes to rest. Ex. A hammer striking a steel pipe. With forced vibration it is the continuous vibration that occurs when a force or several forces are applied repeatedly to an object. Ex. A hammer striking a steel pipe continuously.

What are three main factors that determine the amount of current flow passing through a human body? 1)The body's resistance along the current path. 2)The current capacity of the source. And 3)The voltage of the source.

What should be done to aid a person that has been electrocuted? There are several steps that must be taken in order to protect the victim and the rescuer. View surroundings for hazards. If you are not sure what happened to the victim, take every precaution possible. Get someone else to help you, to notify help, and so that someone else is aware of victim's condition. Determine the cause of electrocution and remove power if possible. If unable to remove power, use an insulated object to move victim away (do not touch victim if they're in contact with the conductor. Check the condition of the victim (breathing, pulse, burns, eyes, etc.) If not breathing or no heartbeat, perform CPR or emergency breathing. Watch for any responses from the victim and continue care until medical help arrives.

What are six ways to produce voltage?

1) Magnetism- relative motion between a conductor and a magnetic field.
2) Chemical reaction- such as in a battery
3) Light- photoelectric cells use light to produce voltage
4) Heat- thermocouples use heat to produce voltage.
5) Pressure- pressure applied to certain crystals can produce voltage.
6) Friction- (static electricity) rubbing two materials together can generate voltage.

What is a psychrometer? A psychrometer is an instrument that is used for determining the dry and wet bulb temperature of air.

What is the difference between combustible and flammable liquids? Combustible liquids will ignite at temperatures at or above 100^0F while flammable liquids will ignite at temperatures below 100^0F.

What is piping tax? Piping tax is the compensation for any heat that is lost in a fluid as it flows through a pipe. This could also be found as a pressure loss, because of the internal pipe friction.

What is material yield strength? Material yield strength is the stress that produces 0.2% permanent set in ductile materials. The risk of cracking greatly increases as the stress to the material equals or exceeds the materials yield strength.

What is the brittle to ductile transition temperature? The brittle to ductile transition temperature is the temperature at which metal enters into a ductile state from a brittle state. When metal is brittle it breaks with little bending. When metal is ductile it can bend more without breaking.

What is the difference between wire sizes? Electrical comes in many different sizes depending on the purpose the wire is to be used for. The wire size is based on the maximum amount of electrical load that the wire will carry before becoming overheated and damaged. The term gauge is used to indicate the wire size. The higher the number on the gauge scale the smaller the wire will be and the smaller the amount of electrical current it will carry. A 12 gauge wire (household wiring) can handle up to 20 amps of current at 120 volts without being damaged. A 6 gauge wire can handle up to 60 amps at 240 volts without being damaged. And of course a 22 gauge wire can only handle small amounts of current before becoming damaged.

What is the difference between wire and cable? A cable is considered of two or more wires formed together where as a wire is just a single strand.

What are carcinogens? Carcinogens are chemicals or elements that can cause or have potential for causing cancer.

What is the difference between mutagens and teratogens? Mutagens cause a change in the genetic information in the human reproductive system. Teratogens cause damage to the fetus during pregnancy, producing either birth defects or death to the fetus. Ex. (lead & PCB's)

What is the difference between a primary cell and a secondary cell? A primary cell is a battery that will release energy by a chemical reaction that cannot be recharged. A secondary cell is a battery that will release energy by chemical reaction only after it has been charged (rechargeable battery).

What is a polarized receptacle? A polarized receptacle is similar to a regular outlet. The only difference is that on a polarized receptacle, one of the slots is longer than the other. This design keeps the hot current flowing along the black and/or red wires and the neutral current flowing along the white or gray wires.

What is the wireless power transmission system? The wireless power transmission system is based upon the hertzian waves, an idea that started with Nikola tesla's dream of transmitting electrical power without the use of wire transmission system. There have been several successful experiments where test aircraft where flown by the power that was sent to them through the air from ground control. The idea is to take electricity and convert it into a microwave so that it can be sent to a relay station in space and sent back to some other part of the Earth as a microwave and is then converted back into electricity for consumption. Countries that have little or no electricity could one day be provided with electricity without even building a power plant.

How does recycling paper help our environment? By recycling a ton of paper, 17 trees, 25 barrels of oil, approx. 7,000 gallons of water, and 3 cubic yards of landfill space is saved.

What is the largest human made structure in the United States? The largest human made structure in the United States is the Fresh Kill landfill on Staten Island in New York City. The landfill mass is estimated at 100 million tons of waste with a volume of 250 million cubic feet.

How does a neon light work? A neon light very similar to a fluorescent light where electricity is applied to the ends of a glass tube that is filled with neon gas. As electrons flow through the tube, they collide with the neon atoms. The collision results in the electrons that orbited within the neon atoms are now knocked out of orbit. As the electrons return to their original orbit, extra energy is given off as electromagnetic radiation. It is this radiation that gives off the visible light that you see as a bright red/orange glow. To produce different colors in neon lighting, different gases are used: Helium - yellow, Krypton - light violet, etc. Even a combination of these gases can be used to adjust the color to be seen.

What causes an incandescent light to burn out? An incandescent light will burn out depending on the rate at which the metal filament evaporates, which depends on the operating temperature the light bulb operates at.

What causes a fluorescent light to burn out? A fluorescent light will burn out because of the exhaustion of the oxide coating on the filaments, which cause the tube to fail. This is not including problems with the starting circuit.

Which countries consume the most energy?

1) United States	6) Canada
2) China	7) India
3) Russia	8) United Kingdom
4) Japan	9) France
5) Germany	10) Italy

Courtesy of the EPA

Why is it hard to let go of a wire when being shocked? First we must understand that our muscles only perform one function, to contract when stimulated, that is it. Secondly we must understand that our brain sends signals (electrical impulses) to every part of our body in order to stimulate those muscles. When a person is being shocked, say by grabbing a hot wire, the electricity flowing through the hand causes the muscles in the hand to contract. The electricity is much stronger than the electrical impulses that our brain can generate, so the muscle will respond to the stronger electrical signal being generated during the shock. This is why it is hard to let go of a wire when being shocked. And if you are wet, the shock is even stronger.

What actually happens during the combustion process? There are generally three chemical reactions taking place simultaneously. 1) Carbon from the fuel combines with oxygen to form carbon dioxide, which releases approximately 14,100 BTU's for every pound of carbon burned. 2) Hydrogen combines with oxygen to form water vapor, which releases approximately 61,000 BTU's for every pound of hydrogen burned. 3) Sulfur combines with oxygen to form sulfur dioxide, which releases approximately 4,000 BTU's for every pound of sulfur burned. A total of 79,000 BTU's is released from being one pound of carbon, hydrogen and sulfur.

What is in the air we breathe? There are two elements that are normally abundant in the air we breathe, one is oxygen at 21 to 23% and the other is nitrogen at 77 to 79%. There also some trace gases in the air that make up less than one percent of the total percentage of air.

How can you find out other countries electrical setup when traveling? There is a web site called Voltage Valet www.voltagevalet.com . Here you can find all the information you will need when traveling to other countries. At this site they even sell the products to help you convert the different power source so that your hair dryer doesn't become a flame thrower. This is very important because electrical equipment will not last very long when using inappropriate voltage ratings let along at different frequencies. And if you have any further questions please email them at hybrinet@voltagevalet.com, they have some wonderful people there to help you with your electrical conversion problems. (Hybrinetics, Inc.)

Does the nuclear industry share information with each other? YES, the nuclear industry does share information about their individual site events. When the nuclear industry first started, sharing information was never thought of…until Three Mile Island. TMI opened the eyes of the nuclear industry and determined in the benefit of mankind that information must be shared between the nuclear sites in order not to repeat the lessons learned from other sites.

What is the difference between an elevated zero and a suppressed zero? Elevated zero is when zero is not the lowest number on the scale. Suppressed zero is when zero is not one of the numbers used on the scale.

Elevated Zero				**Suppressed Zero**			

Measured Variable Range

-25   0   100

Measured Variable Range

20   60   100

Range	Lower Range Value	Upper Range Value	Span	Range	Lower Range Value	Upper Range Value	Span
-25 to 100	-25	100	125	20 to 100	20	100	80

What is radiolysis? Radiolysis is the dissociation of molecules by radiation. It is the cleavage of one or several chemical bonds resulting from exposure to high-energy flux. For example water dissociates under alpha radiation into a hydrogen radical and a hydroxide radical, unlike ionization of water which produces a hydrogen ion and a hydroxide ion. The chemistry of concentrated solutions under ionizing radiation is extremely complex. Radiolysis can locally modify redox conditions, and therefore the speciation and the solubility of the compounds.

Of all the radiation-chemical reactions that have been studied, the most important is the decomposition of water. When exposed to radiation, water undergoes a breakdown sequence into hydrogen peroxide, hydrogen radicals and assorted oxygen compounds such as ozone which when converted back into oxygen releases great amounts of energy. Some of these are explosive. This decomposition is produced mainly by the alpha particles, that can be entirely absorbed by very thin layers of water. The current interest in nontraditional methods for the generation of hydrogen has prompted a revisit of radiolytic splitting of water, where the interaction of various types of ionizing radiation (α, β, and γ) with water produces molecular hydrogen. This reevaluation was further prompted by the current availability of large amounts of radiation sources contained in the fuel discharged from nuclear reactors. This spent fuel is usually stored in water pools, awaiting permanent disposal or reprocessing. The yield of hydrogen resulting from the irradiation of water with β and γ radiation is low (G-values = <1 molecule per 100 electronvolts of absorbed energy) but this is largely due to the rapid reassociation of the species arising during the initial radiolysis. If impurities are present or if physical conditions are created that prevent the establishment of a chemical equilibrium, the net production of hydrogen can be greatly enhanced.

Another approach uses radioactive waste as an energy source for regeneration of spent fuel by converting sodium borate to sodium borohydride. By applying the proper combination of controls, stable borohydride compounds may be produced and used as hydrogen fuel storage medium.

If radioactive gas and liquid are released to the environment, what are the legal limits and/or where can they be found? The radioactive gas and liquid waste is processed before it is released into the environment thus minimizing the impact to the environment. This release is **always** below the limits set forth in the Code of Federal Regulations.

What are some sources of liquid radioactive waste? There are several sources and does not include all of them: regenerative heat exchanger drains, safety injection tank systems, primary loop drains, quench tank drain, reactor coolant pump relief valve drains, leak-off from containment valves and reactor coolant pump vapor seal leak-off.

What does the waste processing system look like?

RC Waste Processing System Flow Diagram

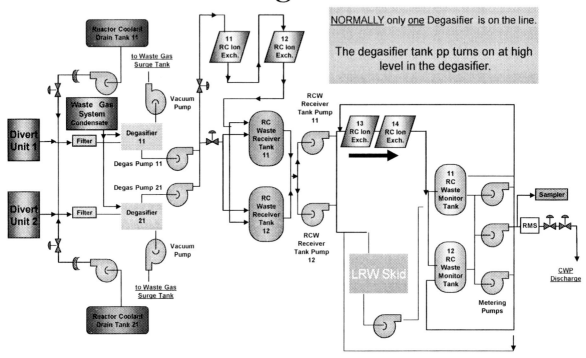

When did nuclear operators begin needing a license to operate nuclear reactors?

The Nuclear Regulatory Commission licenses the individuals who operate the controls of a nuclear power plant. There are two categories of licenses, a reactor operator and a senior reactor operator. A senior reactor operator is a supervisory position overseeing the work of the reactor operators. The license is issued after the individual passes both a written examination and an operating test. The senior reactor operator examination also measures the ability of the individual to direct the activities of licensed operators.

Section 107 of the Atomic Energy Act of 1954, as amended, requires the NRC to determine the qualifications of individuals applying for an operator's license, to prescribe uniform conditions for licensing those individuals, and to issue licenses as appropriate. Additionally, Section 306 of the Nuclear Waste Policy Act of 1982 directed the NRC to promulgate regulations, or other appropriate guidance, for training and qualifying nuclear power plant operators, including requirements governing the administration of requalification examinations and operating tests at nuclear power plant simulators. These statutory requirements are implemented by the NRC's regulations located in Part 55, "Operators' Licenses," in Title 10 of the Code of Federal Regulations (CFR). Detailed NRC policies, procedures, and guidelines that pertain to those regulations are published in NUREG-1021, "Operator Licensing Examination Standards for Power Reactors."

Where does the miscellaneous waste processing system come in? It provides controlled handling, temporary storage and disposal of various liquid wastes from the reactor and sub systems. It processes the liquid waste prior to disposal so that the release is minimized and the concentration of the effluent is below the limits set forth in the Code of Federal Regulations.

Miscellaneous Waste Processing System Simplified Functional Diagram

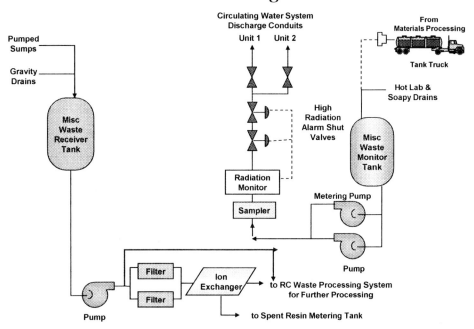

What is vitrification? A process by which uranium processing wastes are mixed with molten glass to form slag for ultimate burial in a geological formation.

What is critical point? On the curve of pressure against temperature for a gas, the point at which the gas and its liquid form coexist...there is no differentiation between the two phases.

What is critical velocity? The speed at which laminar flow changes to turbulent flow.

What is diffusivity? A measure of how fast heat diffuses through a material, equal to the ratio of the material's thermal conductivity to the product of its specific heat and density (expressed in square meters per second).

What is energy? Energy is the capacity to do work.

How are gaseous radioactive waste handled? A waste gas system is used to provide controlled handling and disposal of radioactive gaseous wastes from the reactor. It stores gases removed from liquid waste and other sources to allow radioactive decay of the short lived isotopes before the gases are released from the plant.

Waste Gas System

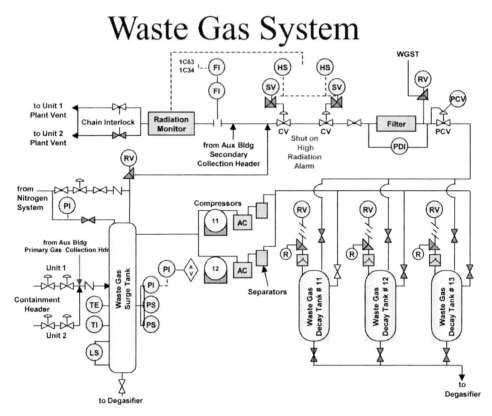

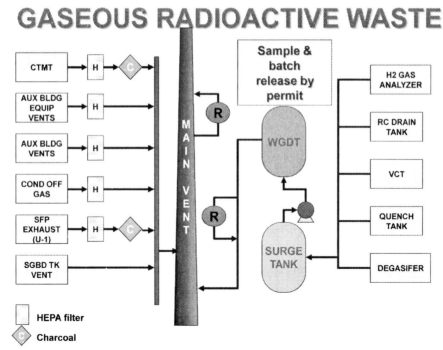

What about solid waste processing?

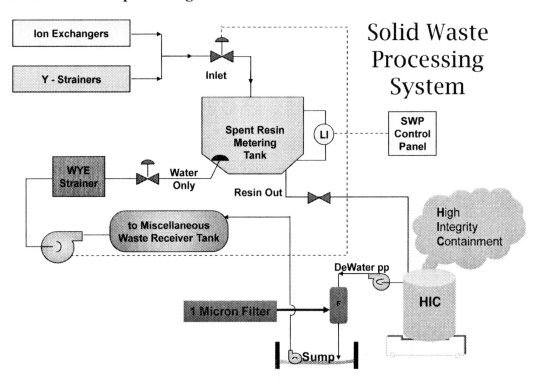

What is the difference between enthalpy and entropy? Enthalpy is the amount of energy in a substance, sensible heat. Entropy is a measure of the unavailability of a system's energy to do work.

What is flux? Flux is the rate of flow of mass, volume, or energy per unit cross-sectional area.

What kind of training does a nuclear operator need? Before the NRC licenses an individual to operate or supervise the controls of a commercial nuclear power reactor, the applicant must complete extensive training and pass rigorous examinations. Once licensed, operators and senior operators must comply with a number of requirements to maintain and renew their licenses. For more details see:

- Process for New Operator Licenses
- Operator License Maintenance
- Operator License Renewal Process
- Operator Licensing Examination Performance Trends
- Operator Licensing Examination Schedules

Process for New Operator Licenses

NRC's four regional offices are responsible for issuing licenses for operators and senior operators of commercial nuclear power plants in accordance with NRC's regulations for "Operators' Licenses" (10 CFR Part 55) and (NRC Form 398). An applicant sends or delivers a completed application (NRC Form 398) to the Regional Administrator having jurisdiction over the plant at which the applicant hopes to work.

A completed application (10 CFR 55.31) describes the applicant's qualifications and requires the facility licensee for which the applicant will work to certify that the applicant has satisfied the facility licensee's training and experience requirements to be a licensed operator or senior operator. Applicants must also undergo a physical examination (10 CFR 55.21) and be certified (NRC Form 396) physically and mentally fit to be an operator.

If the NRC determines that the applicant's qualifications and physical condition are acceptable (10 CFR 55.33), the applicant will be scheduled to take the NRC licensing examination. The examination process begins with a 50-question, multiple-choice written exam covering reactor theory, thermodynamics, and mechanical components; this Generic Fundamentals Examination (GFE) is actually administered early in the applicants' training program and is a prerequisite for taking the site-specific examination. The site-specific examination for reactor operators (10 CFR 55.41) consists of 75 multiple-choice written questions and an NRC-administered operating test (10 CFR 55.45) that includes a plant walk-through and a performance demonstration on the facility licensee's power plant simulator. Individuals who apply for a senior operator's license must pass an additional 25-question written examination (10 CFR 55.43) and a more rigorous operating test. The examinations may be prepared by the facility licensee and approved by the NRC, or the facility licensee may request the NRC to prepare the examinations (10 CFR 55.40). In either case, the examinations are prepared and administered using the guidance in the Operator Licensing Examination Standards for Power Reactors (NUREG-1021).

If the applicant passes both the written examinations and the operating test, the responsible NRC regional office will issue a license (10 CFR 55.51) in a form and containing any conditions it considers appropriate and necessary. The license is only valid to operate the facility for which the applicant applied, and it expires 6 years after the date of issuance or upon termination of employment with the facility licensee or other conditions specified in the regulation (10 CFR 55.55).   Courtesy of the NRC

Who determines operator's license requalification's?

Each license is subject to a number of conditions (10 CFR 55.53) whether or not they are stated in the license. For example:

- Licensed operators and senior operators are required to observe all applicable rules, regulations, and orders of the Commission (10 CFR 55.53(d)).

- Licensed operators and senior operators are required to maintain their proficiency (10 CFR 55.53(e) and (f)) and to complete their facility licensee's requalification training and examination program (10 CFR 55.53(h)) as described in 10 CFR 55.59.

- Licensed operators and senior operators must have a medical examination by a physician every 2 years (10 CFR 55.21). If, during the term of the license, an operator or senior operator develops a permanent physical or mental condition (10 CFR 55.25) that may adversely affect the performance of their duties, the facility licensee must notify the Commission within 30 days of learning of the diagnosis. When appropriate, the NRC may issue a conditional license in accordance with the requirements in 10 CFR 55.33(b).

- Licensed operators and senior operators are prohibited from using, possessing, or selling illegal drugs (10 CFR 55.53(j)) and from performing licensed duties while under the influence of alcohol or any prescription, over-the-counter, or illegal substance that could adversely affect their performance. They are also required to participate in their facility licensee's drug and alcohol testing programs (10 CFR 55.53(k)) established pursuant to 10 CFR 26.

Operator License Renewal Process

If an operator or senior operator applies for renewal at least 30 days before the expiration date of the existing license, the license does not expire until the NRC determines the final disposition of the renewal application (10 CFR 55.55). The renewal process (10 CFR 55.57) requires the applicant to complete NRC Form 398 and submit it to the applicable NRC regional office with the following information: written evidence of the applicant's experience under the existing license, a certification from the facility licensee that the applicant is a safe and competent performer who has satisfactorily completed the requalification program for the facility (10 CFR 55.59), and certification on NRC Form 396 that the applicant's medical condition and general health are satisfactory (10 CFR 55.23). The NRC regional office will renew the license if, on the basis of the application and certifications, it determines that the applicant continues to meet the regulatory requirements (10 CFR 55.57(b)).

How different are nuclear submarine reactors? Naval reactors are of the pressurized water type which differ from commercial reactors producing electricity in that:

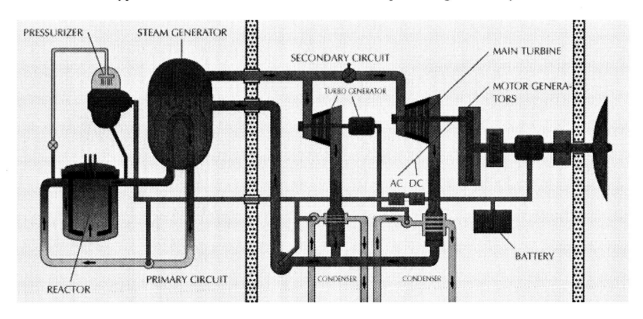

- they have a high power density in a small volume; some run on low-enriched uranium (requiring frequent refueling), others run on highly enriched uranium (>20% U-235, varying from over 96% in U.S. submarines (They do not need to be refueled as often[1] and are quieter in operation from smaller core[2]) to between 30–40% in Russian submarines to lower levels in some others),
- the fuel is not UO_2 (uranium oxide) but a metal-zirconium alloy (circa 15% U with 93% enrichment, or more U with lower enrichment),
- the design enables a compact pressure vessel while maintaining safety.

The long core life is enabled by the relatively high enrichment of the uranium and by incorporating a "burnable poison" in the cores which is progressively depleted as fission products and minor actinides accumulate, leading to reduced fuel efficiency. The two effects cancel each other out. One of the technical difficulties is the creation of a fuel which will tolerate the very large amount of radiation damage. It is known that during use the properties of nuclear fuel change; it is quite possible for fuel to crack and for fission gas bubbles to form.

Long-term integrity of the compact reactor pressure vessel is maintained by providing an internal neutron shield. (This is in contrast to early Soviet civil PWR designs where embrittlement occurs due to neutron bombardment of a very narrow pressure vessel.)

Reactor sizes range up to 55 MW in the larger submarines and surface ships. The French *Rubis*-class submarines have a 48 MW reactor which needs no refueling for 30 years.

The Russian, U.S. and British navies rely on steam turbine propulsion, while the French and Chinese use the turbine to generate electricity for propulsion (turbo-electric propulsion). Most Russian submarines as well as most American aircraft carriers since CVN-65 are powered by two reactors (although *Enterprise* has eight). U.S., British, French and Chinese submarines are powered by one.

Decommissioning nuclear-powered submarines has become a major task for US and Russian navies. After defueling, U.S. practice is to cut the reactor section from the vessel for disposal in shallow land burial as low-level waste (see the Ship-Submarine recycling program). In Russia, the whole vessels, or the sealed reactor sections, typically remain stored afloat, although a new facility near Sayda Bay is beginning to provide storage in a concrete-floored facility on land for some submarines in the Far North.

Russia is well advanced with plans to build a floating nuclear power plant for their far eastern territories. The design has two 35 MW units based on the KLT-40 reactor used in icebreakers (with refueling every four years). Some Russian naval vessels have been used to supply electricity for domestic and industrial use in remote far eastern and Siberian towns.
Courtesy of Wikipedia

How many naval nuclear accidents have there been?

United States

- USS *Thresher* (SSN-593) (sank, 129 killed)
- USS *Scorpion* (SSN-589) (sank, 99 killed)

Both sank for reasons unrelated to their reactor plants and still lie on the Atlantic sea floor.

Russian or Soviet

- Komsomolets K-278 (sank, 42 killed)
- Kursk K-141 (sank, 118 killed)
- K-8 (loss of coolant) (sank after fire, 52 killed)
- K-11 (two refueling criticalities)
- K-19 (two loss of coolant accidents, 27 killed due to one accident)
- K-27 (scuttled)
- K-116 (reactor accident)
- K-122 (reactor accident)
- K-123 (loss of coolant)
- K-140 (power excursion)
- K-159 (radioactive discharge) (sank recently, 9 killed)
- K-192 (loss of coolant)
- K-219 (sank after collision, 6 killed)
- K-222 (uncontrolled startup)
- K-431 (refueling criticality, 10 killed)
- K-320 (uncontrolled startup)
- K-429 (sank twice, 16 killed)
- The Soviet icebreaker Lenin is also rumored to have had a nuclear accident.

What is a LOCA? A LOCA is a Loss Off Coolant Accident, where the main coolant cooling the reactor is lost. This loss of cooling can be large or small.

What is the difference between operating instructions (OI), operating procedure (OP), abnormal operating procedure (AOP) and emergency operating procedure (EOP)? Operating instructions (OI) are system specific directions for equipment manipulation. Operating procedures (OP) are overall plant integrated plan to accomplish power and mode changes. Abnormal operating procedures are pre-planned responses to anticipate operational events. AOPs give directions when to trip the unit. Emergency operating procedures (EOP) are procedures designed to optimize site responses after a post trip.

What is DNBR? DNBR stands for Departure from Nucleate Boiling Ratio. It is the ratio of heat flux to cause departure from nucleate boiling to the actual local heat flux. In a PWR DNBR may also be called critical heat flux (CHF) and for a BWR it could be called dryout. CHF is sometimes called the boiling crisis.

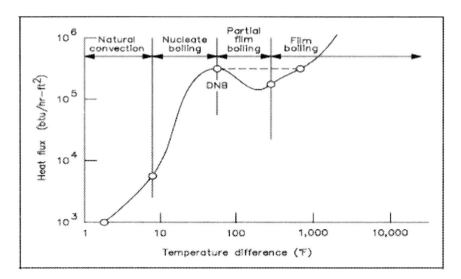

Figure 13   Boiling Heat Transfer Curve

What is PAMS? Post-Accident Monitoring System was developed earlier from previous nuclear events. PAMS allows the control room to monitor different plant systems to determine specific critical variables.

How is liquid & gaseous radioactive waste handled? These systems capture and monitor liquid and gases that are radioactive. These waste materials are monitored and permitted for safe release into the environment. This ensures that the public and the environment are protected from any unsafe releases of radioactive waste. These releases are reported to the NRC and become public record.

What is PV energy? PV energy is the energy possessed by a system by nature of its pressure and volume.

What is meant by internal energy? The energy possessed due to the rotation, vibration and translation of the atoms and molecules of a system.

What is an adiabatic process? A process during which no exchange of heat occurs.

What is an isothermal process? A process during which the system temperature remains constant.

What is an isobaric process? A process during which the system pressure does not change.

What is an isometric process? A process during which the system volume does not change.

What is an isenthalpic process? A process during which the enthalpy of the working fluid does not change.

What is an isentropic process? A process during which the entropy of the working fluid does not change.

What is psychrometrics? A term used to describe the field of engineering concerned with the determination of physical and thermodynamic properties of gas-vapor mixtures. The term derives from the Greek word *psuchron* meaning "cold" and *metron* meaning "means of measurement".   Wikipedia

What is statics? Statics is the study of rigid bodies which are in equilibrium. Bodies in equilibrium are at rest or in uniform motion.

What is the difference between translational equilibrium and rotational equilibrium? Translational equilibrium is when a component is neither accelerating nor decelerating. Rotational equilibrium is when a component has no angular motion.

What is Helmholtz free energy? In thermodynamics, quantity equal to the difference between the internal energy of a system and the product of its temperature and entropy.

Why are diesel generators for emergency power? Diesel generators are used because they are the most reliable, self-sufficient emergency power supplies a utility could have. The maintenance cost is a lot less than if it was a peaking (gas) turbine.

What is Radon? Radon is colorless and the most dense gas known. It is a gas with 20 known isotopes, all of which are radioactive. It is produced as one of the by-products of the radioactive decay of uranium and thorium, and it is the heaviest known gas, being about eight times heavier than air. It is also chemically unreactive.
Radon was discovered by a German physicist Friedrich Ernst Dorn in 1900 while studying the decay products of radium. He named it radium emanation because the gas seemed to come from the radium. William Ramsay and R.W. Whytlaw-Gray, two early investigators of the chemical properties of radon, later changed its name to niton, from the Latin word nitens, which means "shining". Since 1923, however, the element has been called radon.
Radon-222 is the longest lived isotope of radon, with a half-life of 3.82 days, and it is the isotope most generally available and studied. Radon-222 is found in substantial concentrations as a gas in the soil because trace amounts of uranium are present throughout the Earth's crust. The gas diffuses through the soil and into the air. The amount of radon in the air varies from one region to another, but it is considered a potential hazard in many homes in certain areas. The gas can diffuse into the home through basement floors and walls because pressure inside the house is always lower than the outside. Without a constant interchange of fresh air from the outside for air within the

house, radon can build up to dangerous concentrations. This situation is particularly critical during the winter months when windows are likely to be kept shut. Inexpensive detectors are available to measure the amount of radon present in a home.

As early as the 16th century, it was known that uranium miners in Bohemia often died prematurely from diseases of the lung. We now know that the miners suffered from lung cancer caused by radon. As it decays, radon-222 emits alpha particles, which are essentially the nuclei of helium atoms, and simultaneously initiates a decay process that eventually produces lead-210. When radon-222 is inhaled, some of it will decay in the lungs before it can be exhaled. Exhaling removes much of the radon-222, but the decay product, lead-210, is also radioactive and settles in the lungs. Its half-life of 20.4 years is much longer than that of radon-222, and it is not eliminated during breathing. It is this lead-210 that exposes the lungs to radiation for long periods and can produce lung cancer. Health officials in the U.S. have estimated that approximately 10% of all lung cancers are caused by radon.

Smoking a cigarette poses a radiation risk as well as a chemical one. While it is growing, tobacco is subject to contamination by radon from the soil, and the phosphate fertilizers used by planters are rich in uranium. As a result, the broad tobacco leaves become dusted with trace amounts of lead-210, and when this leaf is burned, the inhaled smoke subjects the smoker to levels of radiation 1,000 times higher than those encountered by a worker in a nuclear power plant.

Despite the advances made in radiation therapy through the use of particle accelerators and isotopes such as cobalt-60, radon is still used in many hospitals for cancer therapy. It is usually pumped from a radium source and sealed into tiny glass vials called "seeds" which are implanted in patients at the sites of tumors. Courtesy of Stwertka

What is an electron volt? In physics, the **electron volt** (symbol **eV**; also written **electronvolt**) is a unit of energy equal to approximately 1.602×10^{-19} J. By definition, it is equal to the amount of kinetic energy gained by a single unbound electron when it accelerates through an electric potential difference of one volt. Thus it is 1 volt (1 joule per coulomb) multiplied by the electron charge (1 e, or $1.60217653(14) \times 10^{-19}$ C). Therefore, one electron volt is equal to $1.60217653(14) \times 10^{-19}$ J.[3]

The electron volt is not an SI unit and its value must be obtained experimentally.[4] It is a common unit of energy within physics, widely used in solid state, atomic, nuclear, and particle physics. It is commonly used with the SI prefixes milli-, kilo-, mega-, giga-, tera-, or peta- (meV, keV, MeV, GeV, TeV and PeV respectively). Thus meV stands for milli-electronVolt.

In chemistry, it is often useful to have the molar equivalent, that is the kinetic energy that would be gained by one mole of electrons passing through a potential difference of one volt. This is equal to 96.48538(2) kJ/mol. Atomic properties like the ionization energy are often quoted in electron volts.

For comparison:

- $>10^8$ TeV: The most-energetic known ultra-high-energy cosmic rays.
- 14 TeV: The 'Designed' collision energy of protons at the Large Hadron Collider(as yet not reached).
- 1 TeV: A million million electronvolts, or 1.602×10^{-7} J, about the kinetic energy of a flying mosquito[6]
- 210 MeV: The average energy released in fission of one Pu-239 atom.
- 200 MeV: The total energy released in nuclear fission of one U-235 atom (on average; depends on the precise break up).
- 17.6 MeV: The total energy released in the fusion of deuterium and tritium to form He-4 (also on average); this is 0.41 PJ per kilogram of product produced.
- 1 MeV: Or, 1.602×10^{-13} J, about twice the rest mass-energy of an electron.
- 13.6 eV: The energy required to ionize atomic hydrogen. Molecular bond energies are on the order of one eV per molecule.
- 1/40 eV: The thermal energy at room temperature. A single molecule in the air has an average kinetic energy 3/80 eV.

In some older documents, and in the name Bevatron, the symbol "BeV" is used, which stands for "billion electron volts"; it is equivalent to the GeV.   Wikipedia

Emissions from nuclear plants?

Emissions Avoided by the U.S. Nuclear Industry
1995 - 2008

Year	Sulfur Dioxide (Million Short Tons)	Nitrogen Oxides (Million Short Tons)	Carbon Dioxide (Million Metric Tons)
1995	4.19	2.03	670.60
1996	4.16	1.89	645.30
1997	3.97	1.76	602.40
1998	4.08	1.76	646.40
1999	4.13	1.73	685.30
2000	3.60	1.54	677.20
2001	3.41	1.43	664.00
2002	3.38	1.39	694.80
2003	3.36	1.24	679.80
2004	3.43	1.12	696.60
2005	3.32	1.05	681.92
2006	3.12	0.99	681.18
2007	3.04	0.98	692.71
2008	2.65	0.91	688.72
Total	49.84	19.83	9,406.93

What are the licensing basis – 10 CFR 50 Appendices?

Appendix A – General Design Criteria
Appendix B – Quality Assurance Criteria
Appendix C – Financial Data
Appendix D – (Reserved)
Appendix E – Emergency Planning and Preparedness
Appendix F – Siting of Fuel Processing and Waste Management Facilities
Appendix G – Fracture Toughness
Appendix H – Reactor Vessel Material Surveillance
Appendix I – Nuclear Power Reactor Effluents
Appendix J – Containment Leakage Testing
Appendix K – ECCS Evaluation Models

Appendix L – Antitrust Review of Facility License Applications
Appendix M – Licenses for Construction and Operation
Appendix N – Licenses for Construction and Operation at Multiple sites
Appendix O – Staff Review of Standard Designs
Appendix P – (Reserved)
Appendix Q – Pre-application Early Review of Site Suitability Issues
Appendix R – Fire Protection Program
Appendix S - Seismic

INDEX

References:

American Nuclear Society (ANS) 555 North Kensington Avenue, La Grange Park, Illinois 60526, 800-323-3044 www.ans.org

Ardley, N., Ashurst, G., Backnist, C., Bailey, J., Basham, M., etc…, Random House Dictionary of Scientists, Random House, New York

Babcock and Wilcox, 13024 Ballantyne Corporate Place, Suite 700, Charlotte, NC 28277

Britannica Encyclopedia, 331 North La Salle St., Chicago, IL 60654

City Public Service Energy (CPS Energy) 145 Navarro, San Antonio, TX 78205

Constellation Energy (Calvert Cliffs Nuclear Plant) 1650 Calvert Cliffs Parkway, Lusby MD, 20657   410-495-4647

Department of Energy (DOE) 1000 Independence Ave, SW Washington, DC 20585 800-342-5363, www.energy.gov

Electric Power Research Institute (EPRI) 3420 Hillview Ave, Palo Alto, CA 94304

Energy Providers Coalition for Education (EPCE) 6021 S. Syracuse Way, Suite 213 Greenwood Village, CO 80111 Phone: 303-804-4672 Fax: 303-773-0026 info@epceonline.org

Environmental Protection Agency (EPA)   www.epa.gov

General Physics, 6095 Marshalee Dr., Suite 300, Elkridge, MD 21075, 800-727-6677

Gray, Theodore, The Elements, Black Dog & Leventhal Publishers, New York

http://inventors.about.com/od/timelines/tp/nuclear.htm -timeline

http://inventors.about.com/od/timelines/tp/nuclear.htm -timeline

http://library.thinkquest.org/C005271F/timeline.html - timeline

http://tonto.eia.doe.gov/kids/energy.cfm?page=tl_nuclear - timeline

http://www.aboutnuclear.org/view.cgi?fC=History,Time_Line –timeline

Institute of Nuclear Power Operations (INPO) 700 Galleria Parkway, SE, Suite 100 Atlanta, GA 30339-5943

Jefferson Lab, 12000 Jefferson Ave., Newport News, VA 23606

National Academy of Nuclear Training thru electronic Learning (NANTeL) Email: nanteladmin@inpo.org

Nuclear Energy Institute (NEI) 1776 I Street NW, Suite 400, Washington, DC 20006-3708

Nuclear Regulatory Commission (NRC) U.S. Nuclear Regulatory Commission Washington, DC 20555-0001, 800-368-5642, www.nrc.gov

Rhodes, Richard, 1986, Making of the Atomic Bomb, Simon and Schuster, ISBN 0-671-44133-7.

Roggenkamp, Paul, L. The Influence of Xenon-135 on Reactor Operation

Stwertka, Albert, A Guide To The Elements, Oxford University Press, New York

www.aip.org/history/climate/timeline.htm - timeline

www.allaboutbatteries.com/history-of-batteries.html
www.americanhistory.si.edu/Subs/operating/propulsion/auxiliary/index.html
www.chemcases.com/nuclear/nc-10.html

www.em.doe.gov/publications/timeline.aspx - timeline

www.energy.gov/about/timeline1939-1950.htm - timeline

www.greatachievements.org/?id=3691 - timeline

www.gsinstitute.org/dpe/timeline.html - timeline

www.iptv.org/exploremore/energy/Energy_In_Depth/sections/timeline.cfm - timeline

www.newscientist.com/article/dn9955-timeline-the-nuclear-age.html - timeline

www.NuclearFiles.org – timeline.

www.radiationworks.com

www.radiochemistry.org/history/nuclear_timeline/index.shtml - timeline

www.world-nuclear.org